A Geography of the European Union

SECOND EDITION

The European Union does not exist just at a point or in a vacuum. It is made up of a mosaic of regions of various sizes, shapes and functions. The attributes of each region and its location in relation to other regions are as vital for its existence and development as the way in which it is managed. Geography draws attention to the features and locations of regions and to their implications. People, natural resources and means of production are distributed in complex ways and people are moved from place to place. Increasingly the influence of the Union is being felt in its fifteen Member States and a new geography is emerging, with pressure to reduce regional disparities in living standards by a process of convergence.

The purpose of the book is to identify and assess the problems and prospects of the various regions of the EU and also to view the Union as a system of interdependent regions. The EU has become increasingly important economically and more recently politically in a changing Europe, and its enlargement to take in a considerable number of Central European countries seems only a matter of time. The assessment of controversial issues is frank: problems of unemployment, social stress, ageing and the place of women are covered objectively. Drawing on their extensive research and experience in Europe, Cole and Cole question whether convergence can ever be achieved and whether environmental problems can be controlled over the short term, and argue that the accession of the countries of Central Europe could have a devasting effect on the present financial and structural situation in the EU. Evidence is provided in the form of tables, diagrams, maps and pictures.

Focusing on the activities and influence of the EU in the broader context of Europe and neighbouring parts of Africa and Asia, this book offers all those studying, working or interested in the European Union a greater understanding of its complexities and its prospects.

John Cole is Emeritus Professor and Lecturer in Geography, University of Nottingham. **Francis Cole** is a Conference Interpreter for the European Parliament in Brussels.

A Geography of the European Union

SECOND EDITION

John Cole and Francis Cole

London and New York

First edition published 1993
Reprinted 1994
Second edition published 1997
by Routledge
11 New Fetter Lane, London EC4P 4EE

Simultaneously published in the USA and Canada
by Routledge
29 West 35th Street, New York, NY 10001

Typeset in Garamond by J&L Composition Ltd, Filey, North Yorkshire

Printed and bound in Great Britain by Butler & Tanner Ltd, Frome and London

British Library Cataloguing in Publication Data

A catalogue record for this book is available from the British Library

Library of Congress Cataloging in Publication Data

A catalogue record for this book has been requested

ISBN 0–415–14310–1
ISBN 0–415–14311–X (pbk)

To Isabel and Carol Ann

CONTENTS

PLATES

BOXES

FIGURES

TABLES

PREFACE AND ACKNOWLEDGEMENTS

Great changes have taken place in Europe since the first edition of this book was completed in 1992. As far as possible, changes such as the accession of Austria, Finland and Sweden have been taken into account. There will undoubtedly be further major changes in the next few years. EU publications are numerous and, for UK readers, these can be traced through HMSO.

The authors wish to thank two reviewers appointed by the publisher for their helpful comments and suggestions for improvements. They are also grateful to Rosemary Hoole for typing the text and tables and to Chris Lewis and Elaine Watts, who drew many of the new maps. Several years of undergraduates in the Department of Geography, Nottingham University, who took the courses on the EU, unknowingly also made a major contribution.

The section on 'The costs of multilingualism' (Chapter 2, pp. 58–60) is based on F.J. Cole (1996) 'The lesson of Babel', *European Voice*, 22–28 February, and is used here with permission. The authors would like to thank *Le Monde Editions* for their permission to reproduce the Plantu cartoons on pp. 51, 149 and 369, and the European Commission for permission to reproduce plates 4.4, 4.5, 6.8, 9.1, 10.7, 10.8, 11.2, 11.3, 11.4, 13.5, 14.1 and 14.2. Thanks are also due to the European Commission Direction Généralex, Audiovisuel for the permission to use the photographs on pages 122, 176, 223, 264, 279, 282, 339 and 350.

CONVENTIONS

Countries The English version has been used throughout the book for names of countries. The reader should note, however, that in EU publications, initials for the 15 EU Member States are often used and these start as in the national spelling (e.g. E for España, Spain).

Regions The names of EU regions are given throughout the book in their own languages (e.g. Sicilia rather than Sicily, Corse rather than Corsica, Bretagne rather than Brittany). Readers who intend to use EU regional data sets need to recognise these names.

Italy is subdivided into four main regions: Italia Settentrionale, Centrale, Meridionale and Isole. These have been written with initial capital letters as North, Central, South and Islands. Northern and southern Italy are used in a more loose sense. Western Europe refers to EU and EFTA countries, Central Europe (formerly Eastern Europe) to countries between Western Europe and the former USSR.

Cities For names of cities, the English version has been used, but the reader will obviously meet the names in their own languages in the literature and when travelling. Among EU cities with English equivalents are the following: Germany – Köln (Cologne), München (Munich), Hannover (Hanover); France – Marseille (Marseilles), Lyon (Lyons); Italy – Firenze (Florence), Genova (Genoa), Milano (Milan), Napoli (Naples), Roma (Rome), Torino (Turin), Venezia (Venice); The Netherlands – Den Haag (The Hague); Belgium – Brussel (Dutch); Bruxelles (French). Brussels has been used to avoid choosing one of the two spellings. In Belgium, note also Antwerpen/Anvers and Gent/Gand; Denmark – København (Copenhagen); Greece – Athinai (Athens), Thessaloniki (Salonika); Spain – Sevilla (Seville); Portugal – Lisboa (Lisbon), Porto (Oporto). Note too that Luxembourg refers both to the Grand Duchy (GD) and to a province in Belgium.

TERMS AND ABBREVIATIONS

TERMS AND ABBREVIATIONS

billion	(shortened to bln, bn) is one thousand million
ha	hectare(s)
kg	kilogram
km	kilometre
km/h	kilometres per hour
kW	kilowatt
kWh	kilowatt-hour
Motorway	This refers not only to British motorways but also to Autobahnen, Autoroutes, Autostrade, etc.
sq km	square kilometre
t.c.e.	tonnes of coal equivalent
t.o.e.	tonnes of oil equivalent
tonne	a metric ton

ACRONYMS/INITIALS

ACP	African, Caribbean and Pacific countries (the Lomé Convention)
ASEAN	Association of South East Asian Nations
Benelux	Economic Union of Belgium, the Netherlands and Luxembourg
CAP	Common Agricultural Policy
CCT	Common Customs Tariff
CFC	Chlorofluorocarbon
CFP	Common Fisheries Policy
CFSP	Common Foreign and Security Policy
CIS	Commonwealth of Independent States (the former USSR, excluding the three Baltic States)
CMEA	Council for Mutual Economic Assistance (also Comecon)
Comecon	see CMEA
COR	Committee of the Regions
DOM	Départements d'outremer (French overseas departments Guadeloupe, Guyane, Martinique, Réunion)
EAEC	European Atomic Energy Community
EAGGF	European Agricultural Guidance and Guarantee Fund
EBRD	European Bank for Reconstruction and Development
EC	European Communities (often Community)
ECSC	European Coal and Steel Community
ECU	European Currency Unit (also Ecu, ecu)
EDC	European Defence Community
EDF	European Development Fund
EEA	European Economic Area
EEC	European Economic Community
EFTA	European Free Trade Association

EIB	European Investment Bank		NACE	General industrial classification of economic activities within the European Communities
EMI	European Monetary Institute			
EMS	European Monetary System			
EMU	Economic and Monetary Union		NAFTA	North American Free Trade Agreement
EP	European Parliament			
EPC	European Political Cooperation		NATO	North Atlantic Treaty Organisation
EPU	European Political Union			
ERDF	European Regional Development Fund		NCI	New Community Instrument
ERM	Exchange Rate Mechanism		NUTS	Nomenclature of units of territory for statistics (levels 1, 2, 3 frequently used in EU publications)
ESC	Economic and Social Committee			
ESF	European Social Fund			
EUR 6, 9, 10, 12, 15	European Community according to the joining sequence of Member States			
			OECD	Organisation for Economic Cooperation and Development
Euratom	see EAEC		OPEC	Organisation of Petroleum Exporting Countries
EUROSTAT (also Eurostat)	Statistical Office of the European Communities			
FAO(PY)	Food and Agriculture Organisation (of the United Nations) (Production Yearbook)		OSCE	Organisation for Security and Cooperation in Europe
			PHARE	Poland–Hungary Assistance in the Restructuring of Economies
FRG	Federal Republic of Germany (West Germany)		PPS	Purchasing Power Standard
GATT	General Agreement on Tariffs and Trade (superseded by WTO)		PRB	Population Reference Bureau (Washington)
			R&TD	Research and Technological Development
GDP	Gross Domestic Product (see also GNP)		RSFSR	Russian Soviet Federal Socialist Republic, now Russia
GDR	(former) German Democratic Republic (East Germany)		SEA	Single European Act
GNP	Gross National Product (GNP measures the resources available after the transfer of factor incomes such as interest payments and dividends but unlike GDP is not used at regional level)		SME	Small and medium-sized enterprise
			TACIS	Technical Assistance to the CIS
			TEN	Trans-European Network
			TEU	Treaty on European Union
			TFR	Total fertility rate
GRT	Gross registered tonnage		TGV	Train à grande vitesse
HGV	Heavy goods vehicle		TOR	Treaty of Rome
IGC	Intergovernmental Conference		UK	United Kingdom of Great Britain and Northern Ireland
IT	Information Technology			
JET	Joint European Torus		UN	United Nations
LNG	Liquefied natural gas		UNDP	United Nations Development Programme
Maghreb	Usually refers to Algeria, Morocco and Tunisia			
			USA	United States of America

USSR	Union of Soviet Socialist Republics (also Soviet Union), dissolved in 1991	WEU	Western European Union
		WPDS	World Population Data Sheet (see PRB)
VAT	Value Added Tax	WTO	World Trade Organisation

1

INTRODUCTION

'Oui, c'est l'Europe, depuis l'Atlantique jusqu'à l'Oural, c'est l'Europe, c'est tout l'Europe, qui décidera du destin du monde.'
'Yes, it is Europe, from the Atlantic to the Urals, it is Europe, it is the whole of Europe, that will decide the fate of the world.'

(Charles de Gaulle, President of France, 23 November 1959, Strasbourg)

This book is intended to serve as a basis for the study of the European Union (EU) within a broader European and global context. Since its foundation as the European Communities (EC) in the 1950s the EU has grown into a supranational entity rivalled in its economic scale only by the North American Free Trade Agreement (NAFTA). In the mid-1990s the future of the EU has been very much under debate in a fast-changing continent, and with increasing pressures upon it from other regions of the developed and developing worlds. The geographical approach is used in this book to analyse and to evaluate the major economic, social and political aspects of the EU with particular reference to spatial issues and problems. In order to establish the framework for the book it is appropriate first to put the EU in its European and global contexts.

Membership of the EU is explicitly open only to countries located in Europe. At least in principle the policy of the EU is to welcome eligible countries to membership. It is therefore necessary here to note the traditional or conventional extent and limits of the continent of Europe. General de Gaulle (cited at the start of

the chapter) was prophetic when he emphasised the continent of Europe as a whole and as a single entity, although he overlooked the fact that Russia extends eastwards far beyond the Ural Mountains, indeed to the Pacific coast. However, only since the break-up of the Soviet bloc, the Council for Mutual Economic Assistance (CMEA), in 1990, and of the Soviet Union itself in 1991 has the definition of the limits of Europe been a matter of more than academic interest to the citizens of the EU.

Europe conventionally extends eastwards to the Ural Mountains of Russia, the Ural River and the Caspian Sea. In the south it is separated from Africa by the Mediterranean, and from Asia by the Bosphorus, the Black Sea and the mountains and rivers south of Transcaucasia. Figure 1.1 shows recent variations on the theme of Europe's limits. These limits are traditionally associated with physical features, but they do not run through barriers of great note politically, militarily or culturally.

Early in the fifteenth century the Chinese briefly explored and occupied parts of southeast and south Asia beyond the traditional units of their Empire. It was the Europeans, however,

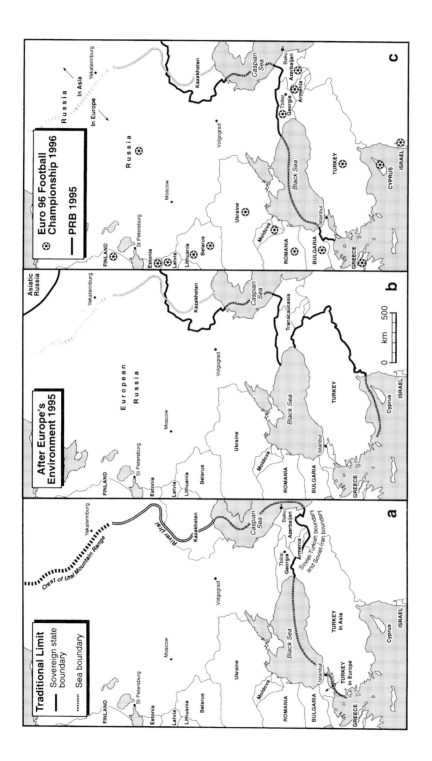

Figure 1.1 Different definitions of the limits of Europe
Map a Traditional limit as described in the text
Map b Stanners and Bourdeau (1995) in *Europe's Environment* include the whole of the Ural economic planning region of the former USSR, which extends both sides of the crest of the Ural Mountains. They exclude a small part of Kazakhstan, between its western boundary and the Ural River, and all three Transcaucasian Republics of the former USSR, but include the whole of Turkey
Map c shows how the latest Population Reference Bureau's *World Population Data Sheet* puts the whole of Russia into Europe but excludes Transcaucasia and Turkey. The football symbols show all the countries in the eastern part of Europe that participated in the EURO 96 Football Championship, with Turkey, Cyprus and Israel deemed to be part of Europe for the purposes of the Championship

who, late in the fifteenth century, began to conquer and colonise lands beyond the limits of their continent. Thus Europe became the first region of the world to establish its influence on a global scale. In 1494 the Pope divided the world beyond Europe between Spain and Portugal in the Treaty of Tordesillas. Christian Europe thus effectively gave itself a mandate to conquer anywhere in the world.

The empires of the Western European countries continued to grow in some parts of the world into the twentieth century, but more than 200 years ago they started to disintegrate with the War of American Independence. Such countries of the EU as France, Belgium and the UK have lost most of their colonies over the last few decades. Ties have, however, been maintained through the Lomé Convention, linking the African, Caribbean and Pacific (ACP) countries with the EU by trade and development cooperation agreements. The British Commonwealth and the Francophone Community (which links France with many of its former colonies in Africa) also help to retain ties between the EU and former colonies.

While Spain, Portugal and, subsequently, other Western European countries were acquiring colonies in various parts of the world from the sixteenth century onwards, Russia was expanding in several directions in the eastern half of Europe, and in the seventeenth century it conquered much of what is now Siberia. In the early 1990s the Soviet Union, inheritor of the vast Russian Empire, was in its turn dissolved, and various non-Russian peoples in the Union readily accepted independence. Such a development has particular significance for the EU, since several former Soviet Republics, after achieving independence from Russia in 1991, have been looking towards the European Union for support. At the time of writing, the three Baltic Republics had already applied for EU membership.

Much of Europe is characterised by great physical and cultural diversity. Many cultural features have origins far back in the past. These affect the organisation and expansion of the EU today, whether they are linguistic, religious or political. Until the sixteenth century Europe was frequently the recipient of migrants, technologies and cultural features from Asia and Africa. At times, as when Greek and Roman conquests extended European power into North Africa and southwest Asia, its own influence spread beyond the continent. Most of the invaders of Europe came from the east or southeast, some from North Africa. As noted above, after 1500 the tide turned, and for the next four centuries the influence of a number of European powers extended to many parts of the world with varying degrees of intensity. Only the Ottoman Turks continued to have control over a part of Europe after the middle of the sixteenth century.

While engaged in conquering other parts of the world, during the last five centuries European powers have also frequently been in conflict among themselves both in Europe itself and elsewhere in the world. Since the Second World War, and during the last stages of the disintegration of their empires, Western European countries have come together in entities for cooperation in defence and trade (see the next section). The creation of the supranational bodies of the EU has brought together former rivals and enemies in a completely new situation. In the European Parliament, politicians of similar political views from the 15 Member States of the EU form transnational groups or alliances. The same politicians are regrouped and allocated to delegations, each vested with the care of the Parliament's relations with the parliament and/or responsible national authorities of a particular country or group of countries elsewhere in the world. Such strange bedfellows would have been unthinkable 60 years ago.

In the present book the EU is considered both as an entity moving, sometimes against historical traditions, towards greater unity, and as a region of the world coming to terms rapidly with a new situation in Europe and likely also to be influenced increasingly by regions outside Europe. Early in the twentieth century, the British geographer Halford Mackinder (1904) argued that future historians might refer to the period 1500–1900 as 'the Columbian epoch':

> Broadly speaking, we may contrast the Colum-bian epoch with the age which preceded it, by describing its essential characteristic as the expansion of Europe against almost negligible resistances, whereas mediaeval Christendom was pent into a narrow region and threatened by external barbarism. From the present time forth [1904] we shall again have to deal with a closed political system, and none the less that it will be one of worldwide scope. Every explosion of social forces, instead of being dissipated in a surround-ing circuit of unknown space and barbaric chaos, will be sharply re-echoed from the far side of the globe, and weak elements in the political and economic organism of the world will be shattered in consequence.
>
> (Mackinder 1904)

We are now well into Mackinder's post-Columbian epoch. If his view of the world has turned out to be correct, then in the future Western Europe's global position may soon be more like its position in the pre-Columbian epoch than it was in the Columbian epoch. From now on there will be pressures from outside: from the USA and Japan to maintain innovative and competitive industry; from Central Europe and the former USSR to help in their economic restructuring; and from the developing countries to be more forthcoming with assistance. There is also growing eco-nomic and social pressure from immigrants arriving from both the former CMEA countries and the developing world.

Since the Second World War, a feature of world affairs has been the emergence of supra-national blocs in various parts of the world. Apart from the recent demise of the CMEA, the process continues, with much of the eco-nomic power of the world concentrated in a few countries or groups of countries, including, for example, the North American Free Trade Agreement, consisting of the USA, Canada and Mexico, with 387 million inhabitants alto-gether, and the Association of South East Asian Nations (ASEAN), with 320 million inhabi-tants. Whether or not existing supranational entities expand and new ones emerge, it must be appreciated that between 1995 and 2025 the combined population of the developing coun-tries of the world is forecast to grow by about 2.5 billion, adding the population of more than six EUs in that time, the equivalent of a new EU every five years.

Throughout its history Europe has never formed a single political unit. During several periods, mostly brief, large portions of the ter-ritory of Europe have, however, been held, at times loosely, in a single political unit (see Box 1.1). The most recent attempts to bring together a number of separate countries of Eur-ope have been the creation of the Council for Mutual Economic Assistance, founded in 1949, and of the European Economic Community itself (EEC), founded in 1957. The CMEA was disbanded in 1990 but the EEC, now referred to as the European Union, has expanded since its foundation in a series of 'steps' (see Figure 1.2). The European Union has appropriated the name of the continent, although in 1995 its 15 Member States con-tained only about one-third of the territory of Europe and had less than half of its population.

It would be premature to speculate if and when the whole of Europe will form a single supranational unit, but with regard to two countries in particular the definition of Europe is unclear. The former USSR and Turkey each has territory in both Europe and Asia. Over two-thirds of Russia lies east of the Urals in Asia. Although almost all of the territory of Turkey is in Asia, Turkey could also be con-sidered for EU membership, in spite of its ambivalent position in relation to the two con-tinents, but at the time of writing, the Eur-opean Commission was not in favour of such a

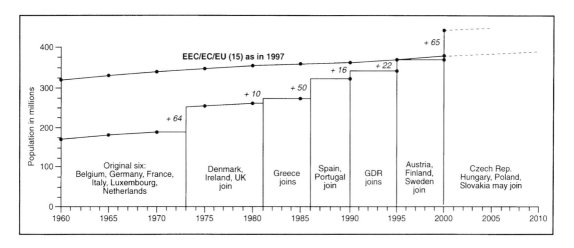

Figure 1.2 The joining sequence of the current 15 Member States of the EU, showing the population added at each stage of expansion, population growth between stages and the total population of all the present EUR 15 countries since 1960

move. The 'rule' that applicants should be in Europe has indeed already been broken with regard to Cyprus, with which negotiations for membership are soon to begin; Cyprus has conventionally been regarded as part of Asia. It would, however, be inappropriate, for example, for New Zealand, South Africa or Uruguay to apply, even though the EU does include several small French Départements d'Outre-Mer, located far from Europe.

ENTITIES IN POST-1945 EUROPE

The Second World War left more than physical scars on a Europe devastated by death and destruction. Citizens and their leaders, stunned by the way in which history had been able to repeat itself, resolved to do their utmost to establish mechanisms and a framework to minimise the risk of armed conflict between European states occurring again. The Conference of Yalta in 1945 established the return of occupied territory, the reaffirmation of most national boundaries and in Central Europe some significant transfers of territory. At the

same time, however, in a matter of months after the war ended the clear demarcation was evident in Europe of two separate blocs created from the Allied powers who had defeated the Axis countries, the first under the control of the Soviet Union and the second under the countries of Western Europe and the United States.

The separation was most clearly evident in the division and occupation of Germany, whose former capital, Berlin, was also divided into Western and Soviet occupied sectors in spite of its location in the Soviet occupied part of the country. In 1949 the Western powers sponsored the formation of the Federal Republic of Germany, which was matched by the German Democratic Republic, set up under the influence of the Soviet Union. Very different visions were held of the future of Europe in political, economic and social terms. The years after the end of the Second World War were thus characterised by the establishment, on the one hand, of entities for economic and cultural cooperation and, on the other, of defence and security organisations designed to protect the blocs in question from the real or imagined threat from the other side.

BOX 1.1 HISTORICAL PERSPECTIVES

Up to its traditional limits of the Ural Mountains and the Ural River, Transcaucasia and the Black and Mediterranean Seas (see Figure 1.1) Europe occupies 10,400,000 sq km, only 7 per cent of the world's land area. Although it is roughly comparable in area to Canada, China, the USA or Brazil, each of which is a single political unit, it has been politically fragmented throughout its history. The great diversity of Europe's geographical features may have contributed to its political fragmentation, with mountain ranges and seas forming obstacles to movement, and major rivers often providing convenient markers at the limits of political units. Linguistic, ethnic and religious differences have also contributed to the separation of peoples. Large parts of Europe have briefly formed states or 'empires', only to disintegrate and later gather again into new combinations.

A review of the periods in the history of Europe when substantial parts of the continent were for one reason or another organised into single entities, while not expected to be more than a rough guide to possible future situations, can serve as a basis for speculation. With the help of the six political maps in Figures 1.3a–1.3f six past 'supranational' versions of Europe will be briefly considered, and each compared with the area of the EU.

(a) The Roman Empire around AD 200 included territory in Asia and North Africa while also extending in places eastwards and southwards in Europe beyond the limit of the present EU, but it did not contain Scotland, most of Germany, the northern Nether-

lands or Denmark. The area of modern Germany proved difficult for the Romans to conquer. Most of the area in the Roman Empire in Figure 1.3a around AD 200 was successfully held and governed from Rome for several centuries.

(b) The medieval German Empire of around 1190 was a loosely organised entity emerging as a major force in Western Europe, bounded in the West by France and in the east by Hungary and Poland. It bore some resemblance to the original EEC of the Six in 1957.

(c) Following its final integration in 1492 with the conquest of Granada and the expulsion of the Moors, Spain became a leading power in Europe in the sixteenth century. Although separated by France from its central European territories, it controlled an appreciable part of Western Europe through the Habsburg Empire.

(d) Around 1810 Napoleon's France, which included territories ruled directly from Paris, states ruled by members of Napoleon's family and other dependent states, resembled the present EU with 15 Member States, except for Portugal, Great Britain and Ireland, and the Nordic countries, but it included the Grand Duchy of Warsaw.

(e) Hitler's hold over Europe at its greatest extent in 1942 consisted of the Grossdeutsches Reich plus a much larger territory occupied by German forces. Of the present EU it did not include the UK, Ireland, Sweden, Spain, Portugal or, notionally at least, Vichy France.

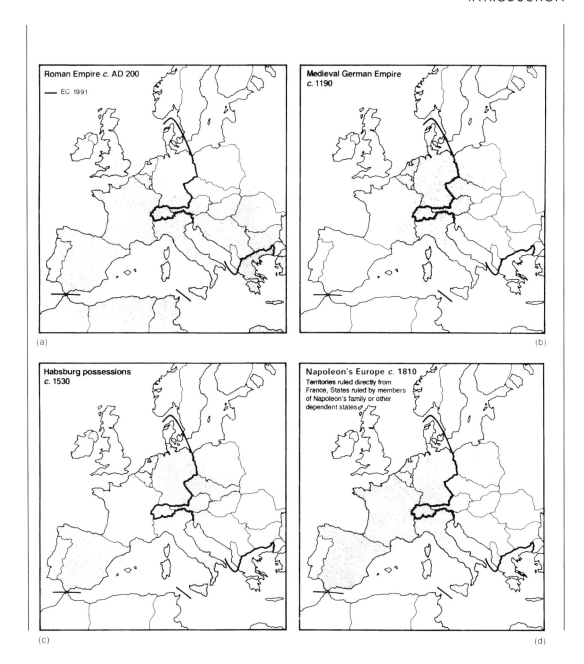

Figure 1.3 Europe under (a) Rome, (b) medieval Germany, (c) Habsburgs, (d) Napoleon (*continued overleaf*)

(f) The USSR superseded Germany as the dominant power in Europe after 1945; around 1950 Stalin's Europe, including the satellites of Central Europe, extended well into Western Europe in the Soviet zone of occupied Germany. The neutrality of several non-Communist countries could be regarded indirectly as a consequence of Soviet pressure. Of the original EEC Six, West Germany formed the front line against perceived Soviet pressure, with NATO forces from other countries stationed there.

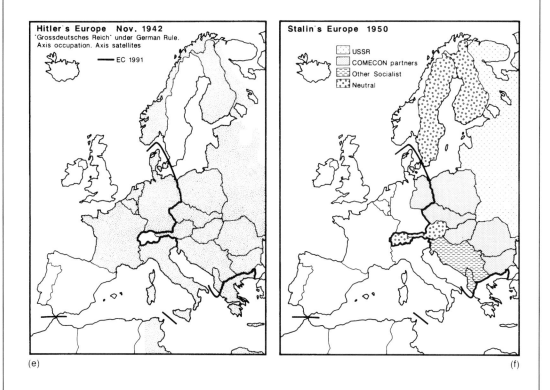

(e)

(f)

Figure 1.3 Europe under (e) Hitler, (f) Stalin
Note that the EU limits shown with a heavy line on each map are for EUR 12, before the accession of Austria, Finland and Sweden

The Treaty of Brussels, signed on 17 March 1948 between Belgium, France, the Netherlands, Luxembourg and the United Kingdom, was created for 'collaboration in economic, social and cultural matters and for collective self-defence'. It became the framework within which the Council of Europe and the North Atlantic Treaty Organisation (NATO) were formed in 1949. Belgium, the Netherlands and Luxembourg also formed a Customs Union in 1948, a precursor to the Benelux Economic Union formed in 1958. The Brussels Treaty Organisation became the Western European Union (WEU) in 1955 after the founder Member States of the European Coal and Steel Community (ECSC) failed in their attempts to create a more concentrated defence core in the form of the European Defence Community (see next section). The WEU was later enlarged to include Italy and the Federal Republic of Germany. At the same time the Council of Europe and NATO extended their membership during the 1950s, and the Nordic Council, a regional grouping of Denmark, Iceland, Norway and Sweden, was formed in 1952. Following the creation of the European Economic Community in 1957, and the subsequent alienation of the United Kingdom from the founder members, another regional economic grouping was formed in 1960, the European Free Trade Association (EFTA), comprising those West European countries not in the EEC, and having broadly similar objectives.

The Cold War was running its course through the 1950s and the Soviet Union was consolidating its sphere of influence in Central and Eastern Europe, first with the creation of the Council for Mutual Economic Assistance in 1949, which brought in Bulgaria, Czechoslovakia, Hungary, Poland and Romania. The purpose of the CMEA was to facilitate and coordinate the economic development of the countries concerned. The same countries, with the addition of East Germany and Albania, formed the Warsaw Pact in 1955, providing for a unified military command and the stationing of Soviet military units on the territory of the other Member States. It soon became clear, however, that the main purpose was to strengthen Soviet control over its satellites, in which it had installed sympathetic governments. Figure 1.4a shows the state of supranational entities in Europe in 1960, with a clear division between Eastern and Western Europe, and each bloc with entities for cooperation in defence and economic affairs.

The following two decades were relatively calm, with no major changes in the membership or scope of these entities. The EEC was gradually taking in additional Member States and the Soviet Union was increasing its military and economic control over its satellites, exemplified in the suppression of the uprisings earlier in Hungary in 1956 and in Czechoslovakia in 1968. The demolition of the Berlin Wall in 1989 and the subsequent break-up of the Soviet Union and its sphere of influence in Central and Eastern Europe heralded several years of political and economic upheavals. In addition to the unification of East and West Germany and the collapse of the Warsaw Pact and the CMEA, the 1990s have also seen the break-up of Yugoslavia and Czechoslovakia and the emergence of fifteen former Soviet Republics as sovereign states. The Council of Europe and the EC have continued to expand their membership and zones of influence as new democracies have looked to the West for future cooperation in cultural and economic areas. EFTA membership has halved as three of its members have joined the EU. The remaining EFTA countries, with the exception of Switzerland, have even joined with the EU Member States in the European Economic Area (EEA), an extension of the EU's Single Market. Russia established the Commonwealth of Independent States (CIS) as an attempt to preserve a degree of economic cooperation between most of the former Soviet republics, but its functions have been limited.

Defence and security structures have remained largely unchanged, with the notable exception of the Organisation for Security and Cooperation in Europe (OSCE), which began as a series of meetings known as the Conference for Security and Cooperation in Europe (CSCE) in 1975 and became institutionalised in 1990. Its role has been further enhanced and it works actively in the areas of democracy, human rights, crisis management and conflict prevention, with 52 participating Member States.

Figure 1.4b shows the state of European entities early in 1996. The main changes since 1960 (contrast with Figure 1.4a) can be seen clearly. The Iron Curtain separating East from

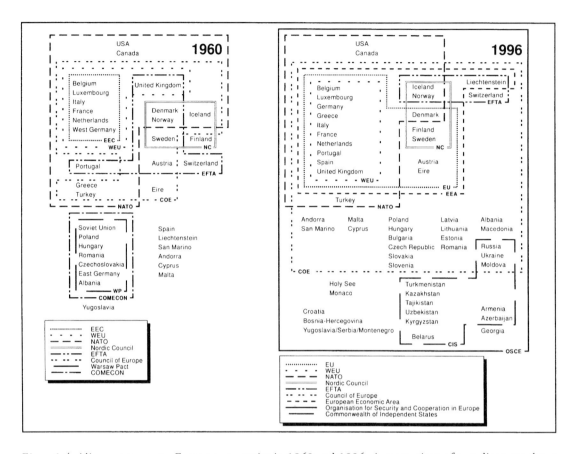

Figure 1.4 Alignments among European countries in 1960 and 1996. A comparison of two diagrams shows the greater complexity in 1996 resulting from both the break-up of the USSR and the emergence of many new countries from the former USSR, Yugoslavia and Czechoslovakia

West has disappeared, allowing many states formerly under Soviet control to make very marked moves towards Western entities. Certain states have disappeared, while many others have emerged or reappeared. The European Union has become much more consolidated in terms of its membership and sphere of influence.

In spite of the numerous political changes in Europe in the 1990s, Western defence entities have, however, remained in place, a clear indication of the view that such structures are still necessary for security, in spite of the end of the Soviet threat, and that it is premature to abolish

them or to merge them with other bodies, such as the EU. The view that NATO is now superfluous, or that the United States no longer has a role to play in the security of Europe, has been refuted by the involvement of NATO with the active participation of US troops in the enforcement of peace in the former Yugoslavia on the basis of the agreement reached in Dayton, Ohio, in November 1995.

Now firmly established as the supreme economic entity in Europe, and the pole of attraction for many former Soviet satellites, the EU still has very clear ambitions in the areas of foreign and defence policy. It remains to be

Table 1.1 The countries of Europe 1996

Name English	Own	Date of membership
European Union		
Belgium	België/Belgique	1952
Denmark	Danmark	1973
Germany	Deutschland	1952 (West) 1990 (GDR)
Greece	Ellas (also Ellada)	1981
Spain	España	1986
France		1952
Ireland	Eire	1973
Italy	Italy	1952
Luxembourg		1952
Netherlands	Nederland	1952
Austria	Österreich	1995
Portugal		1986
Finland	Suomi	1995
Sweden	Sverige	1995
United Kingdom		1973

	Other organisation	Date of application to EU
Non-EU		
Albania		
Armenia	CIS	
Azerbaijan	CIS	
Belarus	CIS	
Bosnia-Hercegovina		
Bulgaria		
Croatia		
Cyprus		1990
Czech Republic		1996
Estonia		1996
Georgia		
Hungary		1994
Iceland	EFTA/EEA	
Latvia		1996
Liechtenstein	EFTA/EEA	
Lithuania		1996
Macedonia (FYROM)		
Malta		1990
Moldova	CIS	
Norway	EFTA/EEA	
Poland		1994
Romania		
Russia	CIS	
Slovakia		
Slovenia		
Switzerland	EFTA	
Ukraine	CIS	
Yugoslavia (Serbia/ Montenegro)		
Non-Europe		
Turkey		1987 (negative opinion)
Morocco		1987 (refused)

seen whether the reluctance of certain Member States to part with sovereignty over such matters will succeed in the face of the increasing desire for further integration in Western Europe. The European Union itself is dealt with in more detail in the next section and in Chapter 2.

INTRODUCTION TO THE EU

The fifteen Member States of the European Union are listed in Table 1.1 in the official alphabetical order used by the institutions of the EU, based on the name of the country in its own language. Other European and certain non-European countries are also included in the table. The location of the countries in Table 1.1 is shown in Figure 1.5. Although the principal subject of the book is the EU, reference will be made, where appropriate, to non-EU countries, in particular to those in the rump of EFTA and to countries in Central and Eastern Europe that are candidates for accession in the next enlargement.

The European Union has its origins in the European Communities, which were founded in the 1950s from a Europe that emerged in 1945 from the turmoil and devastation of the Second World War. Table 1.2 shows a selection of key dates in the formation of the EU. Profoundly influenced by wars dominated by Franco-German conflicts, the founders of the EC saw closer economic and political ties between these two historical protagonists as being the best way of reducing risks of a repetition of such a conflict. This vision, first inspired by Jean Monnet, Commissioner-General of the French National Planning Board, led the 'Father' of the EC Robert Schuman, French Minister for foreign affairs, to propose on the 9 May 1950 that Franco-German coal and steel production should be placed under a joint High Authority in an organisation to which other European nations could belong. At that time, the coking coal of Germany and the iron ore of France complemented each other in the context of geographical proximity and environmental similarities.

The first of the European Communities, the European Coal and Steel Community, was founded in 1952, following the signing of the

Table 1.2 A summary history of the EU

Year	Event
1945	End of Second World War
1950	The Schuman Declaration
1952	ECSC Treaty enters into force
1955	Messina Conference
1958	EEC and Euratom Treaties enter into force
1960	EFTA established
1973	UK, Denmark and Ireland join EC
1979	First direct elections to European Parliament
1981	Greece joins EC
1986	Spain and Portugal join EC
1987	Single European Act enters into force
1990	Unification of Germany
1993	Treaty on European Union enters into force
1993	Completion of Single Market
1994	EEA agreement enters into force
1995	Austria, Finland and Sweden join EU
1996	Start of IGC on reform of TEU

Treaty of Paris on 18 April 1951. The first six Member States were West Germany, France, Italy, the Netherlands, Belgium and Luxembourg, with the four official languages being Dutch, French, German and Italian. The Treaty established the institutions for the management of resources, production and trade in ECSC coal and steel, the High Authority (later the Commission), the Common Assembly (later the Parliament), the Council and the Court of Justice.

Since one of the fundamental objectives of the founders of the EC was to ensure peace and stability in post-war Europe, there was also an attempt to develop a parallel European Defence Community, but proposals foundered when, in 1954, they were rejected by the French Assemblée Nationale, which saw it as a threat to national sovereignty. Proposals for further cooperation in the areas of defence and security were then shelved until the negotiations for the Treaty of Maastricht, which led to agreement on a fledgling common defence policy in the Treaty on European Union in 1993.

The success of the ECSC led the six founding Member States to commit themselves at the Messina Conference in June 1955 to further integration in other sectors. Following two years of negotiations, the Treaties of Rome were signed on 25 March 1957, setting up the European Economic Community and the European Atomic Energy Community. The EEC Treaty laid down a series of objectives in various economic sectors governed by the principle, presented in the preamble, of reducing economic divergences between regions of the Community and improving the living standards of its citizens. The Euratom Treaty dealt specifically with research, production and safety in the nuclear energy sector. Both Treaties contained more detailed provisions governing institutional, financial and administrative procedures for the attainment of their objectives and, together with the ECSC Treaty, they form the corpus of primary law that has governed the activities of the EC over the subsequent three decades.

During the period of the establishment of the EEC, the United Kingdom made attempts to join, but negotiations failed to produce conditions acceptable to it, partly owing to obstacles created by France, led by General de Gaulle. In 1959 a frustrated UK was instrumental in the creation of the European Free Trade Association, signed in the Stockholm Convention in November 1959 with Austria, Denmark, Norway, Portugal, Sweden and Switzerland, and which became operative in 1960. Liechtenstein, which has a customs union with Switzerland, and Finland took associate membership, and Iceland joined in 1971.

Although EFTA was more limited in scope and ambition than the EC, concentrating primarily on the removal of trade barriers between members, it became a significant counterweight organisation. In spite of the fact that the UK, Denmark and Portugal left it to join the EC, it played a leading role in areas of economic and commercial cooperation into the 1990s until the accession of Austria, Finland and Sweden to the EU in 1995. Its more limited membership since 1994 continues in close cooperation with the EU, in particular through the European Economic Area, which entered into force on 1 January 1994, and which is a Single Market comprising the EU Member States and the EFTA members with the exception of Switzerland, which voted in a referendum to remain outside.

Applications for membership of the EU continued and in 1970 the UK, the Republic of Ireland, Denmark and Norway applied for membership. In a closely fought referendum the population of Norway decided to remain outside, but the other three countries joined in 1973, bringing the EC up to nine Member States, with six official languages. Greece applied in 1975 and joined in 1981, followed by Spain and Portugal which applied in 1977 and joined in 1986.

Plate 1.1 Labour Prime Minister Harold Wilson prepares to apply for membership of the EEC in 1967. He is supported by Holland but is unlikely to make the EEC in view of the weight of opposition to the British application by France (Charles de Gaulle), Germany (Konrad Adenauer) and members of his own party, represented by Emmanuel Shinwell
Source: Redrawn from a cartoon in *The Times*, London, 1 March 1967 by Mahood

Plate 1.2 No Entry In January the British Government heard the result of its application to join the Common Market. It was a categorical refusal – inspired by General de Gaulle's jealous protection of France's leadership with The Six
Source: Illingworth 1963

The activities of the EC over its first 30 years were substantial in a wide range of areas such as agriculture, international trade, harmonisation of customs procedures and cooperation in energy policy and in research and development. Nevertheless, it was felt in the course of the 1980s that insufficient progress was being made towards the more fundamental objectives of the Treaties, which included the reduction of social and regional inequalities and the harmonisation of legislation to a Single Market. Following negotiations that coincided with the accession of Spain and Portugal, the Single European Act (SEA) was signed in 1986, entering into force on 1 July 1987. It was the first significant reform of the substantive law of the EC since its foundation.

The SEA set the agenda for the following years of EC activities, establishing a timetable for the completion of the Single Market by the end of 1992 (Art. 8a EEC) and the development of EC policies in such areas as transport and environmental protection. The need for 'economic and social cohesion' (Art. 130a EEC) has been regarded as a central priority for the further development of the EC. Moreover, experience of lengthy negotiations and slow decision-making over previous decades had shown the need for legislative procedures to be speeded up, so the SEA also contained provisions for rendering EC decision-making more efficient and democratic. In particular, it extended qualified majority voting in the Council, strengthened the executive powers of the Commission, and gave a more substantial role to the Parliament (SEA Arts. 6–9).

Following the signing of the SEA there was substantial impetus and a greater sense of urgency in the policy-making and integration of the EC. The Commission's White Paper on the Internal Market, which predated the SEA by eight months, set the objectives for the completion of the Single Market, with 282 directives being stipulated as essential for its legislative and technical framework. During the same period there was also an ongoing process of negotiation for Treaty reform, with initiatives such as the Delors Committee proposals for Economic and Monetary Union (EMU) and the establishment of the Social Charter. This process culminated at the end of 1991 in the completion of the intergovernmental conferences on political union and economic and monetary union leading to eventual agreement on the Treaty on European Union (TEU), signed in Maastricht in 1992, which entered into force on 1 November 1993 following ratification difficulties, in particular in Denmark. The TEU amended the existing EC Treaties as well as establishing new areas for policy-making and cooperation, such as a common foreign and security policy and cooperation in justice and internal affairs. Although the other Treaties and Communities continue to exist, the term European Union has now become the accepted title for the sum of all its constituent parts, and will thus be used in this book wherever appropriate.

Developments in 1989 in Central and Eastern Europe, in particular with the demolition of the Berlin Wall, resulted in the unification of Germany, with the rapid and almost automatic accession of the territory and population of the former GDR to the EC over a period of months. These developments led to a substantial reappraisal of the future direction and scope of the EC. Membership of the Community had become increasingly attractive to other countries in EFTA and Central Europe, leading to a debate on crucial issues of the enlargement and continued integration of the EU. Some Member States talked of further integration into a deeper federal union, whereas others preferred the concept of a broader and shallower confederation of European states in an EU based more on cooperation and decentralised decision-making, with limited central decisions based on consensus, rather than a full political and economic union.

The TEU established a form of compromise in the debate, acknowledging the need for further enlargement, but nevertheless producing a longer term set of objectives for further union, in particular for Economic and Monetary Union and a Common Foreign and Security Policy (CFSP). It was also decided to press ahead with further enlargement to include the applicant countries from EFTA, and on 1 January 1995 Austria, Finland and Sweden became members. In Norway the events of 1972 were repeated as its population decided in a close-run referendum not to join in spite of accession having been successfully negotiated. The EU presently has a total of fifteen Member States, with eleven official and working languages.

At the time of writing, the Member States of the EU were embarking on a further round of negotiations for reforming the Treaties. In the new Intergovernmental Conference (IGC) it is proposed to extend policy-making in certain areas and to make the institutions more effective, efficient and transparent, or open, in their activities, in particular with a view to further enlargements of the EU at the start of the next century. Several countries have applied for membership, and more are expected to do so (see Chapter 13). The present structures and procedures in the EU are felt to be too cumbersome and inefficient to cope with a larger membership.

There is, however, increasing scepticism within existing Member States of the EU over the usefulness of further integration, given the economic crisis of the early 1990s and its apparent failure to solve problems such as poverty and unemployment. Ambitious plans such as EMU are coming up against increasing technical and political obstacles, and the failure of the EU to settle the conflict in the former Yugoslavia without US intervention casts grave doubts over the viability of a Common Foreign and Security Policy. Fears are also being voiced over the impact that the accession of poorer applicants would have on the present financial

Plate 1.4 In Ullapool, the district council of Ross and Cromarty, a county of Scotland until it became part of the the Highland region in the 1960s, shows its international link with Poland and welcomes visitors in Polish, French, German and Russian, the latter presumably including visitors from Russian fishing vessels

Plate 1.3 A roadside sign on entering Montpellier, Southern France, from the north, from the heart of the Massif Central. French-speaking (France, Belgium, Switzerland, Canada) and English-speaking (UK, USA, Canada, Australia) visitors are welcomed. The largest number of non-French visitors to the Montpellier area are actually German. The presence of the many North Africans in Languedoc is recognised in the lower part of the sign, which is in Arabic, but any North Africans entering Montpellier from the north would be local immigrants working there, not tourists. Non-EU countries outnumber the only two EU countries welcomed, the UK and Belgium, showing the image projected by Montpellier as a centre of attraction worldwide

and regional balances in the EU, as well as on industry, social services and sectoral policies such as the Common Agricultural Policy. All these issues will be examined in greater depth in other chapters of this book.

EUROPE IN THE 1990s

What is the end product of the frequent changes in Europe in the last five centuries?

While economic and social changes have generally been gradual, changes in the political map have often come about suddenly, as the result of settlements and treaties following conflicts. When the 'music stopped' at the end of the First World War, substantial changes appeared in the political map of Europe, with for example Alsace/Lorraine transferred from Germany to France, and Poland reappearing as a sovereign state after more than a century in oblivion. There were further major changes in the political map of Europe at the end of the Second World War. The EEC originated in 1957 as a combination of six countries, most of which had themselves taken their present forms less than a century and a half earlier. A comparison of the two diagrams in Figure 1.4 shows how the political situation and alignment of countries have grown in complexity from the Cold War period (1960) to the post-Cold War period (1996).

The changes in the political map of Europe over recent centuries have invariably produced a pattern of boundaries that does not exactly match underlying cultural features such as language, while also failing to produce areal units of reasonably comparable size. Most of the present sovereign states of EUR 15 have ethnic minorities (e.g. Basques in Spain and France) or submerged nations (e.g. Scotland), which aspire to greater autonomy or even complete independence, an issue to be discussed in Chapter 12. The Member States of the EU themselves differ greatly in both population and area size, a feature making political and administrative organisation difficult, particularly in view of the practice of giving greater representation in political bodies in relation to population size to the small countries. This procedure for determining the size of electoral divisions is referred to as degressive proportionality.

Before the outstanding features of the growth and organisation of the European Union are described in the following section it is appropriate to show the Union in its European context. When the USSR broke up in 1991, the term 'near abroad' (*blizhneye zarubezhye*) was adopted by Russians to refer to the newly independent former Soviet Socialist Republics. Figure 1.5 shows the location of the 15 Member States of the EU together with what might be referred to in a similar way as the 'near abroad' of the EU. This area includes four types of country, listed in Table 1.3 and discussed below. Table 1.4 is a supplementary table listing the very small countries.

1 The EFTA countries: Iceland, Norway, Switzerland and Liechtenstein (in Table 1.4). These are already closely associated with the EU, and if and when they apply or re-apply for EU membership they would almost certainly be accepted as full members.

2 Countries like Malta and Cyprus, which at the time of writing were already under consideration for future membership of the EU, or countries that aspire to membership, including Poland, the Czech Republic, Slovakia, Hungary and Slovenia, the three Baltic Republics (Estonia, Latvia and Lithuania), and also Romania and Bulgaria. Albania and the countries of former Yugoslavia, apart from Slovenia, could also in due course be considered.

3 Those former Republics of the USSR that are in Europe (rather than Asia) and therefore in theory are eligible for consideration as candidates for EU membership, including Russia and six other former Soviet Socialist Republics (countries 36–42 in Table 1.3).

4 African and Asian countries with a coastline on the Mediterranean, in the 'wrong' continents, but closely involved with the EU in economic transactions and sharing environmental problems of an increasingly polluted sea. Only Turkey has a territorial foothold inside the traditional limit of Europe.

In Chapter 13 the possible future enlargement of the EU is discussed at greater length, but in the view of the authors many of the

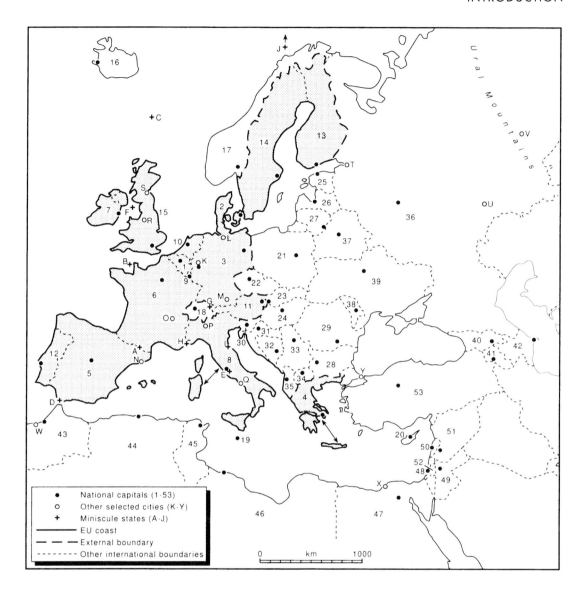

Figure 1.5 The countries of the EU and its 'near abroad' in the mid-1990s. Refer to Table 1.3 for the key to the numbering of the countries 1–53 and to Table 1.4 for countries A–J (the minuscule states)

Forty-eight of the countries took part in the EURO 96 Football Championship, including four from the UK (15) but excluding (32) and (33) of the former Yugoslavia and (43)–(47), North Africa and (49)–(52), Southwest Asia. Three of the minuscule states participated. The cities that are not national capitals are:

K	Cologne	P	Milan	U	Volgograd
L	Hamburg	Q	Naples	V	Yekaterinburg
M	Munich	R	Manchester	W	Casablanca
N	Barcelona	S	Edinburgh	X	Alexandria
O	Lyon	T	St Petersburg	Y	Istanbul

Table 1.3 Who's who in the EU and its 'near abroad' and key to the numbering of countries in Figure 1.5

Country	Capital	(1) Area in 000s sq km	(2) Population in millions mid-1995	(3) mid-1996	(4) expected 2025	(5) Per cent urban	(6) Real GDP per capita in 000s ppp$ 1992
European Union of 1995							
1 Belgium	Brussels	31	10.2	10.2	10.5	97	18.6
2 Denmark	Copenhagen	43	5.2	5.2	5.3	85	19.1
3 Germany	Bonn/Berlin	356	81.7	81.7	76.1	85	21.1
4 Greece	Athens	132	10.5	10.5	10.0	63	8.3
5 Spain	Madrid	505	39.1	39.3	37.1	64	13.4
6 France	Paris	574	58.1	58.4	63.6	74	19.5
7 Ireland	Dublin	70	3.6	3.6	3.5	57	12.8
8 Italy	Rome	301	57.7	57.3	52.8	68	18.1
9 Luxembourg	Luxembourg	3	0.4	0.4	0.4	86	21.5
10 Netherlands	Amsterdam	34	15.5	15.5	17.6	89	17.8
11 Austria	Vienna	84	8.1	8.1	8.3	54	18.7
12 Portugal	Lisbon	92	9.9	9.9	9.8	34	9.9
13 Finland	Helsinki	337	5.1	5.1	5.2	64	16.3
14 Sweden	Stockholm	450	8.9	8.8	9.6	83	18.3
15 UK	London	244	58.6	58.8	62.1	92	17.2
EFTA							
16 Iceland	Reykjavik	103	0.3	0.3	0.3	91	17.7
17 Norway	Oslo	324	4.3	4.4	5.0	73	18.6
18 Switzerland	Bern	41	7.0	7.1	7.5	68	22.6
Central Europe							
19 Malta	Valetta	0.3	0.4	0.4	0.4	85	8.3
20 Cyprus	Nicosia	9	0.7	0.7	0.9	68	15.1
21 Poland	Warsaw	31.3	38.6	38.6	41.7	62	4.8
22 Czech Republic	Prague	79	10.4	10.3	10.7	75	7.7
23 Slovakia	Bratislava	49	5.4	5.4	6.0	57	6.7
24 Hungary	Budapest	93	10.2	10.2	9.3	63	6.6
25 Estonia	Tallinn	45	1.5	1.5	1.4	71	6.7
26 Latvia	Riga	65	2.5	2.5	2.4	69	6.1
27 Lithuania	Vilnius	65	3.7	3.7	3.9	68	3.7
28 Bulgaria	Sofia	111	8.5	8.4	7.5	67	4.3
29 Romania	Bucharest	238	22.7	22.6	21.6	55	2.8
30 Slovenia	Ljubljana	20	2.0	2.0	1.9	50	n.a.

Rest of former Yugoslavia and Albania							
31	Croatia	Zagreb	57	4.5	4.4	4.2	54	n.a.
32	Bosnia-Herzegovina	Sarajevo	51	3.5	3.6	4.5	34	n.a.
33	Yugoslavia	Belgrade	102	10.8	10.2	11.5	47	n.a.
34	Macedonia	Skopje	26	2.1	2.1	2.5	58	n.a.
35	Albania	Tirane	29	3.5	3.3	4.7	37	3.5
Former USSR							
36	Russian Federation	Moscow	17,075	147.5	147.7	153.1	73	6.1
37	Belarus	Minsk	208	10.3	10.3	11.3	68	6.4
38	Moldova (Rep. of)	Chisinau	34	4.3	4.3	5.1	47	3.7
39	Ukraine	Kiev	604	52.0	51.1	54.0	68	5.0
40	Georgia	Tbilisi	70	5.4	5.4	6.0	56	2.3
41	Armenia	Yerevan	30	3.7	3.8	4.3	68	2.4
42	Azerbaijan	Baku	87	7.3	7.6	10.3	54	2.6
Other Mediterranean basin							
43	Morocco	Rabat	447	29.2	27.6	47.4	47	3.4
44	Algeria	Algiers	2,382	28.4	29.0	47.2	50	4.9
45	Tunisia	Tunis	164	8.9	9.2	13.3	60	5.2
46	Libya	Tripoli	1,760	5.2	5.4	14.4	85	9.8
47	Egypt	Cairo	1,001	61.9	63.7	97.9	44	3.5
48	Israel	Tel Aviv	21	5.5	5.8	8.0	90	14.7
49	Jordan[1]	Amman	98	4.1	4.2	8.3	68	4.3
50	Lebanon	Beirut	10	3.7	3.8	6.1	86	2.5
51	Syria	Damascus	185	14.7	15.6	33.5	51	5.0
52	Gaza and West Bank[2]		6	2.4	2.6	6.6	94	n.a.
53	Turkey	Ankara	779	61.4	63.9	95.6	51	5.2

Sources: (1) Area, *UNSYB 1990/91*, Table 11; (2)–(4) *WPDS 1995 and 1996* (PRB 1995 and 1996); (5) *HDR 1995* (UNDP 1995, Table 1)
Notes: 1 Jordan does not have a coast on the Mediterranean; 2 Israeli occupation ceased in 1995 in Gaza and parts of the West Bank
n.a. not available

Table 1.4 Supplementary list of the smallest countries of Europe

	Area in sq km	Population in thousands			Area in sq km	Population in thousands
A	Andorra	453	52	F	Isle of Man	588	64
B	Channel Islands	195	139	G	Liechtenstein	160	29
C	Faeroe Islands	1,399	48	H	Monaco	1	29
D	Gibraltar*	6	31	I	San Marino	61	24
E	Holy See (Vatican)	–	1	J	Svalbard**	62,422	–

Sources: as for Table 1.3 columns (1), (2)
Notes: * British colony
       ** and Jan Mayen Islands, inhabited only during summer period

topics covered in Chapters 2–12 can be more fully appreciated if the reader is aware of what constitutes the rest of Europe. Tables 1.3 and 1.4 are a checklist for the location map in Figure 1.5 of the independent countries of Europe, down to the smallest, as recognised by the United Nations in its Statistical Yearbooks of the early to mid-1990s. All the countries in the list with more than 250,000 inhabitants are numbered in Table 1.3, while the ten smallest, listed in Table 1.4, are each allocated a letter. The reader may care to look first at the map in Figure 1.5 and to try to identify each country and its capital. Learning where places are is no longer an end in itself in geography, but it is helpful at least to know where the EU Member States are in relation to one another, and where the other principal countries of Europe are situated.

Some prominent features revealed by the variables emerge from the data in Table 1.3:

(1) The enormous difference in area size between Russia and the rest. Over two-thirds of the area of Russia is in Asia.

(2)–(4) The great differences in population size. Slow growth of population and even decline in some cases are expected in all the countries of Europe in the next three decades. In contrast, in all the non-European countries (43–53) rapid population growth is expected, with more than a doubling of population in some before zero growth (or a decline) is experienced.

(5) The great differences in the level of urbanisation.

(6) Almost without exception, the EU and EFTA countries (1–18) have much higher GDP per capita than the former CMEA countries and the non-European ones.

GEOGRAPHICAL AND ECONOMIC FEATURES OF THE EU

In spite of the great impact made by Western European countries on the rest of the world in the last five centuries, the area of the European Union (EUR 15) is very small. Its area of 3,242,000 sq km, is a mere 2.4 per cent of the total land area of the world of almost 136 million sq km (excluding Antarctica and Greenland). Its mid-1995 population of 372 million inhabitants was about 6.5 per cent of the world total of 5,700 million in that year. In the mid-1990s the total Gross Domestic Product (GDP) of the EU was about 6,000 billion US dollars, 20 per cent of the world total of 30,000 billion.

Relative to the total population and GDP of the world, the EUR 15 countries have been in decline. Fifty years ago the same countries had over 10 per cent of the total population of the world. Their 2025 population is expected to be roughly the same size as now, but by then they would have only 4–5 per cent of the total population of the world. Their share of total world industrial output and of international trade has also diminished.

With regard to natural resources, in relation to the rest of the world the EU is comparatively poorly endowed. Details will be given in later chapters, but it may be noted here that it has less than 1 per cent of the world's proved oil reserves, about 2 per cent of the proved natural gas reserves, and about 6 per cent of the commercial coal reserves. It has a small or negligible share of almost all major non-fuel minerals. The extent and quality of agricultural land in the EU is more favourable, but agricultural production has risen in recent decades almost entirely through an increase in yields rather than through the growth of the area cultivated. As a result of its high level of industrialisation, the EU has been using up many of its non-renewable natural resources quickly, and it now depends on other regions of the world for about half of its energy requirements, over 80 per cent of the non-fuel minerals it consumes, and also some agricultural products, mainly tropical foods, beverages and raw materials.

The global technological and military superiority of Europe since the end of the fifteenth century was based initially on inventiveness, organisation and indigenous natural resources. Subsequently some other regions of the world have caught up with Europe, while some have natural resources in far greater abundance in relation to their population size. In Mackinder's post-Columbian epoch, Western Europeans are having to work hard to retain their position, and the integration and expansion of the EU is expected to give them greater strength to do so.

As the EU has expanded, two features of geographical relevance have been changing. Enlargement of the territory has resulted not only in greater areal extent but also in greater fragmentation and in a reduction in compactness. The influence of the size, shape and spatial layout of the EU on transport and communications is introduced in the next section and will be discussed further in Chapter 4. Again, in the late 1950s regional disparities existed in the EU, mainly on account of the backwardness of southern Italy compared with the rest of the EEC of the time. With the accession of Greece (1982), Spain and Portugal (1987), and the addition of the German Democratic Republic (GDR, 1990) to Federal Germany, all far below the average EU economic level around 1980, regional disparities have increased. If the EU absorbs some or all of the Central European countries (see Table 1.3), the gap between rich and poor Member States will increase even further. Figure 13.1 (in Chapter 13) gives an idea of the economic abyss to the east of Germany and Austria.

The economic performance of various Member States of the EU and of regional subdivisions within them depends on two different factors of geographical interest, first location within the territory of the Union, and second the availability, quantity and quality of various attributes of each country or region. Location, centrality and peripherality will be discussed in Chapter 3 following the study of the distribution of population in the EU. The prospects and

performance of a country or region are also closely related to its assets and drawbacks, for example heavy dependence on agriculture, declining industry, oil and gas reserves, an attractive climate or landscape. No two regions are identical, although many subsets have features in common. Disparities in economic development are of particular concern to the European Union, some of whose endeavours and resources have been directed to assisting lagging regions.

The attributes of the various regions of the EU will be covered in various chapters, while in Chapters 11 and 12 the problems of regional disparities, cohesion and convergence will be addressed in detail. Figure 1.6 illustrates the subject of cohesion with the example of one sector of the economy, healthcare, exemplified by level of infant mortality, in which marked convergence has taken place in the EU at

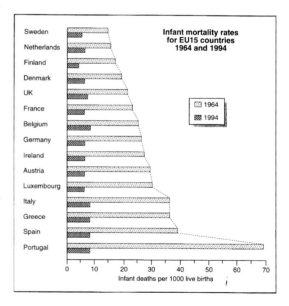

Figure 1.6 Convergence among EUR 15 countries, the example of infant mortality. Note that only six of the 15 countries were in the EEC in 1964. In many other respects, convergence has been much slower and less marked (see Chapter 12 for discussion)

national level over a short period. In many other sectors, however, there has been little change and in some cases even divergence rather than convergence.

THE SHAPE OF THE EU

Western Europe is one of the most irregular and fragmented parts of the earth's land surface. It is not possible to quantify objectively and precisely the effect of the lack of compactness of Western Europe and the resulting extra burden on the provision of an efficient transport network for the EU and its neighbours. In order to give an approximate appraisal, a simple measure of compactness has been applied to each stage of the territorial extent of the EU as it has grown over the decades. The index of compactness used is a simple calculation, arrived at by comparing the area of a given piece of territory with the area of the smallest circle that contains it. The resulting number is multiplied by 100 to produce a score with maximum possible compactness of 100.

As the EU has grown, the addition of new Member States has changed the location of regions relative to the EU as a whole. For example, the relatively peripheral location of Denmark in EUR 12 has been reduced by the accession of Sweden and Finland to its northeast. Similarly, before the EEC came into existence, Strasbourg and its region of Alsace were peripheral in relation to France. When the EEC was formed the same region became central in the context of the whole Community. The accession of Austria, Finland and Sweden to the European Union in 1995 has considerably enlarged its area, although there has been only a small addition to total population and GDP. The presence of Sweden and Finland has increased the already elongated, unwieldy shape of the territory of the EU.

The EU is much smaller in area than any of the six largest countries of the world, Russia,

Canada, China, the USA, Brazil and Australia. Nevertheless, it is large enough for its awkward spatial layout to produce marked differences between central and peripheral areas in terms of accessibility to the whole of the territory. The economic performance of the various regions of the EU can to some extent be accounted for by their locations in relation to parts of or the whole of the EU, a problem acknowledged widely in EU reports, although not quantified with great precision.

In order to illustrate the impact of territorial shape on countries and groups of countries, Figure 1.7 shows the comparative sizes and shapes of EUR 15 and the USA. The greatest distance across contiguous USA (i.e. without including Alaska and Hawaii) is about 4,600 km, between Eastport, Maine, in the east and San Francisco in the west. By coincidence, the greatest distance across EUR 15 is also about 4,600 km, between Lapland in the north of Finland and the Algarve in the south of Portugal. The USA is, however, much larger in area than the EU and therefore more compact. Table 1.5 shows the compactness index of the EU at different stages of its enlargement from the original EUR 6 (1 in Figure 1.8) to EUR 15, with the inclusion of Austria, Finland and Sweden (6 in Figure 1.8). All or part of the five circles forming the basis for the calculation of compactness indices are shown in Figure 1.8. The addition of East Germany (5 on the map) did not increase the longest axis across the EU of the time and its presence therefore slightly increased the index of compactness. The changes in the index of compactness as the EU was enlarged can be seen in the last column of Table 1.5. The indices for three selected single countries are shown for comparison.

The longest axis across the original six EEC countries was between Bretagne (northwest France) and Puglia (the 'heel' of Italy) and the compactness index was 31. The addition of the UK, Ireland and Denmark produced a new longest axis, stretching between northwest

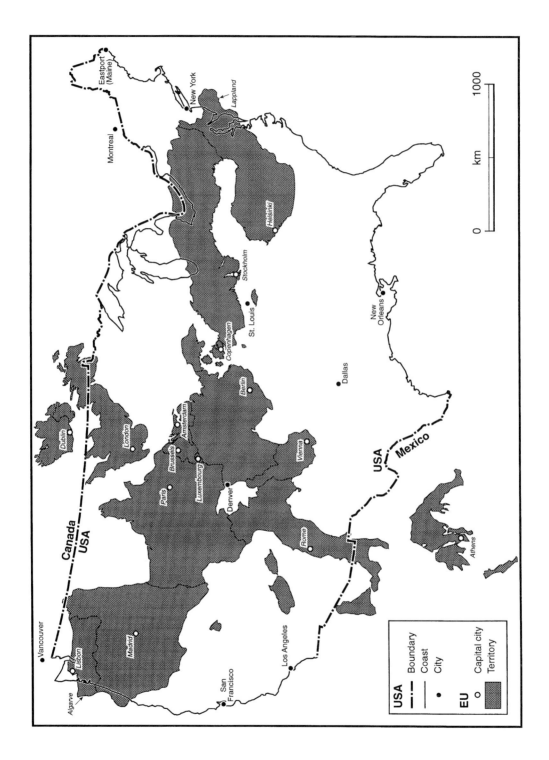

Figure 1.7 Contiguous USA (i.e. excluding Alaska and Hawaii) compared with EUR 15 superimposed and represented on the same scale

Table 1.5 Indices of compactness of the EU and of selected countries

		Area 000 sq km	Smallest circle	Largest axis in km	Radius in km	Compactness index[1]
1	EUR 6	1,175	3,800	2,200	1,100	31
2	EUR 9	1,533	6,160	2,800	1,400	25
3	EUR 10	1,665	9,080	3,400	1,700	18
4	EUR 12	2,260	10,180	3,600	1,800	22
5	EUR 12 + GDR	2,370	10,180	3,600	1,800	23
6	EUR 15	3,240	16,620	4,600	2,300	19
	France	535[2]	950	1,100	550	56
	Norway	324	2,540	1,800	900	13
	USA	7,826	16,620	4,600	2,300	47

Notes: 1 Area of EU or individual country as a percentage of the area of the circle of smallest radius that encompasses the country
2 Excludes Corse (Corsica)

Scotland and the south of Sicilia, reducing compactness considerably. The addition of 'far flung' Greece further lowered the index of compactness, the longest axis extending between northwest Scotland and Kriti (Crete). However, the subsequent inclusion of Spain, Portugal and later East Germany raised the index of compactness. With the accession of Finland and Sweden, projecting far to the northeast of Denmark, compactness was again reduced. For future reference it may be noted that the addition of Switzerland, Norway and any of the countries of Central Europe would increase the compactness index of a further expanded EU, since all these countries would fall within circle 6, increasing the territorial area without requiring a larger circle.

Two matters regarding compactness will now be addressed. First, how compact by comparison are other regions and countries of the world? Second, more importantly, what reservations and qualifications need to be taken into account before an appreciation of the impact of shape and compactness (or lack of it) on EU situations and problems can be of practical use? The comparative aspect will be discussed first.

France is one of the most compact countries in Western Europe, commonly known in French as *l'hexagone*, Norway one of the least

compact. Like France (without Corse – Corsica), contiguous USA is comparatively compact but, like Norway, Chile (South America) is greatly elongated, an atlas maker's nightmare, with an index of compactness of a mere 5 per cent. The inclusion of Corse with mainland France would extend the longest axis without adding much area, thereby lowering the index of compactness considerably. This case is a reminder that, especially with fragmented regions and countries, the inclusion of islands and other separate pieces of territory usually has the effect of lowering an index of compactness appreciably. Thus if Spain's two provinces in the Canarias region in the Atlantic, well to the southwest of mainland Spain, are included in the area of EUR 15, the longest axis of EUR 15 would rise from about 4,600 km to about 6,000 km, lowering the overall compactness index from 19 to 11, since the area of the Canarias is very small. It should be appreciated, also, that different shapes can have similar indices of compactness.

Some aspects of the practical implications of shape and compactness will now be considered. If every small region in the EU was entirely or largely self-sufficient economically, and if its citizens had no reason to travel to other areas, location within the Union would have little influence on the fortunes (or lack of fortune)

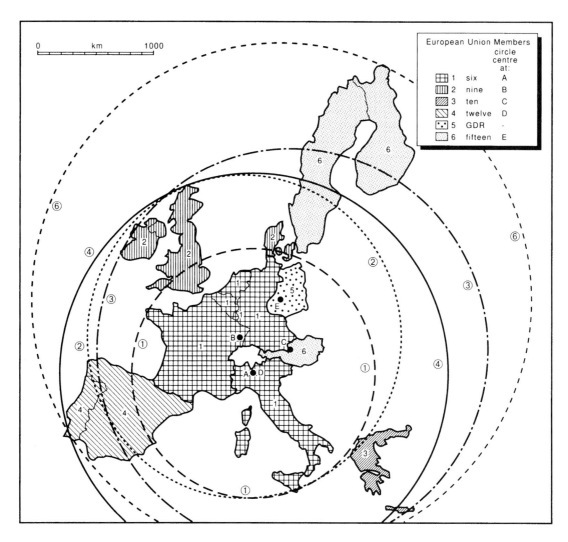

Figure 1.8 The compactness of the EU at various stages in its enlargement. The index used here (see text and Table 1.5) compares the areas of each group of members with the area of the smallest circle that contains it. The circle is the most compact two dimensional form. The centre of each circle is shown on the map by a black dot

of each region. Differences in the economic performance of regions would largely depend on the local natural resources and the endeavours of their respective populations. In practice, every region in the EU depends on other parts of the Union and also on the rest of the world for most of its requirements of food, energy and manufactured goods, and some of its services. Other things being equal, the shorter the distances between a given region and all other relevant regions, especially those with which transactions are greatest, the greater the level of accessibility and the cheaper the transport costs incurred by the transactions. In general, the smaller *and* more compact a given country or group of countries, the lower the accessibility index measured, for example, by total journeys to other relevant places. The subject of accessibility in the EU will be discussed in greater detail in Chapter 4.

THE REGIONAL STRUCTURE OF THE EU

In order to assess the regional situation at the subnational level the Member States of the EU are subdivided into regions at three levels. The regions will be referred to throughout the book and it is therefore appropriate to outline here the structure of the systems used. Before they joined the EU, each of the present Member States had its own system of major and minor civil or administrative divisions, for example the Départements of France and the Länder of West Germany. In some countries these were amalgamated to form *ad hoc* planning regions and special regions for various economic purposes. These systems have remained in existence and are used for internal purposes. The EU has developed in a piecemeal fashion and therefore no attempt has been made to produce a com-

pletely new set of regions, devised with some useful trans-EU criterion in mind, such as uniform area or population size, or economic homogeneity. While, for example, it could be helpful in a few places to override international boundaries and have regions that straddle them (e.g. Northern Ireland (UK) and the Republic of Ireland, Nord-Pas-de-Calais (France) and Hainaut (Belgium)), such a procedure has not been applied. With the establishment of the EU a system of regions and subregions has been created for the use of the supranational bodies of the Community. Statistical data are collected and analysed at subnational level, and decisions regarding the allocation of funds from the EU budget may be influenced by the regional structure.

According to COM 90–609 (1991, Annex 0.1: 5):

> The Nomenclature of Territorial Units for Statistics (NUTS) was established by the Statistical Office of the European Communities, in cooperation with the Commission's other departments, to provide a single uniform breakdown of territorial units for the production of Community regional statistics.

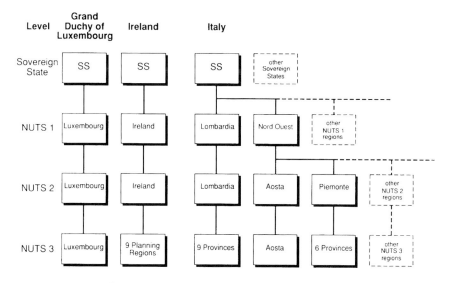

Figure 1.9 The NUTS levels of the regional hierarchy of the EU illustrated by Luxembourg, Ireland and Italy

Three NUTS levels are in use, referred to as levels 1, 2 and 3. Figure 1.9 shows a small part of the hierarchy of NUTS regions. The latest available definitive list of NUTS levels 1 and 2 regions of EUR 15, *Regional Profiles* (1995), has 77 regions at NUTS level 1 and 204 at NUTS level 2. The NUTS level 1 regions are used in Tables A1 in the Appendix and the location of the regions is shown in Figure 1.10.

A confusing feature of the regional system in use at present is that some Member States are not subdivided at NUTS level 1. These are Denmark, Ireland, Luxembourg and Sweden. In effect, continental (mainland) Portugal is not subdivided either, but its two 'overseas' regions are distinguished at NUTS level 1, while mainland Finland is also kept as a single unit but the Ahvenanmaa (Åland Islands) are counted separately, in spite of having only about 25,000 inhabitants. At NUTS level 2, Ireland and Luxembourg remain undivided while in some EU data sets Denmark is not subdivided, although in others it is broken down into three regions. A further complication is the fact that some of the extra-European territories of France, Spain and Portugal are included in the EU. These are described either as 'non-continental' or as 'overseas'.

- France has four overseas territories (Départements d'Outre-Mer): Guadeloupe and Martinique in the Caribbean, Guyane on the mainland of South America, and Réunion in the Indian Ocean.
- Spain has three non-continental units, Islas Baleares in the Mediterranean, Las Palmas and Santa Cruz de Tenerife in the Atlantic, each equal in status to the provinces of the mainland, plus two small territories, Ceuta and Melilla, on the coast of Morocco.
- Portugal has two non-continental units in the Atlantic, the Açores and Madeira.
- Finally, it must be noted that the Isle of Man and the Channel Islands are not in the EU.

In the present book most of the variables mapped at subnational level are broken down only to NUTS level 1 but reference is also made frequently to information about NUTS level 2 units. A minor confusion at NUTS level 2 is the use of the name Luxembourg for both the sovereign Grand Duchy (GD) and a subdivision of Belgium. Similarly, Limburg is the name of subdivisions in both Belgium and the Netherlands.

Some weaknesses of the regional divisions of the EU will now be noted. The many new states of the USA and their county subdivisions that came into existence after the Declaration of Independence were mostly formed in a virtual vacuum, with the US Land Survey having considerable power in shaping them on territory purchased or conquered from France, Russia (Alaska), Spain or the Indians. The Statistical Office of the European Communities (hereafter Eurostat) has been faced with a completely different situation. Each Member State took with it into the Community a complex, long-established set of major and minor civil divisions. One of the advantages of the continued use of existing regions over the creation of a completely new system is that data sets for past periods can be related to current data much more easily and precisely. On the other hand, great disparities in both area and population size are found at each NUTS level, both within and between countries.

At all levels in the EU hierarchy of units, a number of considerations can be expected to carry weight when a system of regions is being created. To avoid friction, protest and possibly even conflict, existing divisions should be used as far as convenient. For example, national groups such as the Welsh and the Basques have been recognised implicitly in the EU system. Administrative units at any level and of any size should, if possible, be reasonably compact rather than elongated or fragmented. Several regions within the EU are unavoidably fragmented because they consist of groups of islands. At any level in the hierarchy, units

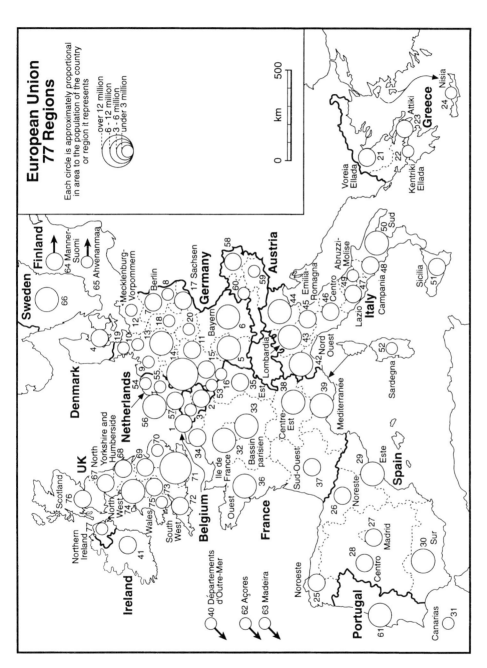

Figure 1.10 The NUTS level 1 system of regions as presented in *Regional Profiles* 1995. The numbering of the regions corresponds to that used in Table A1 in the Appendix, which serves as a key for the regions not named on the map. This map has been used extensively in the book as the base for showing distributions of economic and social features. The circles, roughly proportional to population, have been shaded, rather than the actual areas, since some regions that are very small territorially (e.g. Hamburg) do not show if area is shaded

should for convenience be broadly comparable in area, population and total GDP. These latter criteria are in practice mutually exclusive, except in a hypothetical area with a uniform distribution of population and an identical level of economic development throughout. Perhaps the most important consideration is comparability of population size. In this respect the EU system of regions is far from satisfactory, as shown with examples below.

Sovereign state level

There is a great imbalance in population size. Germany has almost 82 million inhabitants, Luxembourg fewer than 400,000. Blake (1991) alludes to the inconvenience of allowing Luxembourg full status in the EU: a 'luxury' of having a NUTS level 3 region at the top level, likely to be repeated if Malta, Cyprus and Iceland join the Union.

NUTS Level 1

In Germany there are great disparities in population size between the 11 Länder of former West Germany, less between the Länder of former East Germany. The principle applied to the regionalisation of Germany was the retention of the federal structure in the new EU system. In the UK, Scotland, Wales and Northern Ireland are distinct cultural entities, but the eight standard regions of England are no more than convenient groups of counties. Since the Second World War, France, Italy and Spain have reduced the high degree of centralism in their political and administrative organisation and have recognised the aspirations of some of their regions to the achievement of regional autonomy. To some extent, then, the NUTS level 1 regions of the EU take into account traditional features largely overlooked or suppressed in some countries, when each was dominated by a strong centralised administration. In the view of Le Bras and Todd (1981), the appearance of

uniformity was largely illusory in the case of France. In Chapter 12 it will be argued that the development of the supranational EU could actually give new life to 'subnations' in Western Europe, just as the break-up of Yugoslavia and the USSR has done in those countries, but under different circumstances and, it is to be hoped, with less drastic consequences.

NUTS Level 2

At this level, some NUTS level 1 units, such as Lombardia in Italy, Ile de France in France and Hamburg in Germany, are not subdivided. Ireland, Denmark and Luxembourg are still not subdivided. The UK is divided into groups of counties (e.g. Derbyshire and Nottinghamshire) or single counties (e.g. Lincolnshire).

NUTS Level 3

All EU countries except Luxembourg are subdivided at this level, but reference will not be made extensively to them in the book. France (départements), Spain and Italy (provinces) and the UK (counties/regions) are represented by the basic subdivisions with 50–100 units, but the former FRG is much more finely subdivided, with over 300 Kreise, to which may be added over 200 Kreise from the former GDR.

The regional system of the EU has considerable limitations. In any study of data based upon it, variations in size within each level, and relationships between levels, must be looked at carefully. Whatever the region, it should be remembered that its population is an aggregate of individuals, each with a particular type of occupation, level of income and aspirations. Any process of aggregation, whether spatially based on regions, or class based, loses details of individuals and produces averages that cannot do full justice to the complex situation on the ground.

THE AIMS, STRUCTURE AND CONTENTS OF THE BOOK

Since this book is about the geography of the European Union, attention is focused on aspects that vary spatially. Two distinct features of places of various sizes, from a single settlement to a whole country, are their location in relation to other places and their indigenous attributes.

The locational advantages and disadvantages of a place may be influenced by its position in relation to the centre and periphery of the EU. In Figure 3.2 (Chapter 3), circles with a radius of 500 km are centred on four cities in the EU. Over 40 per cent of the total population of the EU is contained within a circle of that radius if it is centred, for example, on Strasbourg, but only a few per cent if it is centred, for example, on Lisbon or Athens.

The attributes of a region include features of climate, relief and soil, and the availability of commercial minerals, as well as features of the population, its educational level and skills, and its activities. The EU contains some relatively backward agricultural regions, other regions that depend heavily on industry, whether long established or newly developed, as well as some of the largest and most sophisticated service-based urban complexes in the world. One of the general aims of the EU is to work towards 'cohesion'. That does not mean that conditions in all regions should be exactly homogeneous, but it does mean that the standard of living, employment opportunities and the quality of life should be broadly similar throughout the Union. In the present book it will be shown that there was still great diversity in the EU in the mid-1990s.

To appreciate spatial distributions and patterns (or lack of patterns), it is, however, necessary also to take into account how the EU and the various parts of it are organised and managed at different levels. There is no objective way of measuring the relative weight of influence and decision-making between the supranational level of the EU, the level of the 15 Member States, and regional and local levels. A rough guide might be the proportion of total public spending accounted for by each level. On this criterion, the EU level accounts for only a few per cent, the national level for around 80 per cent, and the local government level for 10–20 per cent according to country. Although the power of the EU to influence Member States is still limited in most respects, it is prominent in agriculture and fishing, in the coal, iron and steel industries, and in regulating trade between EU Member States and third countries. Again, some types of pollution affect groups of EU Member States, while increasingly trans-European transport links are coming under EU competence.

While most aspects of life in the EU remain in the hands of national governments, the application of the principle of subsidiarity means that many regional and local services, issues and problems are left to local decision-makers. At the other extreme, the EU is not isolated from the rest of the world, but is particularly influenced in its energy policy, in the supply of raw materials and in its policy towards global pollution by events and changes in non-EU countries. Investments from non-EU countries also play a major role in the economic development of some EU regions. Temporary employment or even residence in the EU is the aim of increasing numbers of people from Central and Eastern Europe and from developing countries.

The view of the authors is that the most suitable framework for the study of the geography of the EU is a topic by topic approach rather than a country by country approach. The topics covered will be briefly outlined and justified below. The tables and figures are a vital part of the book and are either for general reference or to illustrate and substantiate particular points in the text. The photographs and boxes are intended to show what conditions are like 'on the ground' and to provide insights into issues and features of the EU that are particularly topical. Examples and case studies are drawn from across the whole of the EU.

Chapter 2 The organisation of the European Union

Prominent features are a gradual transfer of powers to supranational bodies from national governments and a growth in membership from six in 1958 to the present total of 15. The EU is also extending its sphere of influence in economic, political and social matters beyond the original policy areas.

Chapter 3 Population

Little change is expected in the total population of the EU in the next few decades. A gradual increase in average age, moderate internal migration and a possible sharp increase in the rate of migration from extra-EU countries are issues with implications for EU demographic policy in the next few decades.

Chapter 4 Transport and communications

Attention is drawn to the implications of the difficult layout of the EUR 15 countries for the management and improvement of transport links. A fundamental distinction is drawn between the networks and the traffic that uses them. The functions and relative importance of different modes of transport are discussed and the policy of reviving the use of rail transport is critically appraised.

Chapter 5 Energy and water supply

In spite of the development of natural gas and oil reserves, the EU still imports about half of its energy needs. Coal output has declined dramatically but there has been a great increase in the use of nuclear power since the 1960s. The time when renewable, clean forms of energy will replace the present predominance of fossil fuels appears far off.

Chapter 6 Agriculture, forestry and fisheries

There has been little change since the Second World War in the arable area in use, but great increases in yields of crops and livestock. The number of people employed in agriculture has decreased sharply, while mechanisation has increased. A large part of the EU budget still goes to supporting EU farmers through the Common Agricultural Policy.

Chapter 7 Industry

Since the 1970s there has been a gradual decline in the number of people employed in this sector, with a sharp decline in some sectors, but a rise in others. Productivity per worker has increased. Traditional industries in decline include coal, steel and most textiles. Some sectors of engineering have also declined. Investment from extra-EU countries has been beneficial in some EU regions.

Chapter 8 Services

Until the 1990s almost all types of services experienced an increase in the number employed but in recent years there have been considerable job losses in some sectors such as banking and retailing. The traditional public service concept is being re-appraised as healthcare, educational and social security funding are cut.

Chapter 9 The social environment

In spite of the relative affluence of the EU countries, poverty is found among some sections of the population and in some regions. Issues such as the status of women and the elderly are also reviewed.

Chapter 10 The environment

This area has come under scrutiny since the 1970s, with problems of pollution and conservation causing increasing concern. There are major differences between levels of environmental protection in the EU Member States, yet it is clearly a transnational issue.

Chapters 11 and 12 Regional policies and narrowing the gaps in the European Union

Regional differences in the EU at NUTS level 1 are illustrated by disparities in GDP per capita and in levels of unemployment. Causes of regional disparities are discussed and measures to reduce them through the use of EU budget resources are described.

Chapter 13 Enlargement of the European Union

In this chapter the EU is placed in the context of the whole of Europe, and the eligibility and suitability of non-EU countries for future membership of the EU as it enlarges are assessed. Apart from the remaining four EFTA countries, any new members would be much poorer than the present EU average.

Chapter 14 The European Union and the rest of the world

There is a tendency for other blocs to form in the world at least for purposes of trading and economic cooperation if not for the closer kind of union towards which the EU has been moving. The EU is considered in relation both to its immediate 'rivals', North America and Japan, and to various parts of the developing world, including the ACP countries, to the development of which the EU has a special commitment.

FURTHER READING

Clout, H. and Blacksell, M. (1994) *Western Europe: Geographical Perspectives,* Chichester: Wiley. Approach by topics. Post-1989 changes in Central Europe taken into account.

Dawson, A. H. (1993) *A Geography of European Integration*, Chichester: Wiley. A human geography of Europe in the post-Cold War era.

Drake, G. (1994) *Issues in the New Europe*, London: Hodder and Stoughton. Unusual in bringing together Western and eastern Europe. Political, economic, social and environmental issues are covered.

Eurostat (1989) *Europe in Figures*, deadline 1992 (1989/90 edition), Brussels-Luxembourg. London: HMSO ISBN 0–11–972300–X.

Eurostat (1995) *Europe in Figures*, 4th edition, Luxembourg: Office for Official Publications of the European Communities.

Foucher, M. (1993) *Fragments d'Europe*, Lyon: Fayard. Atlas of middle and eastern Europe with useful maps of Europe as a whole.

Masser, I., Sviden, O. and Wegener, M. (1992) *The Geography of Europe's Futures*, London and New York: Belhaven Press. Speculation about future changes in Western Europe, with heavy emphasis on transport and communications.

Miall, H. (1993) *Shaping the New Europe*, London: Pinter.

Murphy, A. B. (1991) 'The emerging Europe of the 1990s', *The Geographical Review*, 81(1), January: 1–17.

Pinder, D. (ed.) (1990) *Western Europe – Challenge and Change*, London: Belhaven Press.

Vujakovic, P. (1992) 'Mapping Europe's myths', *Geographical*, LXIV(9), September: 15–17. The latest changes in the map of Europe.

Williams, A. M. (1994) *The European Community*. Oxford: Blackwell. Account of the development, aims, operations, policies and actions of the EC from its foundation to the 1990s.

2

THE ORGANISATION OF THE EUROPEAN UNION

'The European Union is yesterday's answer to the day before's problem'
(A country called Europe, *Panorama*, BBC1, 15 April 1996)

The purpose of this chapter is to give a brief overview of the structure, objectives, policies and institutions of the European Union, following on from the historical and chronological presentation given in Chapter 1. In view of the legislative and administrative role the EU now plays, with a considerable amount of sovereignty transferred from the Member States to its institutions, together with the wide scope of activities it undertakes in different sectors of the economy and society, its unique nature means that a detailed analysis of all that it does is impossible in one chapter. Individual sectors and policies are therefore covered in separate chapters in the book. This chapter contains a description of the structure and mechanisms that govern the actions of the EU, together with an introduction to the key institutions and policies and the budget through which they are financed. Because of its unique status as a directly-elected multinational assembly, and its particular geopolitical interest, the European Parliament is covered in a separate section of the chapter. Current and future issues of importance are also referred to, with a particular focus on the internal aspects, since enlargement and external relations are

dealt with in more depth in Chapters 13 and 14.

THE OBJECTIVES OF THE EU

The basic objectives of the EU originate in the founding Treaties of Paris and Rome. In their statement of principles and tasks for the EC, the ECSC, Euratom and the EEC Treaties, of which the last is the most comprehensive, all share the same basic aims:

> The Community shall have as its task, by establishing a common market and progressively approximating the economic policies of Member States, to promote throughout the Community a harmonious development of economic activities, a continuous and balanced expansion, an increase in stability, an accelerated raising of the standard of living and closer relations between the States belonging to it.
>
> (Art. 2 EEC)

In order to meet these basic objectives, the EEC Treaty in its next article lists the following priority activities:

a) the elimination, as between Member States, of customs duties and of quantitative restrictions on the import and export of goods, and of all other measures having equivalent effect;

b) the establishment of a common external customs tariff and of a common commercial policy towards third countries;

c) the abolition, as between Member States, of obstacles of freedom of movement for persons, services and capital;

d) the adoption of a common policy in the sphere of agriculture;

e) the adoption of a common policy in the sphere of transport;

f) the institution of a system ensuring that competition in the common market is not distorted;

g) the application of procedures by which the economic policies of the Member States can be coordinated and disequilibria in their balances of payments remedied;

h) the approximation of the laws of Member States to the extent required for the proper functioning of the common market;

i) the creation of a European Social Fund in order to improve employment opportunities for workers and to contribute to the raising of their standard of living;

j) the establishment of a European Investment Bank to facilitate the economic expansion of the Community by opening up fresh resources;

k) the association of the overseas countries and territories in order to increase trade and to promote jointly economic and social development.

(Art. 3 EEC)

As far as the accomplishment of these objectives is concerned, success varied enormously during the first three decades. The main objective of the creation of a 'common market' saw the original deadline of 12 years (Art. 8 EEC) pass by, an unrealistic goal due to continued protection of national interests and cumbersome decision-making procedures. In spite of the fact that the EC was consistently referred to as the Common Market after its establishment, it was only following the Delors White Paper of 1985 and the adoption of the Single European Act in 1987 that the Single Market became a reality in 1993, albeit with certain restrictions still remaining today (see below).

As far as other objectives are concerned, the framework for the common market in the form of the common tariff and customs union was established in 1968 and a common commercial policy means that the EU is represented in entities such as the World Trade Organisation (WTO), formerly the GATT, by the Commission, which negotiates under a mandate on behalf of the 15 Member States. The Common Customs Tariff (CCT) and harmonised customs nomenclature systems mean that imports from third countries are subject to the same duties regardless of where they enter the Union and their identification and description are facilitated through a single nomenclature.

The sectoral policies set out in the Treaties have had mixed results, with the ECSC successfully coordinating the coal and steel sectors of the Member States, in particular in concluding negotiations on the running down of the steel industry through its crisis in the 1980s (see Chapter 7). Without it, there is no doubt that national subsidies would have been used to prolong and aggravate the crisis. Similarly, Euratom has been successful in coordinating strategies and policies in the non-military nuclear energy sphere in those Member States where nuclear power is used (see Chapter 5). The Common Agricultural Policy has also achieved most of its objectives, albeit at a very high cost to the EU budget, due largely to overproduction and surpluses and, it is argued, to the detriment of the world market for agricultural goods (see below and Chapter 6). Transport policy has hardly developed at all since 1958 owing to the fact that many sectors of transport are state-owned and, therefore, subject to strong protectionist tendencies in most Member States. There has been little policy development in the industrial and service sectors, except in the field of research and technological development (R&TD) in new industries, where the EU finances and manages programmes coordinating efforts of Member States. In terms of actual funding, however, the sums involved are small when compared to those provided in the USA or Japan. In social and regional policy areas, the EU has had insufficient funding to make a substantial impact on social and regional imbalances and here, also, the accession of Greece, Spain and Portugal tended to aggravate the problem (see Chapter 11).

In the light of the mixed results described above, in particular the delays in completing the Single Market, a new political impulse was given by the Single European Act, which did not add any actual objectives under Art. 3 EEC, but which provided for new policies in the areas of environment and R&TD, and which established the global objective of 'economic and social cohesion' to attempt to rectify lack of progress made in regional and social policy. The SEA also provided for 'cooperation' at intergovernmental level in the area of foreign policy, and strengthened the call for monetary and economic policy coordination. Its main thrust, however, was in making the decision-making process more efficient through the increased use of directives and qualified majority voting, and more democratic through enhancing the legislative role of the European Parliament. The SEA certainly gave much needed impetus to the completion of the Single Market, but was felt to be inadequate in terms of reforming other areas, leading to calls for new proposals and the intergovernmental conferences on EMU and Political Union, which led to the signing of the Treaty of Maastricht.

The Treaty on European Union, signed in Maastricht in 1992, ran into obstacles before it could enter into force due to its ambitious reforms to the existing Treaties and new policy areas. Not only did the British Government demand and eventually win concessions in the establishment of an 'opt-out clause' for both EMU and the Social Chapter, but the Treaty also met with stiff opposition during the process of ratification by the Member States. Referendums had to be held in Ireland, France and Denmark, with only a close majority in favour in France and the first referendum in Denmark narrowly resulting in a rejection, leading to additional concessions being made to that country before a second referendum came out in favour enabling the TEU to enter into force on 1 November 1993.

The TEU is much more extensive and ambitious in its treaty reforms than the SEA, extending both the powers of the European Parliament and the scope of EU legislation, as well as establishing clear goals for both EMU and the CFSP, including a common defence policy. First, it extends the scope of objectives in Art. 2 EEC with a radical rewording of the article:

> The Community shall have as its task, by establishing a common market and an economic and monetary union and by implementing the common policies or activities referred to in Articles 3 and 3a, to promote throughout the Community a harmonious and balanced development of economic activities, sustainable and non-inflationary growth respecting the environment, a high degree of convergence of economic performance, a high standard of employment and of social protection, the raising of the standard of living and quality of life, and economic and social cohesion and solidarity among Member States.
> (Art. 2 EC as amended by Article G (2) TEU)

When compared with the original article, it can be seen that not only are the environment, employment and social protection given specific reference, but concepts such as convergence, cohesion and solidarity are emphasised, rather than the purely economic objectives from before. As far as the activities are concerned, the original list of Art. 3 (EEC) is amended where appropriate and several significant areas are added:

(j) the strengthening of economic and social cohesion;

(k) a policy in the sphere of the environment;

(l) the strengthening of the competitiveness of Community industry;

(m) the promotion of research and technological development;

(n) encouragement for the establishment and development of trans-European networks;

(o) a contribution to the attainment of a high level of health protection;

(p) a contribution to education and training of quality and to the flowering of the cultures of the Member States;

(q) a policy in the field of development cooperation;

. . . .

(s) a contribution to the strengthening of consumer protection;

(t) measures in the spheres of energy, civil protection and tourism.

(Art. 3 EC as amended by Article G(3) TEU)

The scope of EU activities is thus extended to a large number of economic and social areas, but there is also now a distinction made between what is full or only partial EU competence, shown where appropriate by the words 'a contribution to . . .' rather than 'a policy for . . .'. This principle is further emphasised in a new article 3b, which mentions the principle of subsidiarity, limiting the activities of the Community and its institutions to what is strictly necessary:

> In areas which do not fall within its exclusive competence, the Community shall take action, in accordance with the principle of subsidiarity, only if and in so far as the objectives of the proposed action cannot be sufficiently achieved by the Member State and can, therefore, by reason of the scale or effects of the proposed action, be better achieved by the Community.
>
> (Art 3b. EC as inserted by Article G(5) TEU)

The other major reforms instituted by the Treaty of Maastricht pertain to Economic and Monetary Union (EMU) and the Common Foreign and Security Policy (CFSP). As far as the former is concerned, not only are the single currency and central bank specifically stated as final objectives, but the introductory process and timetable are specifically laid down in Title VI of the Treaty. The CFSP, as well as cooperation in the fields of justice and home affairs (JHA), remain outside Community competence, with the setting of less ambitious

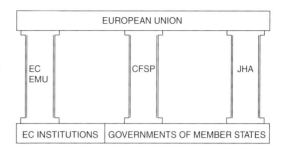

Figure 2.1 The three pillars of the Treaty on European Union

objectives based much more on cooperation than on transfer of powers from Member States to the institutions.

The legal or constitutional structure of the European Union is best visualised, thus, as a structure supported on three pillars, the first of which is the European Community, including EMU, with Community competence and the second and third the CFSP and justice and home affairs in which the Member States retain overall control (see Figure 2.1). The institutional structure and decision-making procedures will be examined in the next section.

THE STRUCTURE AND INSTITUTIONS OF THE EU

As explained in the first section, the founding treaties of the EC, subsequently amended by the Merger Treaty, the Single European Act and the Treaty on European Union, laid down objectives and activities for the European Community and established an institutional framework to perform the tasks required. The EEC Treaty set up the institutions, namely an Assembly, a Council, a Commission and a Court of Justice, assisted by an Economic and Social Committee with an auditing role carried out by the Court of Auditors (Art. 4 EEC).

The Merger Treaty of 1965 merged the Council and Commission for all three Communities, as the 1957 Convention on Common

Institutions had done for the Parliament and the Court, and laid down provisions for their appointment and composition. The powers and roles of the institutions remained largely unchanged through the successive enlargements of the EC until the Single European Act of 1987, which gave some enhanced powers to the European Parliament, the Council and the Commission at the expense of the governments of the individual Member States sitting in the Council of Ministers, and institutionalised the European Council. The Treaty on European Union extended the legislative role of the EP in particular, as well as establishing a clearer delineation of Union competence through the principle of subsidiarity. It also established the Court of Auditors as an institution in its own right and established another advisory body, the Committee of the Regions, and a number of agencies dealing with specific technical subjects with a view to further decentralisation.

The roles of the institutions in the decision-making and legislative process of the EU can be summarised as follows (their location is given in brackets):

- The European Council – (Heads of States and Government) – the supreme political body.
- The Council of Ministers (Brussels) – the main legislative body.
- The Commission (Brussels, some Luxembourg) – the executive and administrative civil service with sole power to initiate legislation.
- The Parliament (EP) (Strasbourg, Brussels and Luxembourg) – the democratically elected assembly with certain legislative powers, more extensive budgetary powers and the role of scrutiny over the other institutions.
- The Court of Justice (COJ) (Luxembourg) – the judiciary which supervises the proper application of EU law by institutions and Member States.

- The Court of Auditors (COA) (Luxembourg) – the external auditor of EU expenditure.
- The Economic and Social Committee (ESC) (Brussels) – an appointed consultative body with members from the social partners and other interest groups.
- The Committee of the Regions (COR) (Brussels) – an appointed consultative body with members from regional and local authorities.
- The European Investment Bank (EIB) (Luxembourg) – provider of loans and guarantees for certain categories of EU projects.
- The European Bank for Reconstruction and Development (EBRD) (London) – provider of loans for technical assistance principally to the former CMEA countries.
- The European Monetary Institute (EMI) (Frankfurt) – the precursor of the Central Bank, responsible for the transition to a single currency.

The European Council

The European Council is composed of the Heads of State and Government of the 15 Member States of the EU, meeting at least twice a year; the presidency is held by each Member State in turn for a six-month period in the order determined by the Treaties (see next subsection). A special arrangement for the Republic of France means that its President and prime minister are both present, hence the reference to the Head of State. Other Member States are represented by their prime ministers.

As the supreme political body, the European Council takes decisions on key political and institutional issues not resolved in the Council of Ministers. Its existence was first formally recognised in the SEA (Art. 2) although it had been meeting periodically since the early 1970s. It usually meets for one or two days every six months in the country of the current holder of the presidency and for one additional session in Brussels every year. The European Council should not be confused with the Coun-

Plate 2.1 The Berlaymont building of the European Commission in Brussels, arguably one of the most photographed twentieth-century buildings in Western Europe, is isolated from the outside world while asbestos is removed from the structure and rehabilitation takes place. Where has the Commission's transparency gone?

Plate 2.2 The new building of the European Parliament in Brussels, currently seating 626 Members of the European Parliament but with an additional capacity of about 150 seats, ready presumably for countries of Central Europe to be represented

cil of Europe, which is a separate organisation located in Strasbourg with 39 member countries covering cooperation in a wide range of social and cultural affairs (see Chapter 1, pp. 8, 10).

The Council of Ministers

The key legislative and decision-making body in the EU, comprising one minister from each of the 15 Member States, is the Council of Ministers. It meets in various forms on a more or less permanent basis, grouping together the respective ministers from the policy areas concerned, for example agriculture ministers for the CAP or transport ministers for matters relating to transport. Ministers alone would not have the time or expertise to deal with all the technical aspects of EU legislation, so preparatory work is done in working groups, committees and COREPER, the Committee of Permanent Representatives, ambassadors from the Member States who meet in almost permanent session to clear the ground for ministerial meetings.

The presidency of the Council of Ministers is held for six months by each of the Member States in an order that was originally based on the spelling of the country's own name in its own official language, but which was then modified to ensure that certain Member States did not have their presidency each time during the first half of the year, which is considered more productive due to fewer holiday periods. The accession of Austria, Finland and Sweden led to the need for further adjustment and the order is now laid down by Council Decision 1 of 1 January 1995 as the following:

1996 Italy/Ireland
1997 Netherlands/Luxembourg
1998 United Kingdom/Austria
1999 Germany/Finland
2000 Portugal/France
2001 Sweden/Belgium
2002 Spain/Denmark
2003 Greece

The outgoing, current and incoming presidencies form the Troika, an unofficial but influential trio of foreign ministers active in particular in foreign policy areas outside EU territory, such as the former Yugoslavia.

The Council has had to concede more decision-making powers to the European Parliament, through the establishment of the cooperation, co-decision and assent procedures, but it still retains the majority share of legislative power among the institutions. The power of Council to enact legislation is generally conditional on the submission of a proposal by the Commission, and there is a large number of policy areas on which it must conduct prior consultation of Parliament, the ESC and the COR, but the influence of such consultation is limited.

Decisions are taken in accordance with Art. 148 of the EC Treaty either by a simple majority, a qualified majority or unanimity, depending on the activity to which the legislative act applies. For simple majority decisions each Member State has one vote. Where the Council acts by a qualified majority, the following weighting of votes applies:

Belgium	5	Luxembourg	2
Denmark	3	Netherlands	5
Germany	10	Austria	4
Greece	5	Portugal	5
Spain	8	Finland	3
France	10	Sweden	4
Ireland	3	UK	10
Italy	10		

Total 87

For the adoption of an act, 62 votes in favour are required where a proposal from the Commission is required by the Treaty, and 62 in favour cast by at least ten members in other cases. Most decisions taken are by qualified majority, although unanimity is still required in some areas, such as treaty changes and enlargement, where a national veto still applies. The Council is also one branch of the Union's budgetary authority, the other being the EP; as such it determines budgetary expenditure in more than half of the Community's activities, although its decisions are subject to Parliament's power to reject the draft annual budget as a whole.

The Commission

The EC Commission has over 16,000 permanent members of staff and uses several thousand additional people on a freelance or expert basis. It is based in Brussels, but has several departments in Luxembourg and research establishments and offices elsewhere in the EU. Its structure is a classic civil service type, with 24 directorates-general covering main policy areas and several other services and offices.

The Commission comprises 20 commissioners, two from each of the five larger Member States (Germany, Spain, France, Italy and the UK) and one from each of the remainder. The commissioners are appointed by the common agreement of the governments of the Member States and, by tradition, the larger countries appoint one from the political party in government and one from the main opposition party. Under the terms of Art. 157 of the EC Treaty commissioners must be 'completely independent in the conduct of their duties', in particular from national or political interest, although this does not preclude their maintaining allegiance to their own political party.

The powers of the Commission can be divided into the categories of initiative, supervision and implementation. It is the only EU institution with the right to initiate legislation, an important power both in terms of the speed and the direction in which EU legislation moves. Its supervisory powers lead to the Commission being called 'The Guardian of the Treaties', supervising the implementation of EU law in Member States and bringing action against states or other entities in case of infringement. The Commission is also responsible for supervising the implementation by national administration of the common policies of the EU, such as the CAP and the structural funds, as well as more direct management of joint programmes and actions such as research and technological development (R&TD), development aid and vocational training. In addition the Commission is responsible for the negotiation of trade and cooperation agreements with third countries and for maintaining relations with other international organisations (Arts. 229–31). In the World Trade Organisation (WTO), formerly the GATT, the Member States of the EU do not participate individually, but speak 'with one voice' through the Commission, which is given a mandate on their behalf.

Although the Commission's role has expanded considerably as the EU has extended its policies into many other areas, its power to act is now more firmly governed by the principle of subsidiarity established by the TEU and described above. Commissioners are responsible for specific areas of EU activity in portfolios shared out at the beginning of its term of office, which they administer using a small advisory cabinet and the overall structure of the directorate-general for the area concerned. In the initiation process for legislation there is a substantial amount of consultation of committees of national government officials, the social partners and other interested parties, and on most days the Commission holds in excess of 50 multinational and multilingual meetings.

The Commission is an institution based on collective responsibility, and decisions taken in the pursuance of its tasks are adopted in the weekly meetings of the commissioners on a collegiate basis. The new Commission is appointed every five years under the TEU in order to coincide with the elections to the European Parliament in June of the preceding year. The President of the Commission and the individual commissioners appear in hearings before the newly-elected Parliament in order to present their priorities and credentials before being sworn in. The Commission is answerable to the Parliament, and commissioners regularly attend its standing committees meeting in Brussels and plenary sittings in Strasbourg and Brussels to answer questions from MEPs. In the last resort the EP has the power to sack the entire Commission with a two-thirds majority in favour (Art. 144 EC) although such action has not been taken to date.

The Court of Justice

The Court of Justice, located in Luxembourg, is the EU institution responsible for the interpretation and correct application of EU law, both in disputes between different parties, such as institutions, Member States or individuals, and in preliminary rulings on questions of EU law referred to it by national courts. The Court consists of 15 judges, one appointed from each Member State, for a term of six years with partial replacement, of eight and seven alternately, every three years. They are assisted in their tasks by nine Advocates-General, to be reduced to eight after 6 October 2000. As the independent EU judiciary, the Court has a vital role in ensuring that EU law is uniformly interpreted and applied throughout the entire territory of the internal market, and that the institutions respect the provisions of the Treaties and the general principles of law common to the Member States, particularly with regard to the protection of individual rights. The Court of First Instance was established on 17 July 1989 to assist the Court in its work, and is responsible for settling disputes between the institutions and its officials, competition cases and matters pertaining to the payment of damages. Its rulings are applicable subject to the right of appeal to the Court of Justice itself.

The Court of Auditors

This EU institution, governed by Articles 188a–188c of the EC Treaty, is responsible for the external auditing of all revenue and expenditure in the EU budget. It consists of 15 members appointed for a six-year term by the Council after consultation of the European Parliament. Situated in Luxembourg, it has substantial investigative powers and produces a weighty annual report as well as special reports on individual issues such as the payment of food aid to developing countries.

The Economic and Social Committee

This institution is based on the French model of a 'Conseil Economique et Social', created in the founding Treaties as a consultative body grouping together 'representatives of the various categories of economic and social activity' and with 'advisory status' (Art. 193 EC). The 222 mem-

bers of the committee are appointed every four years by the Council from the Member States based on the following geographical distribution:

Belgium	12	Luxembourg	6
Denmark	9	Netherlands	12
Germany	24	Austria	12
Greece	12	Portugal	12
Spain	21	Finland	9
France	24	Sweden	12
Ireland	9	UK	24
Italy	24		

They sit in three main groups, representing employers, trade unions and other interest groups such as farmers, consumers, SMEs and cooperatives. The Committee must be consulted and issue an opinion on a wide range of EU legislative activity (Art. 198 EC) and may also produce own-initiative opinions. Its opinions are, however, purely advisory and often merely noted by the Council and Commission.

The Committee of the Regions

Established under the TEU, in particular as a way of applying the principle of subsidiarity, the Committee of the Regions consists of representatives of regional and local bodies and is based in Brussels. It is made up of 222 appointed members with the same geographical composition as that of the Economic and Social Committee. Its role is also advisory, and it must be consulted on five areas with particular regional emphasis: education, culture, public health, trans-European networks and economic and social cohesion.

The European Investment Bank

This is an autonomous public financial institution, based in Luxembourg and established by the Treaty of Rome 'to contribute, by having its own recourse to the capital market and utilising its own resources, to the balanced and steady development of the common market in the interest of the Community' (Art. 198e EC). The EIB's capital is entirely subscribed by the Member States. In order to fulfil its tasks of granting loans and giving guarantees in a wide range of EU activities, such as projects for developing less-favoured regions, for specific actions of assistance in third countries and projects of common interest to several Member States, the Bank normally borrows on the capital markets and re-lends on a non-profit-making basis.

The European Monetary Institute

Established under Stage II of Economic and Monetary Union, the EMI is located in Frankfurt and is responsible for the technical preparations for the introduction of a single currency, in particular the coordination of the monetary policies of the Member States. Under Stage III, it will become the European Central Bank within the European System of Central Banks. EMU is described in more detail later in this chapter.

Other Agencies and Offices

In order to conduct its work in a more efficient and decentralised basis, the Commission has an extensive network of Offices for its activities not only in Member States but also in over 110 countries and several international organisations throughout the world. There are also a number of decentralised Community Agencies responsible for specific technical or vocational areas, situated throughout the territory of the Member States before the most recent accession:

- European Centre for the Development of Vocational Training (Cedefop) – Salonika (Thessaloniki) (Greece)
- European Foundation for the Improvement of Living and Working Conditions – Dublin

- European Environment Agency – Copenhagen
- European Agency for the Evaluation of Medicinal Products (EMEA) – London
- Office for Harmonisation in the Internal Market (trademarks and designs) – Alicante (Spain)
- European Training Foundation – Turin (Italy)
- European Monitoring Centre for Drugs and Drug Addiction – Lisbon
- Translation Centre for bodies of the EU – Luxembourg
- European Agency for Safety and Health at Work – undecided
- Community Plant Variety Office – undecided
- European Investment Fund – Luxembourg

EU Law and the Legislative Process

There are three principal sources of EU law: Treaty provisions; primary legislation adopted by the Council and Commission acting under Treaty provisions; and secondary legislation adopted by the institutions, normally the Commission, to implement primary legislation. The first source of EU law, the Treaties as amended by the SEA and the TEU, covers a wide range of economic, commercial and social activities. The best examples of such primary provisions and their direct impact on EU economic activity are the rules on competition (Arts. 85ff. EC) and state aids (Arts. 92ff. EC) which are directly applied by the Commission and, if needs be, by the Court of Justice, to ensure that there is as little as possible distortion of competition between industries and markets. Secondary legislation is a more complex and dynamic area, with EU legal acts divided into different categories: regulations, directives, decisions and recommendations.

- *Regulation*: a law which is binding and directly applicable in all Member States without any implementing national legislation.

Both Council and Commission can adopt regulations, used in many areas such as management of agricultural markets, transport policy and commercial policy.

- *Directive*: a law binding on Member States as to the results to be achieved, but to be implemented into national legislation in the form each Member State sees appropriate. Most of the Single Market legislation is in the form of directives.
- *Decision*: an act binding entirely on those to whom it is addressed, with no national implementing legislation required.
- *Recommendation*: no binding effect, can be adopted by Council and Commission.

The first two categories are the most important and widespread of EU secondary legislation and are adopted in accordance with the various procedures laid down in the reforms of the Treaties through the SEA and the TEU, which can be summarised as follows:

- *Co-decision*: the TEU establishes powers of co-decision for the Parliament with Council in specific areas including the Single Market, education, research, and certain environmental programmes (Art. 189b EC).
- *Cooperation*: the cooperation procedure was established by the SEA in order to give increased influence to the Parliament in legislative matters and it is faster than the co-decision procedure. It is applicable in matters relating to the internal market, social policy, economic and social cohesion and R&TD (Art. 189c EC).
- *Consultation*: the EP is consulted, along with other bodies, but has little influence on the final outcome.
- *Assent*: in the specific area of the enlargement of the Union and of association agreements with third countries, the EP has full co-decision powers through the assent procedure, whereby the Council may adopt such an agreement only if there is an absolute majority of the EP's members in favour (Arts. 237 and 238 EC).

The co-decision, cooperation and consultation procedures are illustrated in graphic form in Figure 2.2.

THE EUROPEAN PARLIAMENT

Originally a consultative assembly with appointed members, in 1979 the European Parliament became the world's first international parliament to be elected by direct universal suffrage, following the implementation of the Act of 1976. After the accession of Austria, Finland and Sweden, the number of members of the EP reached its present level of 626, distributed as follows between the Member States:

Belgium	25	Luxembourg	6
Denmark	16	Netherlands	31
Germany	99	Austria	21
Greece	25	Portugal	25
Spain	64	Finland	16
France	87	Sweden	22
Ireland	15	UK	87
Italy	87		

Elections to the EP take place every five years. There is no single electoral system for the election of MEPs, but a variety of proportional representation and constituency-based systems is found among the Member States, and often the political balance in national delegations is very different from the domestic political situation. For the formation of political groups, 29 members are needed for a single nation group, 23 for a two nation group and 18 for a group comprising three or more nations. At the beginning of 1996, MEPs from over 100 national parties formed the following political groups in the EP, including the nationalities represented (in brackets):

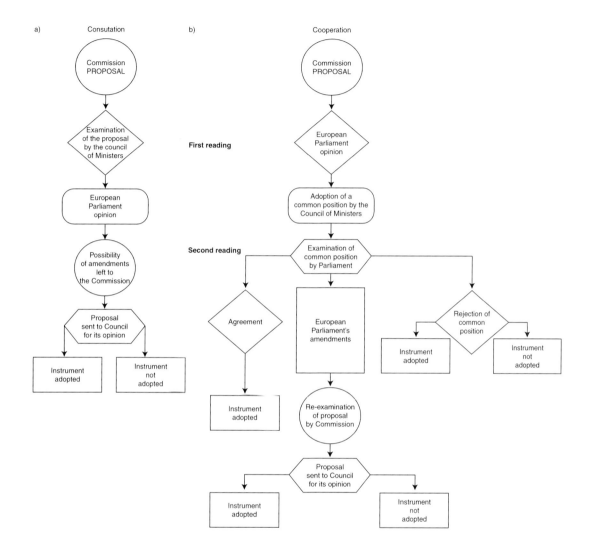

Figure 2.2 Decision-making procedures in the EU

- *PES – Socialists –* 214 members (all countries)
- *EPP – Christian Democrats (including Conservatives)* – 181 (all countries)
- *UE – Union for Europe (Gaullist/Forza Italia)* – 57 (France, Italy, Ireland, Greece, Portugal)
- *ELDR – Liberal, Democratic and Reformist* – 43 (all countries except Greece; Germany and Portugal)
- *GUE/NGL – United Left/Nordic Green Left (former Communist)* – 33 (Spain, France, Italy, Greece, Portugal, Sweden and Finland)
- *V – Greens* – 28 (Germany, Belgium, Italy, Netherlands, Sweden, Finland, Austria, Ireland, Luxembourg and France)
- *ERA – Radical Alliance (Regionalists)* – 20 (France, Belgium, Italy, UK, Spain, Luxembourg)

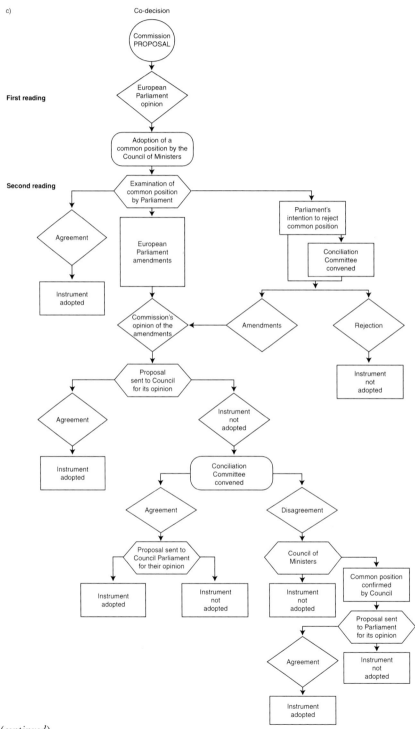

Figure 2.2 (continued)

- *EN – Europe of Nations (anti-integrationists) –* 19 (France, Denmark, Netherlands, UK – Ulster Unionists)
- *Non-attached – (mainly far Right) –* 32 (France, Italy, Austria, Belgium, UK)

The European Parliament has its own administration, with over 4,000 officials working in Brussels and Luxembourg divided into seven directorates-general and a legal service. The treaties establishing the EC originally granted the Parliament supervisory and consultative powers in the legislative process which were extended through the SEA and further through the Treaty of Maastricht. The EP now enjoys active participation in the legislative process through the cooperation and co-decision procedures, which apply to a number of areas of legislation, as well as the assent procedure, all of which are described above (pp. 47–8). The supervisory powers of the EP are such that the Commission and Council are bound to come before the House in order to present proposals and account for action or lack of it. MEPs are able to address oral or written questions to the Commission or to the Council Presidency-in-Office and actively exercise this right in particular in the light of increased media coverage of its activities. The EP also has the power of investiture of the incoming Commission, to hear the candidate Commissioners and to censure the Commission, thereby forcing it to resign. This process, described above (p. 43), has not, to date, ever happened.

The European Parliament conducts its business very much like a national Parliament, in spite of two major handicaps: the first, described in Box 2.1, relates to its three places of work; the second, discussed at the end of this chapter, relates to its 11 working languages. A normal month in the life of the EP consists of one week of political group meetings and two weeks of meetings of standing committees and delegations, including a short plenary part-session in Brussels, and one week of plenary in Strasbourg. Most MEPs are members or substitute members of several committees and delegations and tend to be actively involved in a range of different policy issues and geographical areas. There are 20 standing committees and 3 subcommittees covering areas as varied as Agriculture, Transport and Tourism, Human Rights and Social Affairs. There are also 10 delegations to Joint Parliamentary Committees, principally for candidate countries and the EEA; 21 Interparliamentary delegations for relations with parliaments from individual or groups of third countries such as the USA, South Africa, Transcaucasia and South America; 70 MEPs are also members of the Joint Assembly of the ACP-EU Agreement countries. Although not official organs of the Parliament, there is also a large number of 'intergroups' that deal with more specific areas of interest such as animal welfare or minority languages.

The European Parliament has seen its institutional role increase over recent years and it remains to be seen whether its own calls for further powers following the next IGC will be heeded. Not only is it competing with the other institutions, but its true legislative role can be strengthened only at the expense of the national parliaments, who themselves are opposed to the continual erosion of their sovereignty and influence over the EU and its activities.

THE BUDGET OF THE EU

The budget of the EU must first be seen in proportion to the GDP of the individual Member States. The total 1997 draft budget, 90 billion ECU, is only approximately 1 per cent of total EU GDP, or around 200 ECU per citizen. When the wide range of activities and the ambitious goals of the EU are borne in mind, it becomes apparent that the resources available are quite limited. The sources of revenue for the budget and the destination of

BOX 2.1: THE TRAVELLING PARLIAMENT (See Figure 2.3)

The European Parliament, unique in terms of its directly elected composition, is also unlike other Parliaments in that it has three places of work. The EP meets in plenary form for one week every month in Strasbourg, holds most of its political group and standing committee meetings in Brussels, at a distance of 450 kilometres, while much of its secretariat is based in Luxembourg, half-way between the two. This inefficient system originates from the precursor to the EP which was an appointed consultative body with little direct influence over EC affairs. It seemed appropriate that it should meet in Strasbourg, symbolic regional capital at the interface of French and German traditions, and it was less important that its activities should be conducted close to the main institutions in Brussels.

As the influence and powers of the EP have expanded, many of its activities have been shifted, to be closer to the Council and Commission, with which it cooperates in legislation and over which it exercises parliamentary scrutiny. The powerful French and Luxembourg lobbies have resisted the efforts to create a single location for the EP in Brussels, and obtained, at the Edinburgh Summit in 1994, the undertaking that 12 plenary weeks a year would continue to be held in Strasbourg, in exchange for agreement on the location of a number of new EU agencies. For France in particular, having the EP in Strasbourg is not only a matter of prestige, but also one of financial interest, since the hotel and catering trade in the city benefits enormously from the thousands of members, officials and visitors who descend on it once a month, almost all on expenses. For the EP, there are also the huge costs of maintaining offices and meeting facilities in all three places, and the costs of ferrying staff and the contents of their offices as well as tons of documents backwards and forwards most weeks.

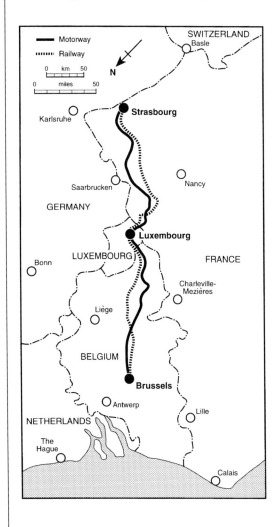

Figure 2.3 The three 'capitals' of the EU

Plate 2.3 The travelling parliament
Source: 'Le douanier se fait la malle', cartoon by Plantu in *Le Monde Éditions*, Paris, 1992, p. 91

The conflict between Brussels and Strasbourg has escalated with the construction of a huge parliamentary complex in each city, with Brussels encouraging the holding of additional plenary sittings in Belgium and the French defending their established right. The cost of the travelling parliament is huge, both to the EU taxpayer, who pays for the administrative costs involved through the EU budget, and to the French and Belgian taxpayers who indirectly finance the construction costs of the EP premises. According to Conradi (1995) the cost is estimated at 180 million ECU out of total running costs of the EP of 800 million ECU. Until pragmatism and realism prevail, however, the expense and inefficiency of the present system are bound to continue, much to the chagrin of many staff and members, and to the delight of Strasbourg restaurateurs and the building industry.

expenditure are shown in Tables 2.1 and 2.2 and are discussed below.

Income

The EU is financed by its own resources, which are made available to it in the amounts and under the conditions laid down in a series of decisions of the Council subsequently ratified by the Member States. The difference between income (85 billion ECU) and expenditure (90 billion ECU) is explained by the fact that the EU may commit itself to expenditure to be financed later. The income comes from the sources shown in Table 2.1.

Table 2.1 Breakdown of the EU budget for 1997 (preliminary) by type of revenue

Type of revenue	Millions of ECU	%
Agricultural and sugar levies	2,016	2.4
Customs duties	12,203	14.4
VAT	34,588	41.0
4th resource (GNP)	35,022	41.5
Miscellaneous	616	0.7
	84,445	100.0

Source: COM(96) 300, p. 36

Table 2.2 Breakdown of the EU budget for 1997 (preliminary) by subsections of expenditure

Subsection	Millions of ECU	%
European Agricultural Guidance and Guarantee Fund, 'Guarantee' Section (EAGGF)	42,305	47.0
Structural operations, structural and cohesion expenditure, financial mechanism, other agricultural and regional operations, transport and fisheries	31,729	35.3
Training, youth, culture, audiovisual media, information and other social operations	748	0.8
Energy, Euratom nuclear safeguards and environment	193	0.2
Consumer protection, internal market, industry and trans-European networks	880	1.0
Research and technological development	3,450	3.8
External action	5,807	6.5
Common foreign and security policy	50	0.1
Guarantees, reserves, compensation	541	0.6
Administrative expenditure (all institutions)	4,293	4.8
Total	89,996	100.0

Source: COM(96) 300, p. 17

- *Agricultural levies*: import levies charged at the external EU frontiers in order to bring the price of imported foodstuffs up to the higher EU price level; sugar and isoglucose levies charged on EU producers aimed at limiting surplus production.
- *Customs duties*: all duties charged on products imported from third countries.
- *VAT*: a proportion of the VAT base of each Member State, set at 1.4 per cent, with a ceiling of 55 per cent of GNP on the VAT base so as not to penalise those countries in which private consumption represents a relatively high proportion of national wealth (Portugal, Greece, Ireland and the UK).

- *GNP resource*: contributions from Member States based on their GNP at a rate set in the annual budget.
- *Miscellaneous*: other minor sources of income include EU taxes on the salaries of EU staff, who are exempt from paying national income tax, the revenue from fines imposed under EU law, and the activities of the Office for Official Publications.

Expenditure

The expenditure of the EU budget has been dominated by spending on agriculture through the Common Agricultural Policy. In spite of

Table 2.3 Contributions from and receipts to Member States from the EU budget

	(1) Contributions		(3)	(4)	(5)
	Bn ECU	%	ECU per capita	Receipts %	Receipts to contributions ratio
Belgium	3.4	3.8	333	2.0	53
Denmark	1.6	1.8	308	3.0	167
Germany	26.3	29.3	322	13.0	44
Greece	1.3	1.4	124	7.0	500
Spain	5.7	6.3	145	14.0	222
France	15.9	17.7	272	14.0	79
Ireland	1.0	1.1	278	4.0	364
Italy	10.3	11.4	180	17.0	149
Luxembourg	0.2	0.2	500	0.2	100
Netherlands	5.3	5.9	342	8.5	144
Austria	2.4	2.7	296	2.4	89
Portugal	1.4	1.5	141	4.0	267
Finland	1.3	1.4	253	1.4	100
Sweden	2.3	2.5	261	2.5	100
UK	11.8	13.1	201	7.0	53
EU	90.2	100.0	242	100.0	100

Sources: COM(96) 300, p. 90; Cole and Cole 1993: 39, from which receipts of EU funds have been updated to give estimates of the authors for 1997. See also *The Economist*, 23 November 1996, for their estimates of net 'winner' and 'loser' countries in the allocation of funds from the EU budget, the 'generosity stakes'

reductions obtained through reforms, it can be seen from Table 2.2 that this continues to be the case, severely limiting spending on other crucial policy areas and regional development. The principal categories of expenditure are as follows:

- *EAGGF Guarantee*: the section of the European Agricultural Guidance and Guarantee Fund that pays for guaranteed agricultural prices to EU farmers.
- *Structural Funds*: these are aimed at improving regional economic and social development in the EU and include the European Regional Development Fund, the Guidance section of the EAGGF and measures for transport and fisheries.
- *Other internal policies*: these include R&TD, training, youth, culture, energy, consumer protection, industry and trans-European networks.
- *External action*: this category includes EU

development assistance towards the Mediterranean, Asian and Latin American countries as well as special measures for the development of Central European countries under the PHARE programme and the former Soviet Union through TACIS. The European Development Fund for development aid to countries under the Lomé Convention (an Association Agreement between the EU and most African, Caribbean and Pacific countries) is funded directly by Member States and is not part of the EU budget.

- *Administration*: this category covers the administrative costs of running the EU and its institutions, 40 per cent of which are due to the policy of full multilingualism and the 11 official and working languages.

Table 2.3 shows estimates for net contributors and beneficiaries among Member States. Table 2.4 shows projected spending under the financial perspectives from 1995–9. It can be

Table 2.4 EUR 15 budgetary financial perspective, 1995 and 1999

		Billions of ECU	
		1999	*1995*
1	Common agricultural policy	37.9	43.4
2	Structural operations	26.3	34.6
	Structural Funds	24.1	31.7
	Cohesion Fund	2.2	2.8
	EEA financial mechanism	0.1	0.0
3	Internal policies	5.1	6.1
4	External action	4.9	6.5
5	Administrative expenditure	4.0	4.5
6	Reserves	1.1	1.2
	Monetary reserve	0.5	0.5
	Guarantee reserve	0.3	0.3
	Emergency aid reserve	0.3	0.3
7	Compensation	1.5	0.0
Total appropriations for commitments		80.9	96.2
Appropriations for payments (% of GNP)		1.20	1.24

Source: COM(96) 300, p. 9

seen that the main trend is increased spending under the Structural Funds as well as in internal operations and external actions. It is unclear what the governments of Member States will agree to for future budgets for the next millennium, in particular in the light of the economic recession of the mid-1990s. It is clear that the continued predominance of agricultural spending will limit what can be spent on other policies and that the next enlargement with the accession of a number of net beneficiaries will put further pressure on resources that are already stretched. Any major increase in the EU budget will require substantial efforts on the part of taxpayers in the richer Member States and it remains to be seen whether they would be prepared to pay for policies that are of direct benefit to citizens of a country geographically and culturally far removed from their own.

CURRENT ISSUES IN THE EU

There now follows a more detailed analysis of four subject areas that are crucial to the future development of the European Union and which, although not exhaustive, are worthy of particular attention.

Enlargement and the 1996 IGC

There is a firm commitment on the part of the European Union to continue enlarging its membership, in particular to include in the medium term most of the countries of Central Europe. The detailed aspects of enlargement are addressed in Chapter 13, but it is to be noted here that it involves the greatest challenge facing the European Union in its 40-year history. The candidate countries are mostly far poorer than the EU average and are only recently established democracies and market economies. They also entail increased difficulties in institutional terms, because the EU institutions were designed to operate with a far more limited number of Member States and, in spite of certain changes to the decision-making process, any substantial enlargement will inevitably slow the EU down in terms of its structures and actions.

The Intergovernmental Conference has therefore been called for 1996 in order to propose reforms to the institutional framework of the EU to prepare for an eventual membership approaching 30 Member States. Negotiations are expected to last for over a year and, given that issues of sovereignty and subsidiarity are at stake, they are likely to be difficult. Larger Member States fear an erosion of their influence in the EU institutions while smaller Member States are reluctant to concede any of theirs either. Several governments are also opposed to the idea of conceding their right of veto over certain areas of EU legislation and national parliaments are wary of the transfer of any more of their sovereignty to the European Parliament.

These are the principal issues to be addressed in the work of the IGC, which have been prepared by a Reflection Group in a report dated 10 November 1995 for the Madrid Summit in its introductory section entitled 'A Commitment to the Future' and in its section IV 'An Efficient and Democratic Union'. The report also covers two other essential areas for EU reform, in section II 'The Citizen and the Union: an area of Freedom and Security' and section III 'External Union Action'. These two subjects are analysed in more depth below.

The Completion of the Single Market and the Third Pillar

Although the completion of the Single Market officially occurred on 1 January 1993, there are still in practice many obstacles to the freedom of movement of people, goods, services and capital between the Member States. Some categories of products and services are still excluded from free movement, including, for example, the purchasing and registration of automobiles and the establishment of certain professions. The failure on the part of the Member States to adopt common rules on immigration, visas, asylum, refugees and other issues that affect citizens from third countries means that there are still obstacles to the free circulation of people within the EU.

The Schengen Agreement (see Box 2.2), which comprises seven Member States, Belgium, Germany, Spain, France, Luxembourg, Netherlands and Portugal, is a significant step towards the creation of a passport-free zone. There have, however, been difficulties over maintaining it, with France reimposing border controls as part of its fight against fundamentalist terrorism, and disputes between other members over extradition. The United Kingdom and Ireland have stated their clear intention to remain outside, since their geographical location and island status renders membership difficult. Other Member States have expressed their intention to join, but populations remain sceptical about the unrestricted movement of persons, in particular since it also facilitates free movement of organised crime, drugs, illegal immigrants and terrorists. With high unemployment in many EU Member States, local populations are less enthusiastic about the uncontrolled influx of job-seekers from other countries and sceptical about the EU's ability to control such issues.

The IGC has therefore set as one of its objectives the strengthening of cooperation in the Third Pillar of Maastricht, namely justice and home affairs, in order to improve public confidence in the EU's ability to provide increased security for its citizens. Certain governments are, however, reluctant to transfer sovereignty over these areas to the EU, preferring to maintain their own traditions and policies in criminal and justice affairs.

Economic and Monetary Union

One of the major objectives for the European Union introduced by the Treaty of Maastricht is the introduction of a single currency through Economic and Monetary Union (EMU). The Treaty laid down a timetable for its completion as well as a set of economic criteria to be met by

BOX 2.2 SCHENGEN AGREEMENT

On 14 June 1985 the countries of France, Germany, Belgium, Luxembourg and the Netherlands signed an agreement in Schengen, Luxembourg, on the gradual abolition of controls at their common borders. The Schengen Agreement applied mainly to the free movement of people whereby checks for crime, terrorism and illegal immigration would be reinforced instead at the external borders of the countries concerned. Article 140 of the Agreement stipulated that all Member States of the EU may join, although only nine participated, the original five plus Italy, Greece, Spain and Portugal, by the time the Agreement entered into force in March 1995 for a trial period.

The Agreement, an expression of intent by the participating Member States to comply fully with the requirements of the Single Market, is in the form of an international convention, and has no legal force within the EU framework. Its role at EU level is limited due to the fact that the UK and Ireland are opposed to the mechanism for removing border controls due to their geographical position as islands. In addition, Denmark, Sweden and Finland are members of the Nordic Passport Union in which their citizens, together with those of Norway and Iceland, can travel freely without passport controls. The three Nordic Member States of the EU state that they could not join Schengen if it endangered freedom of movement between the Nordic countries which has existed since the 1950s. Sweden, in particular, points out how impractical it would be to introduce border controls along its 1,500 kilometre border with Norway. One solution now envisaged is that Norway also be allowed to join the Schengen Agreement, even though it was originally stipulated that only EU Member States may be members.

The Schengen Agreement has also encountered difficulties among the founding members, with France unilaterally suspending application of the Agreement in 1995 as part of its fight against Fundamentalist terrorism and problems over extradition of ETA terrorist suspects between Belgium and Spain. A true Single Market for the free movement of persons, despite the existence of the Schengen Agreement, remains an ideal still to be fulfilled in the European Union, and any enlargement of the EU will make it yet harder to fulfil.

participating countries. Both the UK and Denmark obtained an 'opt-out clause' since their governments were unwilling to commit themselves to joining, and Sweden has more recently expressed its intention to use the same exemption.

The timetable laid down is divided into three phases:

- Phase 1 (1 July 1990–31 December 1993)
  - the completion of the Single Market, in particular the free movement of capital;
  - the introduction of measures for convergence between economies of Member States.
- Phase 2 (1 January 1994–1 January 1997 or 1 January 1999)
  - the creation of the European Monetary Institute responsible for technical preparations for Phase 3 and coordination of national monetary policies;
  - Member States avoid excessive public deficits;

- the freezing of the composition of the ECU.
- Phase 3 (From 1 January 1997 or, if a majority of Member States do not meet the convergence criteria, by the latest on 1 January 1999 with those Member States that do meet the criteria)
  - irrevocable fixing of exchange rates and the introduction of the Euro as the common currency;
  - the phasing out of use of national currencies;
  - the introduction of a European System of Central Banks to replace the EMI.

The convergence criteria to be met by participating Member States are as follows:

- Inflation rate within 1.5 percentage points of the average rate of the three States with the lowest rate.
- Long-term rate of interest within 2 percentage points of the average rate of the three States with the lowest rates.
- National budget deficit must be below 3 per cent of GDP.
- National debt must not exceed 60 per cent of GDP.
- National currency must not have been devalued for two years and must have remained within the 2.25 per cent fluctuation margin of the European Monetary System (EMS).

There are clear advantages to the introduction of a single currency. First, it would lead to a consolidation of the Single Market, improving the competitiveness of EU industry and services. Second, transaction costs and hedging costs incurred by travellers and businesses would be removed. In spite of the benefits for EU industry and its citizens, EMU is facing a difficult struggle. There is now increasing doubt over the ability of most Member States to meet these criteria even by the final deadline for Phase 3, and Member States have called for the timetable or criteria to be made more flex-

ible. The Commission has, however, stated that any postponement or loosening of conditions would threaten the whole exercise, since economic confidence is required for the Single currency to be successful. There are also psychological obstacles, with many citizens unwilling to countenance the disappearance of their own currency, which is a part of their own national identity, and fearing a deterioration of their own financial situation without the control of their governments over monetary policy. Criticism is also increasing over public sector finance cuts being introduced by Member States in their efforts to meet the convergence criteria, since inevitable job losses are entailed at a time of already high unemployment. The future of EMU remains uncertain, therefore, and is likely to dominate public debate over the EU as the Phase 3 deadlines approach.

Common Foreign and Security Policy

The Treaty of Maastricht in its 'second pillar' introduced the objective of a Common Foreign and Security Policy leading to the formulation of a Common Defence Policy, although opposition from certain Member States meant that it still remains outside the sphere of EU policy as such, based on cooperation and coordination between national policies. The break-up of the Soviet Union and the dissolution of the Warsaw Pact gave rise to ambitions in EU circles for the creation of a truly European Security entity, with some calling for the disbanding of NATO. Recent events, in particular the conflict in former Yugoslavia, have shown how difficult it is to coordinate the policies of Member States, which have such different foreign policy traditions ranging from the former colonial powers of France and the UK, which remain strongly independent, to countries that have been neutral for decades, such as Sweden, Austria and Ireland. There are also different geopolitical interests that govern the foreign policy attitudes of Member States, the position of Greece over Cyprus and Turkey a particularly striking example of this.

The modest achievements obtained so far in the move towards a CFSP have led to calls for greater impetus to be given to the process, and the 1996 IGC will address the question of how to give more identity and influence to an EU foreign and security entity. Proposals include the establishment of an international legal personality for the Union in this sphere and the setting up of an analysis, forecasting and planning unit for a common foreign policy. On defence, there are proposals for greater coordination between the EU and the Western European Union (WEU), whose members are all EU Member States, with a gradual integration of the WEU into the EU, which would provide a clearer EU security and defence entity. The establishment of a European defence and peace corps is also mooted, as well as closer cooperation in armaments production and procurement.

Foreign, security and defence policy is, however, one of the most sensitive areas of national identity, with very different traditions between the Member States. In spite of the continuing need for a common response to international conflicts and issues there is too much resistance in many national capitals to the transfer of the considerable amount of sovereignty required to give the CFSP and CDP the substance they require to succeed.

THE COSTS OF MULTILINGUALISM

The European Union is faced with a permanent dilemma over its own policy towards languages, on the one hand wishing to promote multilingualism out of respect for national and regional identities and cultures and, on the other, struggling with the technical and financial difficulties of conducting its own business with full cover for official and working languages. With further enlargement now only a matter of time there is talk of a crisis looming, with the addition of new languages threatening to lead to the chaos of Babel in the Old Testament.

This latter-day Babel finds its origins in the founding Treaties of the European Communities in the 1950s, in which equal status was given to the official languages of the original six Member States, creating the EC with four official and working languages. Subsequent enlargements led to the addition of another seven languages, giving the present total of 11 for 15 Member States. Each enlargement was negotiated with little more than lip service paid to the possibility of rationalising the policy of full multilingualism, and the most recent accessions even saw particular attention paid to the importance of maintaining it. Not only did Sweden and Finland insist on equal treatment, but the negotiations with Norway made express provision for the use of both official forms of Norwegian. It was clearly felt, given Scandinavian scepticism over European integration as shown so clearly in the Danish referendum on the Treaty of Maastricht, that it would be a public relations gaffe to suggest that the EFTA countries enter the EU with restrictions placed on the use of their languages. To the relief of certain officials responsible for already overstretched language services, in 1994 the population of Norway voted, as they did in 1972, to remain outside the EU.

The political importance attached to official and working languages in the European Union goes beyond the function of language as a symbol of national identity. The EU as a multinational entity is unique in the world since it enjoys sovereignty transferred from Member States in a wide range of areas. Its institutions are responsible for legislation that is directly applicable on the territory of the Member States and are accountable to varying degrees to the individual citizens for their actions. It is essential, therefore, that communication among the institutions as well as between them and the Member States should be conducted in the official languages of those states. It would thus be inconceivable for a candidate for the European Parliament to be expected to use a language other than his or her own either to get elected or to perform his or her duties once in office. It is this political reality that has thwarted any serious attempts to introduce restrictions on the number of languages covered, and that led to the outcry against the mere suggestion in December 1994 by French Minister Lamassoure that the number of languages should be reduced.

There are also, however, technical and financial realities at stake. The EU institutions constitute by far the largest recruiter of translators and interpreters in the world. The translation service of the Commission translates well over a million pages of text per year. The plenary week of the European Parliament in Strasbourg each month requires the services of over 450 conference interpreters. A single simultaneous 11-language meeting in the Council of Ministers uses 33 interpreters, three in each booth. It is estimated that around 40 per cent of the EU's administrative budget is spent to cope with multilingualism, or around 2 billion ECU per annum. In addition to salaries, the cost of equipping meeting rooms with interpretation systems and providing computer systems for translators, there are also the hidden costs such as office space, and personnel and administration.

The above financial and technical realities have not, however, led to insurmountable difficulties. Anyone attending meetings in the institutions will admit that multilingualism works in purely technical terms. The increasing number of language combinations means that there is ever greater recourse to the use of relay interpretation, with the use of 'core' languages acting as servers for the other booths. Although often criticised for delays, with the old adage of the Danes or, these days, the Finns, laughing at jokes ten seconds later than other delegates, relay interpreting is usually accurate and enables the more economical use of interpreters. Similarly, the EU budget has always provided financing to purchase new equipment and to cope with the increasing workload of language staff at the institutions. Often the problem has been to find enough linguists with the necessary skills capable of meeting the high professional standards required, and the institutions must continue to invest in training and recruitment in order to maintain their services even with the present number of languages.

In the future the political importance of maintaining full multilingualism will still outweigh the financial and technical problems it entails. In spite of the possibility of another dozen languages being added in forthcoming enlargements, the Reflection Group's report for the 1996 IGC stated that 'The instruments of the EU will have to respect the linguistic and cultural diversity of the Union. Respect for transparency and for greater participation by national parliaments must be the criteria governing the Union's treatment of official languages.'

No radical solutions can be expected to the problems facing the Union over multilingualism. A *lingua franca* such as Latin or Esperanto is unrealistic in the short or medium term as a replacement for the present system; bilingualism in all politicians and civil servants is an impossible dream; restrictions on the number of languages used is politically unacceptable. Rationalisation is possible with reductions in the number of languages used for certain purposes, but the domains of legislation and democratic accountability, principally the responsibilities of Council and the Parliament, will require full multilingualism regardless of how many Member States there may eventually be. A foretaste of things to come could be savoured at the Madrid Summit in December 1995, where interpretation for 19 working languages was provided at the special session involving the heads of government of candidate countries. Perhaps multilingualism, and the vital services provided by the thousands of interpreters and translators who work every day for the EU institutions, should be viewed, as Jacques Delors did in 1991, as 'the price to pay for a democratic Europe'.

FURTHER READING

Bennett, R. J. (ed.) (1993) *Local Government in the New Europe*, Chichester: Wiley.

Blacksell, M. and Williams, A. M. (eds) (1994) *The European Challenge: Geography and Development in the European Community*, Oxford: Oxford University Press.

Cecchini, P. (1988) *The European Challenge 1992*, Aldershot: Wildwood House. Made an impact in the late 1980s but later regarded as simplistic and in places suspect.

European Parliament (1994) *Fact Sheets on the European Parliament and the Activities of the European Union*, Directorate General for Research, Luxembourg: Office for Official Publications of the European Communities. Exhaustive compendium of EU organisation and activities.

Evans, R. (1993) 'Contesting the Euro vision', *Geographical*, LXV (1), January: 21–3. A troubled future ahead predicted as the Single Market comes into being.

Evans, R. (1994) 'Sidelined by history', *Geographical*, LXVI (3), March: 32–5. Crisis in NATO. With the end of the Cold War it has lost its *raison d'être*.

Holland, M. (1993) *European Community Integration*, London: Pinter.

Laffan, B. (1992) *Integration and Cooperation in Europe*, London: Routledge.

Lodge, J. (1993) *The European Community and the Challenge of the Future*, London: Pinter.

Miles Arnold Ltd (ed.) (1995) *The 1995 European GIS Yearbook*, Oxford: Blackwell. Reference work on geoinformation.

Nicoll, W. and Salmon, T. C. (1994) *Understanding the New European Community*, Hemel Hempstead: Simon and Schuster.

Scott, A., Peterson, J. and Millar, D. (1994) 'Subsidiarity: A "Europe of the Regions" v. the British Constitution', *Journal of Common Market Studies*, 32 (1), March.

Weigall, D. and Stirk, P. (1992) *The Origins and Development of the European Community*, Leicester and London: Leicester University Press.

3

POPULATION

'Italy's birth-rate is the lowest in the world: since 1988, total fertility rate has hovered somewhere between 1.2 and 1.3 children. A leading Italian demographer, Professor Antonio Golini, is so concerned about what he calls Italy's 'demographic malaise' that he envisages a situation where fertility decline and population ageing reach a point of no return, endangering the long term survival of Italy's population.'

(Arkell 1994)

- The population structure of every Member State of the EU is such that for the next few decades little change in the total is likely.

- Population is spread very unevenly over the territory of the EU, its present distribution reflecting in particular the availability of agricultural land and mineral resources, especially coalfields.

- In the last two centuries the population of what is now the EU has become increasingly urbanised.

- Almost a quarter of the population of the EU lives in areas regarded as rural and on this basis eligible for considerable amounts of economic assistance.

- Almost one-twentieth of the population of the EU is foreign born and in the decades to come it can be expected that immigration into the EU will exceed emigration to other parts of the world.

During the eighteenth and nineteenth centuries most parts of Europe experienced fast population growth through natural increase. As a result of considerable net out-migration from various parts of Europe to the rest of the world until the 1930s, and high mortality rates of infants and children (see Langer 1972), population growth was, however, usually only between 1 and 2 per cent per year even when growth was fastest. In contrast, much faster rates of growth are now experienced in many developing countries, mostly in the range of 2–4 per cent per year. Greatly increased yields in agriculture, widespread industrialisation, increased literacy, emigration, and customs such as late marriage and inheritance laws favouring the oldest son, are considered to have contributed to the reduction in the rate of population growth this century in Europe. The increasing acceptance and use of contraceptive methods has also been an influence. The annual increase of population in the EU is now very small and, as a result, the share of total world population in the Union is declining, and will continue to do so for many decades.

THE POPULATION STRUCTURE OF THE EU

In 1995 the total population of the 15 Member States of the EU was 373 million, compared with 316 million for the same countries in 1960, an increase of only 12 per cent. During 1960–95 the total population of the world grew by over 90 per cent, from about 2,980 million to 5,700 million. In 1960 the countries of the present EU had over 10 per cent of the world's population, in 1995 only 6.5 per cent. The EU's share is expected to drop below 5 per cent by 2020. Columns (1)–(3) in Table 3.1 show the total population, birth-rate and death-rate of the countries of the EU. Of the five largest EU countries in population, only France and the UK are expected to grow in population, although not quickly, whereas the populations of Germany, Spain and Italy are expected to decrease.

There are several possible reasons why the natural increase of population in the EU has declined. First, there has been a sharp decline in total fertility rate (columns (5)–(7) in Table 3.1), with every country now having a level below 2.1, regarded as essential for a population to stay at a given level. A contributing factor affecting this trend has been the increase in the mean age of women at the birth of their first child. Even during 1980–90 it rose in all EUR 12 countries by between one and two years (*The European*, 28 Jan.–3 Feb. 1994: 6). Second, the process of urbanisation, which started in the nineteenth century, or even earlier in some countries, has continued, concentrating more people in cities, where family size tends to be smaller than in rural areas and among agricul-

tural populations. Third, the use of contraceptives (see column (8) in Table 3.1) is widespread in most Member States of the EU, including Italy, Spain and Portugal, all three officially almost entirely Roman Catholic. The impact of net immigration also influences the population size and growth in the countries, with increasing numbers in the 1990s entering the EU from Central Europe and the former USSR as well as from northwest Africa.

The recent population structure of nine countries is shown in Figure 3.1. The mothers of the decade 2010–20 were mostly already in existence around 1990. Assuming no marked change in female mortality rates, the reduced numbers in the youngest cohorts in Italy and Germany in particular confirm the prospect of decline anticipated in Table 3.1. If fertility rates remain very low in Italy, in a few decades time the shape of the population pyramid of Italy could (appropriately) be that of a 'cornetto'. Although different from each other in detail, the population structures of Algeria, Turkey and Azerbaijan are broadly similar in still having a pyramid shaped profile, pointing to substantial increases in population size in those countries at least for several decades to come, one reason being the small number of people in older age groups, in which most deaths occur.

Rapid population growth in developing countries is regarded by many commentators, although not all, to be a serious problem. Whether or not growth is a problem, the countries of the EU no longer face that situation, but instead the prospect that an increasing proportion of the population will be elderly. As the composition of the young and elderly dependent elements of the total population changes, the need to support a large number of children is being replaced by the need to support a large number of elderly citizens. Resources have to be transferred from education to healthcare and pensions.

When the EEC came into being in 1957 its

Table 3.1 Aspects of population change

	(1) Popn 1995 mlns	(2) Birth rate per thousand	(3) Death rate per thousand	(4) Annual natural increase %	(5) Total fertility rate 1950	(6) 1975	(7) 1995	(8) % using contraceptives
Belgium	10.2	12	11	0.1	2.3	1.7	1.6	81
Denmark	5.2	13	12	0.1	2.6	1.9	1.8	63
Germany	81.7	10	11	−0.1	2.1[1]	1.5[1]	1.3	78
Greece	10.5	10	9	0.0	2.6	2.3	1.4	n.a.
Spain	39.1	10	9	0.1	2.5	2.8	1.2	59
France	58.1	12	9	0.3	2.9	1.9	1.7	81
Ireland	3.6	14	9	0.5	n.a.	3.4	2.0	60
Italy	57.7	9	10	−0.0	2.5	2.2	1.2	78
Luxembourg	0.4	13	10	0.4	n.a.	1.6	1.7	68
Netherlands	15.5	13	9	0.4	3.1	1.7	1.6	76
Austria	8.1	12	10	0.1	n.a.	1.8	1.4	71
Portugal	9.9	12	11	0.1	3.0	2.6	1.5	66
Finland	5.1	13	10	0.3	3.2	1.7	1.8	80
Sweden	8.9	13	12	0.1	2.3	1.8	1.9	78
UK	58.6	13	11	0.2	2.2	1.8	1.8	81
EUR15[2]	372.6	12	10	0.2	2.61[3]	2.05	1.6	73[4]

Sources: WPDS 1995, (PRB 1995); PRB 1987
Notes: 1 West Germany only
2 All means for EUR 15 are unweighted
3 Mean for 12 countries
4 Mean for 14 countries
n.a. not available

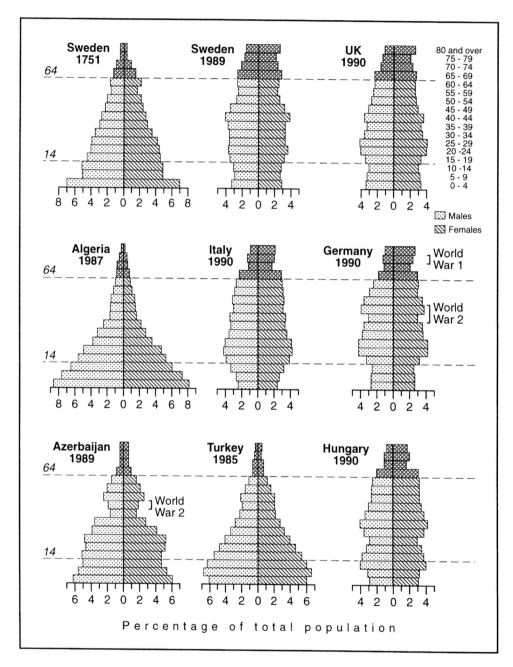

Figure 3.1 A comparison of the population structure of selected EU and non-EU countries. The present structures of population of Sweden, the UK, Italy, Germany and Hungary (Central Europe) are broadly similar, but contrast sharply with those of Sweden in the eighteenth century and the predominantly Islamic countries of North Africa and Southwest Asia. The population aged over 64 years is shaded heavily to draw attention to differences in the proportion of the elderly

six founder Member States had a population of about 165 million. By comparison, excluding the GDR, in 1995 the same countries had 223 million. Figure 1.2 in Chapter 1 shows how the population of the EU has increased as a result of the accession of new Member States to join the original six. If the 'rump' of EFTA (Norway, Switzerland and Iceland), together with Malta and Cyprus, join the EU, these five countries have a combined population of only about 13 million and their impact would therefore be modest. Figure 1.2 shows the effect of the addition of the 65 million people in four other possible future members of the EU. In numbers the impact would be relatively less marked than the addition of the UK, Ireland and Denmark in 1973, but the Central European countries are economically very different from the present EU and are much poorer than most. While the former GDR was 'adopted' by the FRG and given substantial support from the rest of the EU, it is doubtful if much development assistance could be provided for four times the population of the GDR.

THE DISTRIBUTION OF POPULATION

Among various ways of showing the distribution of population on a given territory, two are frequently used. First, symbols such as dots can be used, each representing a given number of people. Second, administrative areas can be coloured or shaded according to the number of people per unit of area, such as a square kilometre, giving a picture of the density. The dot method will be used in the rest of this section and the density method will be used in the following section.

Figure 3.2 shows the distribution of population in EUR 15 using 100 dots, each representing 1 per cent of the total population of the EU in 1990. The total population in *Regional Profiles* (1995) for EUR 12 including East Germany was 344 million, to which the combined population of 21.5 million of Austria, Finland and Sweden was added, making each dot equivalent to 3,650,000 people. The placing of the dots was determined on the basis of the NUTS level 2 population of administrative divisions. The exact positioning of the dots depends on the way population was grouped, but the general structure of the distribution is correct and, given the virtually static demographic situation in the EU, is changing only very gradually in the 1990s.

The dot distribution in Figure 3.2 allows rough calculations to be made of the number of people within given distances of any place in the EU, a calculation in particular of economic interest and applicability. For purposes of calculating the potential market within a given distance of a place, the procedure could be refined to weight the dots according to the GDP per capita in different countries or even individual regions. For example the dots in the former West Germany might carry double the weight of those in Greece or Portugal.

A number of features of the distribution of population in the EUR 15 can be seen on the map. The centre of gravity of the distribution of 100 dots can be calculated by finding the means of their distribution on two orthogonal axes (which do not have to be east–west and north–south). The mean centre of the population distribution is shown on the map (refer to key). In terms of an imaginary model, if each dot is a weight placed on a weightless flat 'tray' corresponding to where it is on the map, the tray would balance on that point. The point is located in the Département of Vosges, in France, near the small town of Bourbonne-les-Bains. The inclusion of Austria, Finland and Sweden in 1995 pulled it a few tens of kilometres to the northeast from its position in EUR 12 (see Cole and Cole 1993: 56) near Dijon.

In Figure 3.2 a circle centred on ⊞ has been drawn to include 50 dots, half of the EUR 15 population. This 'inner' half of the population

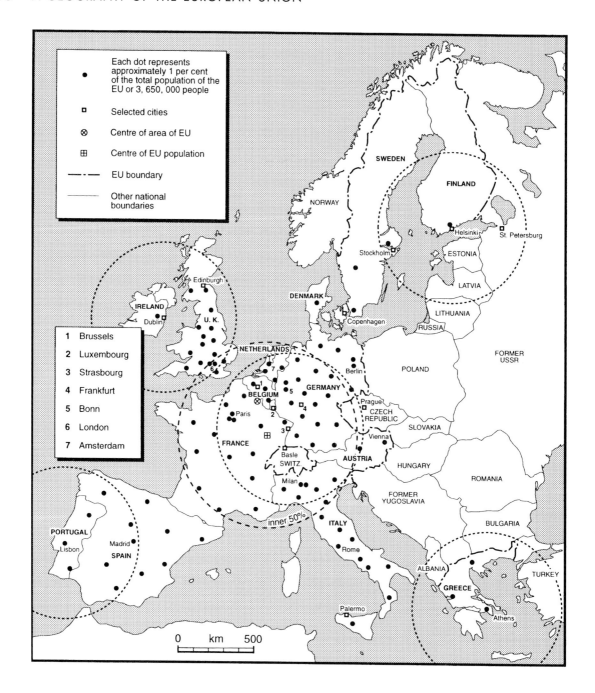

Figure 3.2 Distribution of population by 100 dots. Each dot is placed at the centre of 1 per cent of the total population, calculated on the basis of NUTS level 2 regions. The exact placing of the dots is subjective but the general picture is valid

Table 3.2 EU population within 500 km of selected places

Rank	Place	Population %	millions	Rank	Place	Population %	millions
1	Strasbourg	41	150	12	Dublin	16	58
2=	Brussels	38	139	13	Edinburgh	15	55
2=	Bonn	38	139	14	Rome	14	51
4=	Basle	37	135	15	Madrid	13	47
4=	Amsterdam	37	135	16	Copenhagen	11	40
6=	Luxembourg	36	131	17	Vienna	9	33
6=	Frankfurt	36	131	18	Lisbon	8	29
8	Paris	32	117	19	Palermo	7	26
9	London	29	106	20	Stockholm	4	15
10	Milan	23	84	21	Athens	3	11
11	Berlin	21	77	22	Helsinki	2	7

is located within about 600 km of ⊞ and includes the populations of the Benelux countries, of most of France and Germany, and of northern Italy, southeast England and western Austria. The whole of Switzerland, not an EU Member State, also falls within the circle. With Switzerland, about 190 million of the most affluent people in the world, accounting for over 10 per cent of the total GDP of the world, are concentrated on an area of about 1 million sq km, about 0.7 per cent of the world's land area.

The exact location of the centre of gravity of population is not of particular significance except perhaps to a few villages vying for the prestige and the tourism potential of having it in their place. It does, however, provide one basis for determining central and peripheral parts of the EU. More specific and practical information about the size of the EU market or particular parts of it within a given distance of various places can be calculated by placing circles (or other shapes) of appropriate diameter and counting the number of dots falling within that space. Such circles, each 500 km in radius, have been drawn on the map in Figure 3.2. They are centred on selected places to illustrate the great contrasts in accessibility to the rest of the population of the EU. Thus, for example, about 150 million EU citizens (41 per cent)

reside within a distance of 500 km of Strasbourg in eastern France. In contrast, within the same distance of Helsinki there are about 7 million EU citizens. The Strasbourg circle also includes about 12 million non-EU citizens in Switzerland and the Czech Republic, while Helsinki takes in Latvia, Estonia and St Petersburg. The only EU citizens within 500 km of Athens are some 10 million Greeks. Lisbon's circle reaches out to only about half of the population of Iberia, while Dublin's includes almost all of the population of Ireland and the UK. Table 3.2 shows the number of EU citizens residing within 500 km of 22 selected places in the European Union, including those already referred to. All the places are shown on the map, but only five of the relevant circles are drawn.

The method described above serves only as a rough guide to the relative accessibility of different parts of the EU. For particular purposes, circles with a radius less than 500 km could be of interest. Thus, for example, within a distance of 100 km of the centres of London and of the Rhein–Ruhr conurbation (Düsseldorf) there are about 15 million people, and within that distance of Paris and Milan, about 10 million people. Again, in any of the five circles centred on cities in Figure 3.2, the speed of surface travel varies according to direction. Thus only about 5 million people can be reached directly by land from Dublin, another 53 million in Dublin's 500 km circle only after a sea crossing. Helsinki's 500 km 'catchment' includes the area around Stockholm, again across a sea barrier. To reach the rest of the EU by surface transport from Greece the choice is between a lengthy ferry crossing to Italy and a long, and in the early 1990s hazardous, road journey across the former Yugoslavia or more deviously, to the east of Yugoslavia. The matter of journey times and accessibility will be considered in more detail in Chapter 4.

If and when more European countries are admitted to EU membership, the mean centre of population is bound to shift towards the east, as there are no countries in Europe to the west of it not already in the EU. If, for example, the four Central European countries, Poland, Hungary, the Czech Republic and Slovakia, join, then their combined 65 million inhabitants, centred on Katowice in Poland, would pull the centre of gravity of the EU population eastwards. The distance between the present population centre of EUR 15 and Katowice is about 1,200 km. The relative weights at the two population centres are 365 against 65. The eastward shift would therefore be about 1/7 of 1200 km, putting the new centre roughly at Karlsruhe in Germany, close to the Franco-German border. It took 200 years for the centre of gravity of population of the USA to move from near Washington (DC) in 1790 to a location in the state of Missouri in 1990 after massive population growth and westward migration. In Western and Central Europe, no such process is at work, and with little growth of population expected the population might be described as spatially inert. Even the enlargement of the EU to take on board all of the rest of Central Europe would still leave the centre of population somewhere in southern Germany, while even with the inclusion of Turkey it would move only into western Austria.

THE DENSITY OF POPULATION

A population density map of the EUR 12 countries was published in 1987 (see Commission of the EC 1987a) on a scale of 1:4 million, using NUTS level 3 administrative divisions, except for Germany. A more recent density map, for 1992, covering all of Europe except the former USSR at NUTS level 2, was published by the European Commission (Commission of the EC 1994a). In this section a density map produced by the authors will be described.

While the dot distribution discussed in the previous section can be manipulated and further interpreted, the density map is more a broad picture of the demographic situation. Figure 3.3 has been drawn to highlight differences in density more sharply than is customary on conventional density maps. The average density of population in EUR 15 is 114 persons per sq km. In the compilation of the map in Figure 3.3 data for NUTS level 2 units of EUR 12 have been used, together with the major administrative divisions of the three newest Member States. Five levels of density of population have been distinguished, two above the mean of 114 per sq km and three below it. Solid and open circles have been used to indicate divisions with above average density, to contrast with the minus sign and two shadings for divisions with below average density.

The larger the number of divisions used, the more detailed and precise a density map will be, while the contrast between the highest and lowest densities will be correspondingly sharper. The particular set of administrative divisions used in the countries of the EU at NUTS levels 1 and 2 is intended to reflect and serve local government traditions and needs, but is not necessarily the most appropriate set of areas for mapping spatial variations. Thus in Germany, for example, Hamburg and Bremen are city Länder at NUTS level 1, but are also used at NUTS level 2. They contain a large number of people in very small areas and give a

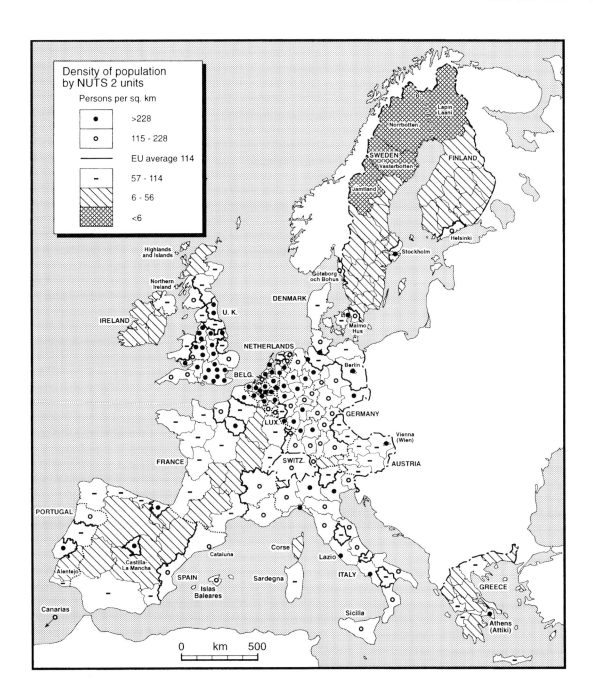

Figure 3.3 Density of population in the EU at NUTS level 2. Dots highlight the regions with above average density, shading and a minus sign the regions with below average density

Plate 3.1 The urban scene: since the death of Franco, Barcelona, like other Spanish cities, has caught up in sophistication with the rest of Western Europe

Plate 3.2 The rural scene: like many other villages in the Massif Central of France, St Enimie in the Tarn gorge is difficult to reach, but its attractiveness to tourists has ensured that it has not declined in population

misleadingly high density of population. In contrast the populations of the cities of Munich and Stuttgart, comparable in size to Hamburg and Bremen, are diluted over administrative areas of considerable extent, the density of which is comparatively low, as a result of the presence of extensive rural areas.

Until the eighteenth century, the density of population in different regions of Western Europe was largely related to the availability and quality of cultivated land, pasture and wood-land to ensure a food supply, water and fuel-wood, as well as fodder supplies to maintain a large number of work animals. In the eighteenth and nineteenth centuries much of the population became concentrated in industrial centres, especially in and near the coalfields of the northern countries. In the twentieth century the proliferation of services has supplemented agricultural employment in many rural areas and small towns and at the same time has resulted in the further concentration of people in relatively few large cities and clusters of cities.

Density of population is one of the variables included in basic data sets provided for EU policy-makers and it is related to at least one of the objectives of the allocation of European Structural Funds (see Chapter 12). The above problems of interpretation and the factors influencing the present distribution of population should therefore be taken into account when density is being considered. Unless the whole picture of a distribution of population is kept in mind it is easy to reach simplistic conclusions about the significance of different densities. Three administrative divisions in Sweden and one in Finland (see Figure 3.3) have densities of population less than one-twentieth the EU average of 114. Here, it might seem, is space for more population. In practice, there is virtually no cultivation or manufacturing in these

regions and the population depends variously on forestry, mining and, in the extreme north, reindeer herding. The problem in Finland, Sweden and also neighbouring Norway is how to maintain even the very small existing population in the northern regions. The general tendency in the EU has indeed been for many regions with already high densities to grow in population while regions with low densities have been losing out relatively and even absolutely in the process of population change.

Table 3.3 shows the density of population in a selection of the NUTS level 2 regions of the EUR 15 countries, the eight with the highest densities, eight with low densities and a selection of eight from several different countries with densities near the EUR 15 average of 114 persons per sq km.

- Four of the eight regions with the *highest density* are UK metropolitan counties,

Table 3.3 Density of population at national level and in selected regions of the EU

Density of population of EUR 15 countries in persons per sq km above average		below average		Eight highest densities at NUTS level 2	
Netherlands	367	Portugal	107	Brussels	5,922
Belgium	328	France	105	Greater London	4,360
UK	237	Austria	93	Berlin	3,876
Germany	225	Greece	77	Vienna	3,694
Italy	188	Spain	77	West Midlands	2,924
Luxembourg	150	Ireland	51	Merseyside	2,222
Denmark	120	Sweden	21	Hamburg	2,209
		Finland	15	Greater Manchester	1,996

Selected near EU average densities persons per sq km			Four lowest densities in Sweden/Finland and other selected low densities		
Zeeland	Netherlands	118	Castilla-La Mancha	Spain	22
Abruzzi	Italy	116	Alentejo	Portugal	20
Namur	Belgium	114	Highlands, Islands	UK	9
Northern Ireland	UK	113	Vasterbotten	Sweden	5
Niederbayern	Germany	106	Norbotten	Sweden	3
Oberpfalz	Germany	105	Jamtland	Sweden	3
Bretagne	France	105	Lapin	Finland	2
Lincolnshire	UK	100	Guyane	South America	2

Main source: Commission of the EC 1994a, Table A.27

specifically designed to contain large urban agglomerations but little rural land, while four are large urban centres elsewhere in the EU, situated within very small administrative districts.

- In all the eight regions with near *average densities*, industry and agriculture are both present. All are poorer than the average level in their respective countries, but not among the poorest in the EU.

- Two of the regions with the *lowest densities* are located in the driest part of Iberia, while mountains and poor soils have restricted settlement in northwest Scotland, and cold conditions and poor soils have kept densities very low in the northern part of the Nordic countries. The French territory of Guyane in South America, an integral part of the EU, consists almost entirely of tropical rain forest. The system of divisions at NUTS level 2, which forms the basis on which density is calculated, gives a 'coarse' breakdown of some features, thus failing to highlight, for example, small areas of very low density in such areas as the high Alps and the Pyrenees.

The overall picture of population density, clearly shown in Figure 3.3, is characterised by two outstanding spatial features. First, a northwest–southeast zone of mainly above average density of population (the circles) extends from England through the Netherlands and Belgium to Italy; at either end of this zone, in Scotland, Ireland and Greece, are sizeable areas of below average density. Second, a greatly elongated southwest–northeast zone of mainly below average density (the minus signs and lined shading) extends from southern Portugal to northern Sweden and Finland. Low density of population in this zone is over-ridden where it overlaps the high density northwest–southeast zone. The low density zone contains several 'islands' of high density, for example Madrid in Spain, Paris and Haute Normandie in France, and Stockholm in Sweden. In eight of

the 15 EU countries, including the more peripheral ones (see Table 3.3), most of the area is below the average density for the EU.

Although the range of density in the EU, extending from almost 6,000 to 2 persons per sq km, is enormous, the range is even greater in the world as a whole. At the high end, for example, the population of the whole of Hong Kong is about 5,140 persons per sq km, but on Hong Kong Island (over a million people on 80 sq km) there is a density of about 15,600 persons per sq km and in Kowloon (over 2 million people on 43 sq km) a density of about 47,600 per sq km. At the other extreme, the very low density of the whole country of Iceland, with 2.5 persons per sq km, compares with that in single regions of northern Sweden, while the US state of Alaska has a density of about 0.3 persons per sq km, the Northern Territory of Australia 0.1 and the northwest Territories of Canada 0.02.

While the wide variations in the density of population among the regions of the EUR 15 are a geographical fact, relating density to level of economic development and other aspects of human activities, and to life in general, is not straightforward. At both national and EU level, areas with a high density of population form obstacles to the movement of traffic between pairs of other places. In areas with a very low density of population it is costly to provide adequate services such as healthcare and education. In general, far more is spent per capita from EU funds to assist development in comparatively poor regions with a low density of population than in large urban areas, much of it on improving transportation links, especially in peripheral areas. While making such areas more accessible to the rest of the EU, the links also make penetration from elsewhere more easy and, in the past more than now, have made emigration more easy from the very areas at which assistance was aimed. Since larger, more innovative, and expanding enterprises tend to flourish in central areas, greater case of movement in the EU as a whole could indeed affect the periphery negatively.

URBAN POPULATION

For at least two centuries the proportion of urban to total population has gradually increased in most parts of Western Europe. This trend broadly correlates with a decline in the proportion of agricultural employment in the economically active population. In practice, however, it is not possible to make a universally consistent subdivision of settlements into rural and urban types. Urban (as opposed to rural) population is determined largely according to three, to some extent mutually exclusive, criteria: the population size of a settlement, its function (agricultural or non-agricultural) and the definition of the administrative district in which it is situated. Many nucleated settlements in parts of southern Europe with over 10,000 inhabitants still contain many agricultural workers, while in other parts of the EU many districts defined as urban contain villages and farmland. On the other hand, persons employed in local services, together with commuters to nearby cities, outnumber agricultural workers in many areas defined as rural.

At national level there is considerable divergence in different sources as to the percentage of total population defined as urban at any given time. In the mid-1990s the differences between countries were, however, very marked, and the divergence far greater than any disparities in definition would produce (see Table 3.4). Belgium and the UK are two of the most highly urbanised countries in the world whereas Portugal is much less highly urbanised than many developing countries, especially those of Latin America. In all 15 countries of the EU the proportion of total population defined as urban has increased this century. In almost all the countries, however, the proportion of non-agricultural population greatly exceeds the proportion defined as urban, underlining the presence in areas defined as rural of many people employed in services and in industry. For the formulation of EU policy, information about the level of urbanisation of regions is of limited value in view of the inconsistency in definition. On the other hand, the location of the larger cities is of great relevance in shaping the human geography of the EU.

In the mid-1990s the EU had 36 cities with over one million inhabitants, together with several others that by a somewhat more extended definition would also qualify. The 36 cities are listed in order of size in Table 3.5. The numbering in the table corresponds to the numbering on the map in Figure 3.4. A large number of other cities with fewer than

Table 3.4 Urban and rural population, and employment in non-agricultural sectors in EUR 15

Rank		Percentage Urban		Rural	In non-	Rank		Percentage Urban		Rural	In non-
		1960	1995	1995	agric.			1960	1995	1995	agric.
1	Belgium	92	97	3	98	9	Italy	59	68	32	94
2	UK	86	92	8	98	10	Spain	57	64	36	91
3	Netherlands	85	89	11	97	11	Finland	38	64	36	93
4	Luxembourg	62	86	14	98	12	Greece	43	63	37	78
5	Germany	76	85	15	96	13	Ireland	46	57	43	88
6	Denmark	74	85	15	96	14	Austria	50	54	46	95
7	Sweden	73	83	17	97	15	Portugal	22	34	66	86
8	France	62	74	26	96						

Sources: WPDS 1995 (PRB 1995) *and FAOPY 1994* (FAO 1995)

Table 3.5 The 36 largest cities of EUR 15 in the mid-1990s, population in millions

Rank size		Country	Population	Rank size		Country	Population
1	Paris*	France	10.7	18	Stockholm	Sweden	1.5
2	London*	UK	10.3	19	Munich	Germany	1.5
3	Rhein-Ruhr*	Germany	9.2	20	Liverpool	UK	1.4
4	Milan*	Italy	5.1	21	Brussels	Belgium	1.3
5	Madrid*	Spain	4.9	22	Lisbon	Portugal	1.3
6	Berlin*	Germany	4.7	23	Oporto	Portugal	1.3
7	Athens*	Greece	3.5	24	Glasgow	UK	1.3
8	Rome	Italy	3.1	25	Sheffield	UK	1.3
9	Barcelona	Spain	2.7	26	Lyons	France	1.2
10	Naples	Italy	2.6	27	Marseilles	France	1.1
11	Birmingham	UK	2.6	28	Frankfurt	Germany	1.1
12	Greater Manchester	UK	2.6	29	Newcastle	UK	1.1
13	Leeds/ Bradford	UK	2.1	30	Genoa	Italy	1.0
14	Vienna	Austria	2.0	31	Lille	France	1.0
15	Copenhagen	Denmark	1.7	32	Valencia	Spain	1.0
16	Turin	Italy	1.6	33	Rotterdam	Netherlands	1.0
17	Hamburg	Germany	1.6	34	Amsterdam	Netherlands	1.0
				35	Dublin	Ireland	1.0
				36	Helsinki	Finland	1.0

Notes: * Numbering of notes corresponds to rank
    1 Ile de France
    2 Greater London plus parts of contiguous counties
    3 Regierungsbezirke of Düsseldorf and Köln
    4 Provincie of Milano, Como and Varese
    5 Comunidad de Madrid
    6 West plus East Berlin
    7 Development region of Attiki
    11 West Midlands
    13 West Yorkshire
    20 Merseyside
    25 South Yorkshire
    29 Tyne and Wear

one million inhabitants are also shown on the map. They are not chosen explicitly according to population size but are based on a map from the European Commission (Commision of the EC 1994a: 111) on which are located 194 centres in Western and Central Europe considered to be of economic importance in the early 1990s. The authors have relocated a number not correctly sited on the map (e.g. Brno in Slovakia, Łódź in Poland), have reduced an excess in Hungary, and have added others that in their view merit inclusion. As might

be expected, the distribution of centres correlates broadly with the distribution and density of population (see Figures 3.2 and 3.3).

Of the 36 'million plus' cities of the EU, 14 are national capitals (Berlin to be, Luxembourg too small to qualify). In addition to the six largest urban centres, each with over 4 million inhabitants, two very large clusters of cities, one in north central England (11, 12, 13, 20, 25 and several other sizeable centres) contains about 10 million inhabitants, and one in the extreme north of France, Belgium and the

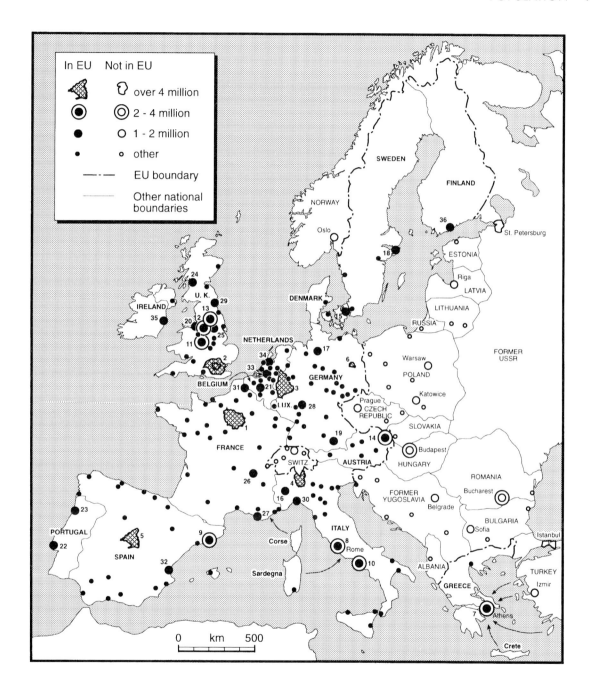

Figure 3.4 Cities of the EU with over about one million inhabitants, and other selected cities. The key to the numbering of unnamed cities is in Table 3.5

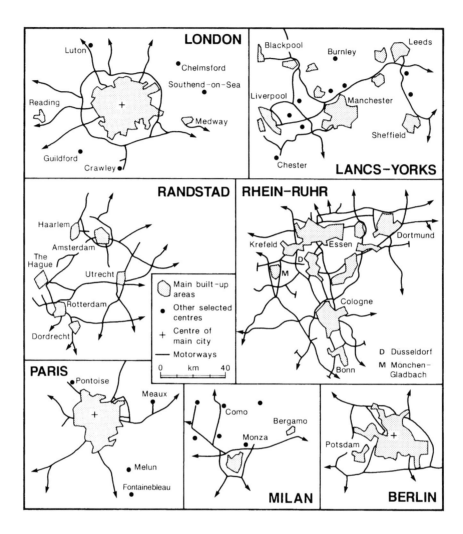

Figure 3.5 Seven major urban agglomerations of the EU. The scale is the same for all seven maps

Netherlands (31, 21, 33, 34) contains about 8 million inhabitants (see Figure 3.5). The triangle Paris (1), Frankfurt (28), Manchester (12) includes five of the eight largest concentrations of urban population in the EU. Four of the EU's busiest airports (or groups of airports), London, Paris, Frankfurt and Amsterdam are here, as well as the most prominent financial centres. In European Commission of the EC (1994a: 100–1), ten major 'islands' of science-based innovation in the EU are identified: Greater London (2), Rotterdam/Amsterdam (33, 34), Ile de France (1), the Ruhr area (3), Frankfurt (28), Stuttgart, Munich (19), Lyon/Grenoble (26), Turin (16) and Milan (4). Outside the general area extending from northern England to northwest Italy there are secondary centres of national importance, the largest mostly being capital cities (e.g. Stockholm, Copenhagen, Vienna, Athens, Madrid), all within one to two hours by air of the largest cities but more remote from the population 'core' of the EU by land travel.

In Chapter 13 the possible expansion of the EU into Central Europe will be discussed. In anticipation it can be noted on Figure 3.4 that there are several million plus cities not far from the eastern side of the EU, for example Prague, situated roughly mid-way between Berlin and Vienna, all of them much closer to the core of EU population than Athens, Lisbon or Helsinki. Waiting in the wings for a chance to become more closely linked are St Petersburg near the border of Finland, and Istanbul, near to Greece.

RURAL POPULATION

The particular importance of the larger cities to the economic life of the EU was stressed in the previous section. In a different sense, however, the rural population also has an important role in the EU, although its contribution to the total GDP is well below its share of total EU population. Somewhere between 96 and 97 per cent of the total area of the EU can be considered rural according to an estimate made by the authors, described below.

In the mid-1990s, according to the definition of urban and rural population in each of the Member States of the EU (see Table 3.4), about 286 million people (76.8 per cent of the total) were urban dwellers, leaving 86.5 million (23.2 per cent of the total) as rural dwellers. Without attempting to take into account whether urban refers to the size, function or status of settlements, it can be calculated that if on average 2,400 urban dwellers of the EU occupy 1 sq km, then 286 million urban dwellers would occupy about 120,000 sq km, which is about 3.7 per cent of the total area of 3,256,000 sq km. There are, of course, great differences in the proportion of urban area to total area in different countries, with a much larger proportion, for example, in the Netherlands, Belgium and

England than in Finland or Portugal. The density of 2,400 was arrived at by taking the average density for ten urban administrative areas. Even if a much lower density is used to calculate the area occupied by the urban population, the overwhelming part of the territory of the EU is still rural.

As noted in the previous section, however, it is mistaken to assume that in the EU of the 1990s agriculture is the main occupation of the rural population. According to *FAOPY 1994* (FAO 1995), in 1994 the agricultural population of the EU was only 18 million (of which 8.5 million or 47 per cent were economically active) out of the total rural population of 86.5 million, that is little over 20 per cent, the majority being in industry and services.

The interaction between rural and urban populations and conditions has been the theme of much literature and the concern of policymakers and planners. In many parts of the EU the long-standing net influx of rural dwellers into urban centres has now been replaced by an exodus from the cities, especially the larger ones, into the surrounding countryside, with commuters prepared to travel tens of kilometres between home in the country and work in the city. Features of the countryside found in urban centres are confined to parks and private gardens. In contrast, the urban population increasingly pushes into the countryside, either permanently in commuter housing added to existing villages, or briefly in outpourings of vehicles and hikers, swarming over areas of particular natural beauty or historical interest. Transport and communication links break up the continuity of the rural landscape and buildings connected with industry or services are widely dispersed over the countryside.

Although living standards are generally somewhat lower in rural than in urban areas in the EU, almost every rural dweller now has access to an electricity supply, running water and paved roads, the latter not only facilitating the movement of farm vehicles and machinery,

{Plate 3.3} Pasture, forest and steep slopes combine to produce a picturesque landscape at Mayerhofen in Austria

{Plate 3.4} In the Republic of Ireland, the west of the country remains largely rural, a long-standing tradition, or so it is claimed, in Mayo

Plates 3.3 and 3.4 In spite of the growth and spread of cities in EU countries this century, large areas remain little affected by urban development

but allowing visitors to penetrate in their cars to the remotest corners of the countryside. Education and health services are universally available, if in practice often more difficult to reach than in cities.

The contrast between urban and rural life is generally much less marked in the EU than it was even at the end of the nineteenth century, when many people had never left the villages in which they were born. Nevertheless, according to Collett (1996), the Rural Development Commission has intensified its campaign for better services in the countryside:

> The latest survey from the Commission, a government agency that suggests ways of improving life in the countryside, shows that nearly a third of English villages have no daily bus service, only half have a post office and 83 per cent lack a permanent doctor . . . . Without public transport, recruitment of staff is a huge difficulty for businesses. People have to rely heavily on voluntary car schemes. There is also little housing at reasonable prices in the countryside and this makes the recruitment worse.

The presence in the EU of a large rural population and the fact that this population is settled in more than 95 per cent of the total area of the EU, is apparently not of sufficient concern to the Commission or to the European Parliament for there to be an explicit reference either to rural or to urban population in *Europe in Figures* (Eurostat 1995a) or *Fact Sheets on the European Parliament and the Activities of the European Union* (European Parliament 1994b). On the other hand, as will be shown in more detail in Chapter 12, much of the EU budget is allocated to supporting and improving the living standards of the rural population, particularly the fifth that is economically active in agriculture. CAP funds (see Chapter 6) benefit both well-off farmers in areas with large farms and good bioclimatic conditions, and small farmers in areas with difficult conditions. Much of the population of the lagging regions of the EU (Objective 1) is rural, as is also

virtually all of the populations of the thinly populated Objective 5 regions and the special Objective 6 regions of the north of Sweden and Finland.

From what was said at the beginning of the section on urban population about the definition of urban and rural it is evident that there is no universal, objective, clear-cut way of dividing settlements into two distinct classes. Rather there is a continuum from places that are indisputably urban to places that are equally distinctly rural. What has been changing in Western Europe particularly in the last two centuries of industrial growth and so-called post-industrial development is the relative preponderance and dominance of urban and rural types. Many parts of the countryside of the EU, with its rural population and traditions, is gradually being transformed into a grey area that is neither urban nor rural, but which reaches well beyond the conventional suburbs of the larger cities. Purely agricultural settlements are becoming a feature of the past and areas in which natural conditions are virtually untouched, in spite of attempts to protect them, are shrinking as mechanisation allows steep slopes to be ploughed, marshlands are drained, peat is cut for power stations and suburban gardens, and trees are felled to make way for roads or for their timber.

Two centuries ago urban areas in Western Europe were small patches on an almost uninterrupted countryside. In the next century the genuinely rural areas will be patches of various size, some explicitly 'protected', among large, sprawling urban and semi-urban patches, with tentacles reaching outwards over great distances. Nuttall (1995) describes a measure used to assess the extent and loss of areas of tranquillity in the English countryside. Such areas are defined as 'peaceful and unspoilt places, typically between one and three kilometres from roads, four kilometres from a power station, beyond large settlements and the noise of military or industrial activity'.

For what it is worth, areas of tranquillity now cover little over half of the area of England (some 70,000 sq km out of 130,000 sq km) and between the early 1960s and the early 1990s alone about 20,000 sq km were lost, while the average size of such areas has diminished greatly.

NATURAL CHANGE OF POPULATION

In the early to mid-1990s there was little change in the total population of the EU through natural increase (or decrease). On the other hand, net in-migration from outside the EU has increased, but its impact on population change varies greatly between countries and between regions within them. Population change through natural increase will be discussed in this section, migration in the next section. In Table 3.1 columns (2)–(4) show that in the mid-1990s there was not much difference between the number of births and deaths in most countries. Between 1985 and 1990 the population of EUR 15 grew from 359.6 million, to 365.8 million, or by 1.7 per cent (0.17 per thousand). Between 1990 and 1995 the total increase was 1.8 per cent. Some of that increase was the result of migration, so the rate of natural change was very low by standards in Western Europe only a few decades ago and even more markedly by current world standards. According to Commission of the EC (1994a Table A.27), population change in EUR 12 was 0.3 per cent per year during 1981–91.

In 1995, there was virtually no difference between births and deaths in Germany, Italy and Greece, while even Ireland, the Netherlands, France and Finland, with the highest rates of natural increase, only recorded 0.3–0.5 per cent per year. At national level it is not realistic to explain the modest differences by cultural influences, such as the policy of the Roman Catholic Church on the use of contraceptives, or by economic performance. At regional level, however, more marked differences occur within countries. For example in Italy in 1992 (ISTAT 1993 Table 2.11) there were 561,000 births, 541,000 deaths, a difference of 0.04 per cent. In North-Central Italy, however, there was a decrease of −0.19 per cent whereas the Mezzogiorno (the south) experienced an increase of +0.42 per cent. At NUTS level 2 in Italy the disparity was even greater: Liguria, which includes Genoa, experienced a decrease of −0.66 per cent, contrasting with an increase of +0.64 per cent in Campania, which includes Naples.

While the natural change of population in a given period of time is the difference between births and deaths, total fertility rate measures the number of children born on average to each woman. Rates are specific to the female population of child-bearing age or to specific subdivisions by age (eg five-year age cohorts) within that child-bearing period, whereas birth rate and death rate are expressed as proportions of total population. A drawback with the use of fertility rate data is that the calculations can be made only from an appropriate sample of women who have already passed their child-bearing limit. Only the child-bearing *intentions* of women still below that threshold can be known. An advantage of the data for total fertility rate (TFR) is, however, that the index calculated for a given year is not only a guide to the birth rate that year but also to births to be expected in 2–3 decades time. The reason is that the number of potential child-bearers is already known and, assuming only limited mortality among females (around 1–2 per cent) between birth and age 20 and limited migration, the number of births that can be expected, given particular fertility rates in the future, can be estimated.

The demographic structure of all EUR 15 countries has been characterised by a sharp drop in TFR since the Second World War. In 1950, TFR was above the replacement rate of about 2.1 per cent in all the countries of EUR

15, although by world standards of the time it was already low in West Germany, the UK, Sweden and Belgium. By 1975 TFR remained above replacement level in only five out of the present 15 Member States. In 1995 TFR had dropped to 1.6 for EUR 15 as a whole, with even Ireland now below replacement level and Italy, Spain and Germany far below it. Elsewhere in the world, comparably low rates are very rare (e.g. Hong Kong 1.2, Slovenia 1.3, Russia 1.4) among the main developed countries of the world, but Japan (1.5), Canada (1.7) and the USA (2.0) show a similar trend.

Even from the data for three years in Table 3.1 (columns (5)–(7)) it is evident that TFR has declined from initially different levels in 1950, but at different rates during 1950–75 and again during 1975–95, in different countries. In Spain, for example, there was a considerable increase in TFR during 1950 and 1975 (improvements following the end of the Civil War in 1939) but a very sharp decline after 1975. In some of the northern Member States of EUR 15, the rate seems to have stabilised during the last two decades a little way below replacement. The evidence points to the prospect of a decline in population in many regions of EUR 15 in the next few decades. While there is no immediate prospect that declining family size will cause serious concern in the EU, in a few decades time the proportion of elderly to total population is likely to be an increasing problem. The question of ageing and support for the elderly will be discussed in Chapter 9.

MIGRATION

Like the term urbanisation, the term migration is not a process that is clearly defined. In principle, migration applies to any permanent change of residence, although a move to the next street would hardly merit inclusion. Information about some types of migration is difficult to obtain, and good information may be available only from census material collected every ten years. Migration flows between distinct administrative areas are the flows normally recorded.

In the context of the EU, three levels of migration are of interest: flows between regions within each Member State, flows between EU Member States, and flows between any part of the EU and any region outside it. Migration between two regions includes flows in both directions. For practical purposes it is usually adequate and appropriate to refer to the direction of the net flow, the difference between the number of people flowing in each direction. Within each of the countries of Western Europe there have not normally been restrictions on inter-regional travel and migration since the Second World War. With the gradual removal of barriers at the national boundaries between EU Member States, restrictions on migration between countries have now in principle also been reduced. In practice it is not always easy or possible for people to migrate to take up employment anywhere in the EU. For the most part immigration from outside the EU is even more strictly controlled. The three levels of migration will now be considered in turn.

Interregional Migration within EUR 12 countries (Eurostat, 1994a, Table 1.4)

Most internal migration in EU countries is not directly related to EU policy and funding but, even so, it may indirectly reflect changes in the relative fortunes, genuine or perceived, of different regions. A brief summary of migration follows, country by country. The number of people migrating in the year studied is expressed as a percentage of total population.

Germany 1990 (1.3 per cent)

The data do not include the very large number of people moving from East to West Germany in

1989–90. In 1990, the most northerly Länder of West Germany, Niedersachsen and Schleswig-Holstein, were recipients of the largest net inflows of migrants, while the three most southerly Länder, Bayern, Baden-Württemberg and Hessen, together with Nordrhein-Westfalen, were net losers. The net north–south flow characteristic of the mid-1980s was therefore reversed. Were the prospects of southern Germany already deteriorating in the late 1980s?

UK 1991 (1.6 per cent)

The net flow here was out of the southeast, West Midlands and northwest, which include the largest urban concentrations, to the more peripheral regions of East Anglia, the East Midlands, the southwest, Scotland and Wales.

France 1990 (1.1 per cent)

An outflow took place from Ile de France (Paris), Nord-Pas-de-Calais and Lorraine (declining heavy industry) to the southern third of France, especially Provence-Alpes-Côte d'Azur and Languedoc-Roussillon. Climatically southern France is perceived to be more attractive than northern France, but economically there is little to choose.

Italy 1991 (0.5 per cent)

Ever since Italian unification was completed in 1870, the tendency has been for net migration to be from south to north. The industrial regions of Piemonte (Turin) and Lombardia (Milan), as well as the new capital, Rome, in Lazio, were the main destinations of internal migrants, while Italians also emigrated in large numbers to the Americas until the 1920s. After the Second World War, many southern Italians worked in Germany, Belgium and France. In

1991 the pace of internal migration was much slower than earlier this century, but the regions with the largest net outflows remain Sicilia, Calabria, Puglia and Campania, all regions with fertility rates well above the Italian average. Emilia-Romagna and Veneto have joined Lombardia among the preferred destinations, whereas Piemonte attracted comparatively few.

Spain 1991 (0.4 per cent)

Comparatively little interregional migration took place. The País Vasco and Catalunya were the main net recipients of migrants and Andalucía and Valencia the biggest losers.

In the Netherlands 1991 (1.4 per cent) and Belgium 1991 (0.9 per cent) considerable internal migration took place whereas in Portugal 1991 (0.2 per cent) there was very little at all. In Greece, the population of Athens has grown rapidly, drawing widely on migrants from other parts of the country.

Intra-EU Migration between Member States

There are considerable differences among estimates of the foreign born population of the EU countries in the early 1990s. The general consensus is that these comprised between 16 and 16.5 million, about 4.5 per cent of the total population of EUR 15, including Austria, Finland and Sweden before they joined. Table 3.6 shows numbers according to two different sources. The numbers do not include children born in the EU of migrant parents. The impossibility of having precise data for migration at any given time is illustrated by the estimate in *Ethnicity in the 1991 Census* (British) published by the Office of National Statistics, in which the Irish born population in Britain (i.e. excluding Northern Ireland) is somewhere between the spuriously precise limits of

Table 3.6 Foreign born population in the countries of the EU in the early 1990s

	(1) Thousands	*(2)* % of total popn	*(3)* Thousands	*(4)* % of total popn	*(5)* % of labour force
Belgium	905	9.1	921	9.1	8.3
Denmark	161	3.1	189	3.6	1.9
Germany	5,518	6.9	6,878	8.5	8.8
Greece	229	2.3	n.a.	n.a.	n.a.
Spain	493	1.3	430	1.1	0.5
France	3,597	6.3	3,597	6.3	6.2
Ireland	88	2.5	94	2.7	3.0
Italy	781	1.4	987	1.7	n.a.
Luxembourg	116	30.1	125	31.1	38.6
Netherlands	692	4.6	780	5.1	3.9
Austria	n.a.	n.a.	690	8.6	9.6
Portugal	107	1.1	n.a.	n.a.	n.a.
Finland	n.a.	n.a.	56	1.1	n.a.
Sweden	n.a.	n.a.	508	5.8	5.1
UK	2,429	4.3	2,001	3.5	3.6
EU	15,116*	4.3			

Sources: (1), (2) 'Migrants in Europe', (1993) *The European*, 26 Nov.–2 Dec., p. 7
(3)–(5) Martin and Widgren 1996: 25, using OECD (SOPEMI) *Trends in International Migration*, 1994
(Paris: OECD), p. 27.
Notes: * EUR 12
n.a. not available

Table 3.7 Regions of origin of EU immigrants in thousands, 1992

Region of origin	Immigrants in thousands
Other EU country	4,907
EFTA country	409
Central/Eastern Europe	747
Rest of Europe	3,304
Africa	2,763
Americas	799
Asia	1,565
Australia/Oceania	89
Unknown	113
Total	14,696

Source: Dynes and Brock 1995

837,464 and 1,089,428. About one-third, or 5 million immigrants in EU countries were from other EU countries and about 400,000 were from EFTA countries, three of which became EU Member States in 1995. Table 3.7 shows the regions of origin of the EU foreign born population.

Since migration within the EU is almost exclusively economic rather than political, the direction of movement is largely from poorer to richer regions. Portugal, Greece, Spain, Italy and Ireland are the principal sources of migrants, while they receive very few migrants either from the 'northern' countries of the EU or from each other. A combination of factors influences the choice of destination of migrants seeking work in other EU countries: distance, cultural conditions, especially language, the type of employment available or thought to be available, preference for places with a large population in a small area and therefore a large turnover of jobs. There are therefore preferences in destinations. Portuguese and Spanish migrants mostly go to France, Belgium and Luxembourg. Those from the Republic of Ireland move to the UK (except for Northern Ireland). Italians have traditionally moved to Switzerland (not in the EU), Germany (about half a million), Belgium and France, while Germany has about a quarter of a million Greeks.

In theory the exchange of population between EU Member States should continue as young people extend their education in 'another' country, the elderly retire and aspire to end their days in pleasant environments and, above all, the search for jobs in other countries grows. While international migration within the EU is facilitated and even encouraged by closer ties between Member States, immigration from outside the EU has increasingly become a controversial issue, regarded by many as potentially the cause of serious problems, such as unemployment, pressure on housing and crime.

Migration between EU Countries and Countries outside the EU

After the Second World War and especially after the break-up of the overseas empires of France, Britain, the Netherlands, Belgium and

BOX 3.1 FOREIGNERS IN FRANCE

France is among the EU Member States with the highest proportion of foreign born population. In detail the situation in France differs from that in Germany, the Benelux and the UK, but since the 1980s the opening up of Eastern Europe and the serious economic crisis in the EU itself have highlighted and sharpened awareness of the problem of migrants seeking work or political asylum throughout the EU.

The presence of a large foreign born population is not new in France. In 1931 there were 2.7 million foreigners in the country, 6.6 per cent of the total population, 90 per cent from other countries of Europe. The proportion dropped in the 1940s, but in the 1980s it was

again between about 6 and 7 per cent, this time numbering 3.6 million people. Ideologically, France's Christian tradition and strong socialist leanings have tended to foster a liberal attitude towards migrants. This has combined with the fact of the low fertility rate among the French themselves, and the need for workers in sectors of the economy in which the French themselves have been unwilling to participate. Following the end of the Second World War, for almost three decades the French economy was expanding and migrant workers were actively recruited. By the 1970s France was one of the most prosperous countries in Europe. In 1974 a tougher policy towards immigration was adopted.

(continues)

In the early 1990s, about 40 per cent of foreigners in France were from other EU countries (Portugal 24 per cent, Italy and Spain each 6 per cent). Roughly another 40 per cent were from the Maghreb countries and Turkey (Algeria 17 per cent, Morocco 13). Most of the remainder were from more distant former French colonies, mainly in Africa.

The foreigners in France are distributed very unevenly in relation to the population of the various regions and departments of France. About one-third live in and around Paris, while in Rhône-Alpes, about half a million are concentrated around Lyon-Grenoble. Provence-Côte d'Azur (including Corse) account for another 400,000. The high concentration of foreigners in comparatively few regions reflects the availability in those regions of jobs in industry, construction and services such as healthcare. There are very few foreign born in northwest France (Bretagne) and west central France. Thus the advantages and drawbacks of having a large foreign presence seriously affect only about a third of France's population.

The foreign born in France are characterised by an average fertility rate per female of 3.2, which is almost twice as high as the average for French women of 1.8. There is an imbalance between males and females, although attempts have been made to encourage the development of family life among migrants. The foreign born are on average younger than the French, there being 40 per cent under 25 years compared with 36 per cent among the French. The crime rate is higher among the foreign population.

In the heavy industrial areas of France, many earlier Polish and Italian immigrants have been successfully assimilated into the local population after two generations. A similar future seems possible for the more recent migrants from Europe, most of them from countries now in the European Union, and therefore no longer restricted in where they work and live. The problem remains as to how to integrate more than 2 million foreigners of African and Asian origin, most with a Muslim background, cultural characteristics at variance with those in France, and conspicuously different from Europeans in the colour of their skin. An immediate solution to the problem of immigrants and further immigration is hoped for in France but, short of repatriation, which is advocated by the National Front of Jean Marie Le Pen, a politically sensitive issue, assimilation needs at least some decades, certainly not a few years.

Source: based on Lauby and Mareaux 1994

Portugal, many immigrants from the Caribbean, South Asia and Africa settled in Western European countries. Some, like the French from Algeria, and the British from Kenya and Northern Rhodesia, automatically had the citizenship of the countries to which they returned. At first, the indigenous population of the colonies was also accepted in large numbers, working in manual jobs such as construction work, in jobs with unsocial conditions such as hospitals, and in factories working shifts at undesirable hours. Germany lost its African colonies in 1919, so there was no such transcontinental migration there, but following the Second World War and the dismemberment of the country, West Germany received a very large number of forced migrants from territories lost to Poland and the USSR. When that flow ceased completely in 1961 following the construction of the Berlin Wall, Germany accepted large numbers of immigrants from elsewhere and in the late 1980s had about 1.5 million

Turks and over half a million Yugoslavs, kept without the right to permanent residence as Gastarbeiter (guest workers).

Since 1992, with the reduction of Russian influence in Central Europe and the former USSR, migration has increased greatly from that direction. The complex situation in the mid-1990s is difficult to summarise and the future difficult to speculate about with confidence. A number of features and trends in the 1990s may however be noted.

- Even without considering the special situation of Luxembourg, it is evident from the data in Table 3.6 that the 'northern' EU countries (apart from Finland) have been far more attractive to migrants than the 'southern' countries. More recently, Spain (proximity to northwest Africa), Italy (the Albanian and Yugoslav connection) and Greece (proximity to the Balkans) are becoming increasingly attractive.
- The political and economic changes in Central Europe and the former USSR since the late 1980s have made it possible for a large number of people to consider leaving the region. Previously the Iron Curtain was maintained to keep Western influence out. Now a new 'curtain' has been created to control the movement of people moving out of Central Europe into the EU.
- Individual EU governments and gradually the EU as a whole are becoming increasingly discriminating as to whom they allow to immigrate. Economic migrants are not generally wanted, while refugees seeking political asylum are carefully checked. Some people, such as 'Germans' from the former USSR (e.g. Volga Germans) and from Romania (Transylvania) are permitted to settle in Germany. Many others enter illegally in the way that, for several decades now, Mexicans have entered the USA.
- With the inclusion of Austria, Finland and Sweden, the land interface between EU and

non-EU countries has increased greatly. Before 1990 Germany and Italy were the only countries sharing international boundaries with non-EU countries, namely Poland, Czechoslovakia, Austria, and Yugoslavia. In 1995 the Swedish–Norwegian boundary, Finnish–Russian boundary and Austria–Czech–Republic–Hungary–Slovenia boundary were added to the EU land boundaries.

- Some coasts of EU countries have in recent years been used by illegal immigrants to reach the EU, especially the southern coast of Spain by migrants from Morocco and Algeria, and the Adriatic coast of Italy by migrants from Albania. According to Webster (1995c), in order to control the flow of migrants entering Spain's colony of Ceuta on the coast of Morocco, with financial assistance from the EU Spain has built a 'wall' equipped with Berlin-style paraphernalia: 30 closed-circuit television cameras, spotlights and sensory pads to detect anyone crossing a strip of no man's land.
- The conflict, ethnic cleansing and massive destruction of buildings and infrastructure in the former Yugoslavia between 1991 and 1995 have produced a very large number of refugees, many of whom have been accepted in EU countries.
- Under the Schengen Agreement (see Box 2.2), the movement of people between certain EU Member States is no longer checked. It is therefore increasingly easy for illegal immigrants, once inside the EU, to move between countries.
- Fifty thousand Hong Kong citizens have the right to settle in the UK, with dependants making about a quarter of a million potential immigrants, in particular following the return of its sovereignty to China in 1997. Once in the UK there is in theory nothing to stop them from moving to any other EU country

FURTHER READING

Champion, A. G., Fielding, A. J. and Keeble, D. E. (1989) 'Counterurbanization in Europe', *The Geographical Journal*, 155, Part 1, March: 52–80.

Hall, R. and White, P. (1995) *Europe's Population*, London: UCL Press.

Heilig, G., Büttner, T., and Lutz, W. (1990) 'Germany's population: turbulent past, uncertain future', *Population Bulletin* (Population Reference Bureau), 45(4): 33.

King, R. (ed.) (1993) *Mass Migration in Europe: The Legacy and the Future*, Chichester: Wiley. Reviews theory and policy, trends in recent decades and prospects for future especially with regard to Eastern Europe and South-to-North movements.

Martin, P. and Widgren, J. (1996) 'International migration: a global challenge', *Population Bulletin*, (Population Reference Bureau) 51(1), April. Western Europe, North America and the Middle East have been the main destinations for international migration in the early to mid-1990s. The EU situation is placed in a global context, with useful comparisons with other regions.

Noin, D. and Woods, R. (eds) (1993) *The Changing Population of Europe*, Oxford: Blackwell (in Progress in Human Geography series).

North, A. (1996) 'Alien nations', *The Geographical Magazine*, LXVIII(6): 22–3. The author argues that while illegal immigration is a great concern for politicians, young immigrants are necessary for the economy of Western Europe, with its ageing population.

Population Reference Bureau (various years) *World Population Data Sheet*, Washington D.C.

van de Kaa, D. J. (1987) 'Europe's second demographic transition', *Population Bulletin*, 42(1): March.

4

TRANSPORT AND COMMUNICATIONS

'An efficient transport system is a crucial precondition for economic development and an asset in the international competition. Personal mobility for work, study and leisure purposes is considered a key ingredient of modern life. With the integration of markets in Europe, economic growth and higher levels of income, transport is also a major growth sector.'

(Stanners and Bourdeau 1995: 434)

- Although considerable sums have been allocated from EU Funds towards improvements in the road and rail infrastructure of the Union, a comprehensive transport policy has not been implemented.

- The shape of the territory of the EU, with numerous peninsulas, islands and water and mountain barriers results in great distances between extremities and the need for a variety of modes of transport to handle goods and passenger traffic most efficiently.

- Road, rail and air transport compete for passenger journeys of varying lengths.

- Road, rail and waterways (inland and coastal) account for almost all the movement of goods within the EU.

- For environmental reasons, rail transport is regarded as preferable to road and air transport but the latter modes are able to compete mercilessly with rail for shorter and longer journeys respectively.

In almost every part of the world, local economic and demographic self-sufficiency no longer exist. Goods, people and information are increasingly transported or transmitted both between centres in the same country and, on continental and global scales, between different countries of the world. In densely populated and highly industrialised regions of the world like the European Union, in which regional specialisation in different economic activities is already highly developed, and is indeed encouraged, an efficient transport network is essential. The need for an integrated transport structure for the EEC was appreciated by the authors of the Treaty of Rome, but until the mid-1980s little was done to develop a common transport policy. Although transport is part of the service sector, the subject of Chapter 8, it is treated separately here on account of its great relevance to and influence on virtually all other activities.

INTRODUCTION

According to Eurostat (1995a: 322)

> transport for hire or reward accounts for approximately 5 per cent of GDP and 5 per cent of total employment in the European Union. Transport as a whole, both own-account and for hire and reward, accounts for 30 per cent of energy consumption and has a serious impact on the environment.

According to *Transport in the 1990s* (Commission of the EC 1993b: 2) transportation accounted for 7 per cent of the GDP of the EU in the early 1990s and directly employed 5.6 million people in the EUR 12 countries, while the manufacture of transport equipment gave jobs to another 2.5 million people. The disparity between the two figures for GDP may be because the lower estimate does not include own-account transport. The provision and improvement of transport facilities directly affects the energy, agricultural, industrial and service sectors, to be covered in Chapters 5–8, as well as the prospects for the regions of the EU, the features and problems of which are discussed in Chapters 11 and 12.

The transportation sector consists of two distinct aspects, networks and traffic. For modes of transport on the land, networks consist of fixed or 'hard' links (e.g. rail, pipeline), while for modes of transport using sea or air, networks consist of more flexible 'lanes'. Some modes are general purpose while others, such as pipelines, are restricted in application. Traffic can be subdivided into three broad categories: passengers, goods and information.

According to Eurostat (1995a: 84) the current internal aims of the EU's transport and infrastructure policy are to interconnect existing networks (e.g. integrating those of the former East and West Germany), to fill in 'missing links' (e.g. the Channel Tunnel), to unblock bottlenecks between the existing networks, and to overcome the isolation of the peripheral regions (e.g. by improving the road links between Portugal and Spain). The break-up of CMEA and of the Soviet Union has added a new task for the 1990s, the need to improve the quality of networks in Central and Eastern Europe, while the accession of Sweden and Finland in 1995 has added a large new, albeit mostly thinly populated, peripheral region to the territory of the EU itself. EU policy is concerned with road transport, with in particular improvements to cross-border links, with railways, with the construction of high-speed links, and with air transport, with emphasis on a unified system of air traffic control, as well as with other modes. According to *Trans-European Networks* (Commission of the EC 1994c: 4), 'the European Union is prepared to provide up to ECU 20 billion a year during 1994–1999 for the development and improvement of transport and communications'.

With regard to the use of the transport networks of the EU, according to European Parliament (1994b: 249) the development of a common transport policy has two main objectives: to eliminate all forms of discrimination and disparities of services, and to establish a common market in transport services, allowing free movement in all sectors of transport. Already in the 1950s, under the ECSC, discriminatory transport charges in the movement of coal and steel were eliminated and most subsidised rates were abolished. Only with moves to implement the Single Market since 1993 has there been an active policy to eliminate discrimination and allow freedom of services, with harmonised laws. For example, in 1993 international transport and cabotage by road were in theory completely liberalised, replacing quotas. Again, only in the 1990s has the negative impact of traffic on the environment been seriously considered. Implicit in many of the proposals for the future of transport in the EU is the intention, where commercially realistic, to transfer traffic currently using roads to the railways. Unless a comprehensive view is taken of the transport needs of the whole of the EU,

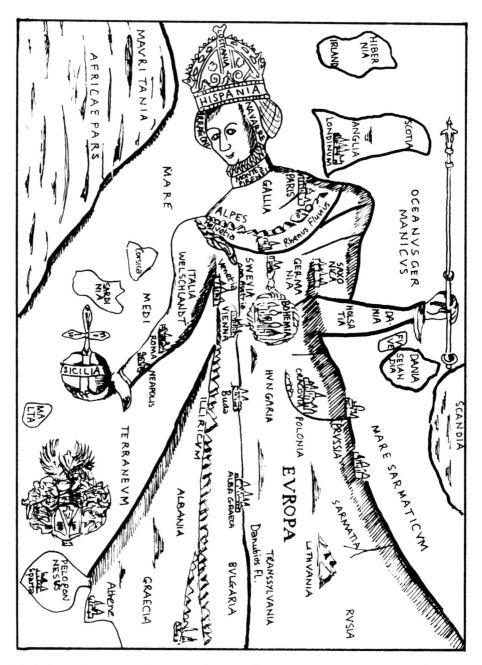

Figure 4.1 A sixteenth-century anthropomorphic map of Europe. The peripheral nature of the British Isles, Italy and Denmark is highlighted. Spain is credited with the greatest influence, while Bohemia is located at the heart of the continent. Note that Iceland and much of European Russia are not shown. Based on the picture on the cover of Foucher 1993. A somewhat different version of the map is reproduced in Vujakovic 1992

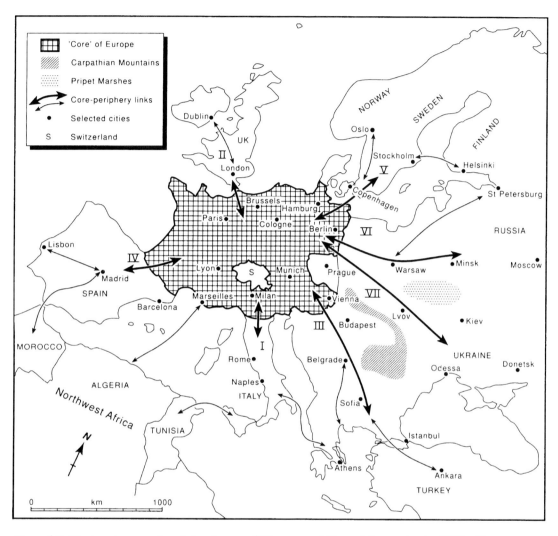

Figure 4.2 The 'core' area and peripheral areas of Western and Central Europe in the 1990s. The supremacy of Spain faded in the sixteenth century. Financial and economic power is focused on the area roughly enclosed by a line through Paris, London, Cologne, Munich, Milan, Lyon, Paris. Switzerland could claim to be the financial centre of Europe. The western part of the Czech Republic (see Prague on the map) is at present the only non-EU contender for a place in the core, apart from Switzerland

increased privatisation of many parts of the system could result in the neglect of links in sparsely populated and peripheral areas, where traffic is limited in scale, although vital for the continuing existence of many settlements.

The implementation of the above proposals for the transport system of the EU would result in yet

another shift of financial resources, power and decision-making from the 15 national governments to the EU level. Should some or all of the countries of Central Europe join the EU, part of the financial resources earmarked for improving the transport system in EUR 15 countries in the years to come would have to be diverted to improving existing road and rail systems outside the EU and constructing new ones there. Alternatively, a completely new extra fund would have to be set up to provide an integrated system for Western and Central Europe, with extensions to parts of the former USSR.

THE LAYOUT OF THE EU

It has been argued that the relative isolation of different parts of the EU, fostered by mountain ranges and arms of the sea, has been responsible for the emergence of distinct ethnic groups and many nation states, and for the great maritime traditions of the past, given that no place is very far from a coast. The shape of Europe, already referred to in Chapter 1 (pp.25–9), was caricatured in the sixteenth century in the anthropomorphic map of the continent shown in Figure 4.1. Figure 4.2 shows the EU in its European setting, a form the reader might visualise as some flying creature, but with Portugal and Spain no longer the head, and various islands and peninsulas as wings.

Figure 4.2 shows a central core area comprising France, Germany, the Benelux, Western Austria, and the northern part of Italy, all in EUR 15, together with Switzerland and part of the Czech Republic. Projecting from this central core area are a number of peninsulas and islands, mostly separated from each other by seas (e.g. the North Sea between the UK and Scandinavia), the main ones numbered I–VII:

    I   Italy
    II   The British Isles
    III The Balkans
    IV Iberia
    V   The Nordic countries
    VI Poland-Russia
    VII Slovakia-Ukraine

Five of the seven protuberances (I–V) are entirely or partly within the EU. Land journeys between places in any pair of these radiating parts of the EU (e.g. between Greece and the UK, Spain and Sweden) must pass through the core, which may be thought of as a great crossover area, carrying not only journeys between places within itself (e.g. Paris to Milan) or between some place within it and some place outside (e.g. Paris to Lisbon) but also 'other people's' journeys (e.g. Dublin to Barcelona or Copenhagen to Rome). Since almost all the journeys made on any transport system are actually over very short distances, it is not implied that long distance road or rail traffic actually places an excessive burden on the transport links in the central area, although heavy air traffic across the core is a serious problem, since most commercial air journeys *are* over considerable distances. France was the 'crossroads' of EUR 12 in the 1980s. The enlargement of the EU since then, particularly with the addition of East Germany and Austria, has made West Germany less peripheral than it was, and any further enlargement of the EU would increase the centrality of Germany in the core area at the expense of France.

The situation described above produces a dilemma for EU policy-makers, one of whose jobs is to allocate funds to improve the transportation system of the Union. The improvement of links in the central part of the EU can be justified, since these are likely to carry international traffic between many pairs of Member States. There is an equally strong case for improving links in peripheral areas, although such developments in relation, for example, to Ireland, Portugal and Finland affect only journeys between one particular Member State and all the others. No through traffic between other pairs of Member States crosses these countries.

The stellar form of the layout of the EU results in the presence of the seven main 'directions' listed above radiating from the central core. The order of numbering I–VII relates primarily to the length of time each direction has been associated with the EU. Italy except the North (I), Bretagne, southwest France and Corse were the main protruberances in EUR 6. The

Plate 4.1 Advertisement in the car-carrying compartment of Le Shuttle. Life at the end of the tunnel?

addition of the UK and Ireland (II) gave an appearance of symmetry, producing a second major protuberance 'opposite' peninsular Italy. The Western Balkans (III) became of direct interest to the EU when Greece joined in 1981, since the new Member State can be reached entirely over land only through a non-EU state. There are also ferry links of prime importance between southern Italy and Greece. The accession of Spain and Portugal (IV) added the Iberian Peninsula to the EU, while the accession of Sweden and Finland (V) brought Denmark into prominence as a key link with the Nordic area of the EU. In due course, if Poland (VI) and the Baltic Republics join the EU, then the eastern side of the Baltic Sea will form yet another prolongation of EU territory. Finally, although hardly a serious contender for

EU membership in the 1990s (see Chapter 13), Ukraine stretches away from Slovakia between the Carpathians and Black Sea to the south and the Pripet Marshes to the north, the latter a thinly populated area, as far as the Sea of Azov and the heavy industrial region around Donetsk.

TRANSPORT NETWORKS AND TRAFFIC FLOWS

The map in Figure 4.2 was designed to focus on the shape of Europe. In Figure 4.3 a general picture, not specifically related to any particular mode of transport, shows the main directions of international traffic flows in two distinct ways. Each of the solid black dots is located at the centre of population of 10 per cent of the total population of the EUR 15 (see Chapter 3). All

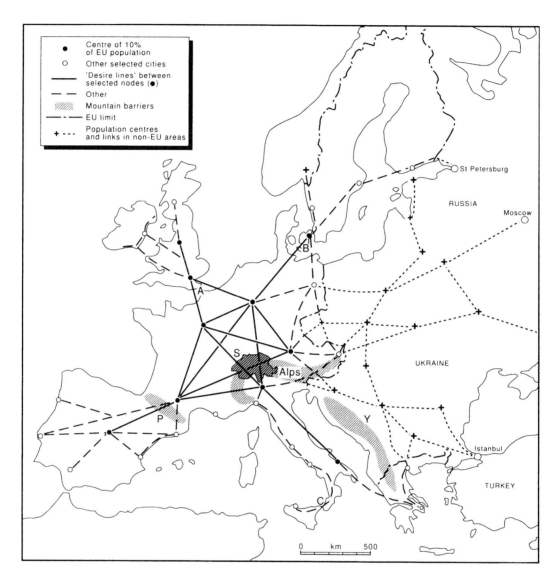

Figure 4.3 Hypothetical flows of longer distance land traffic between major cities or clusters of cities. See text for explanation

ten dots represent the population of two or more countries (e.g. that in Copenhagen takes in all of the population of Denmark, Sweden and Finland and part of the population of Germany). The solid black lines joining some pairs of dots are hypothetical 'desire lines', indicating the general direction of heavy traffic expected in the EU. Each of the five more central dots is joined to all the other four, but the five more peripheral dots are joined only to the central dot or dots closest to them. The open dots in the EU are selected cities, while the broken lines are links between them or with the 10 per cent of population dots.

Beyond the limits of the EU, the removal of the Iron Curtain and the break-up of the USSR and Yugoslavia have brought about rapid changes in the direction of flows of trade and traffic. The nodes in Figure 4.3 are selected major centres of population in Central Europe, the former USSR and Turkey. The links (dotted lines) shown on the map in this part of Europe are only those 'pointing' towards the EU, not those (e.g. Moscow–St Petersburg) linking pairs of places outside the EU. The proposed developments for integrating the EU and Central Europe are discussed in more detail in Chapter 13.

The improvement of transport links is regarded as particularly important in the areas of serious physical obstacles in the EU itself. A number of serious obstacles to the international movement of traffic in the EU are shown in Figure 4.3. They are related either to channels of the sea at present served by ferries but which technically, if not financially, could be crossed by tunnels or bridges (A–C), to mountain ranges (Alps, P – Pyrenees), or to non-EU countries located between EU ones (eg the Czech Republic between the eastern parts of Germany and Austria). Six of them will be described briefly below.

A   The Channel Tunnel itself is now fully operational. Construction costs turned out to be almost double what had initially been estimated. Such an over-run of costs is common with major civil engineering projects (cf. the Hokkaido Tunnel in the north of Japan). For the foreseeable future no dividends will be paid to shareholders. All traffic in the Channel Tunnel is conveyed by rail and consists of four basic types, through passenger and goods trains running between various pairs of cities on either side of the Channel, and Shuttle trains carrying passenger vehicles (cars, buses) or goods vehicles between the Tunnel terminals in Calais and Folkestone. Motorways now link the Channel Tunnel to Paris and London, but a highspeed rail link between Folkestone and London remains to be built.

B   A fixed link is due to be completed between the Danish Island of Sjaelland, on which the capital, Copenhagen, is situated, and the rest of Denmark (Fyn and Jutland). It will carry road and rail transport separately. Two other fixed links are under consideration to link Sjaelland with Sweden (the Øresund link) and with Germany. The completion of the latter two links would supplement if not replace existing train-carrying ferry services and would allow through trains to run between Sweden, Germany and other parts of the EU (see Cole 1993 and Plon 1996).

C   A train-carrying ferry has linked the Italian mainland near Reggio di Calabria to Messina in Sicilia across the Strait of Messina for many decades. Like Scotland, Wales and Northern Ireland in the UK, Sicilia is seen by its inhabitants to be a distinct region, remote from the rest of the country. A bridge across the Strait of Messina is technically possible, although the cost would be astronomical and damage in the future through the impact of a major earthquake a distinct possibility.

P   Motorways and railways pass either end of the Pyrenees between France and Spain without great difficulty. The Pyrenees remain, however, a psychological barrier between Iberia and the rest of Europe, and improvements have been made to road links across the mountains, in particular between Barcelona and Toulouse.

S   The Alps are a problem area in the EU overview of transport. Since the opening of the Mont Cenis and later the Gotthard, Simplon and Kandersteg rail tunnels, the barrier effect of the Alps has been greatly reduced. Since the Second World War, several motorway tunnels have also been opened, eliminating the tortuous and sometimes hazardous crossing of the Alps by ordinary roads. Nevertheless, Switzerland (still not in the EU) and, until 1995, Austria, have been concerned about the increasing amount of heavy goods traffic passing between northern Italy and various destinations in France, Germany and beyond. As a Member State of the EU, Austria now has a period in which to settle

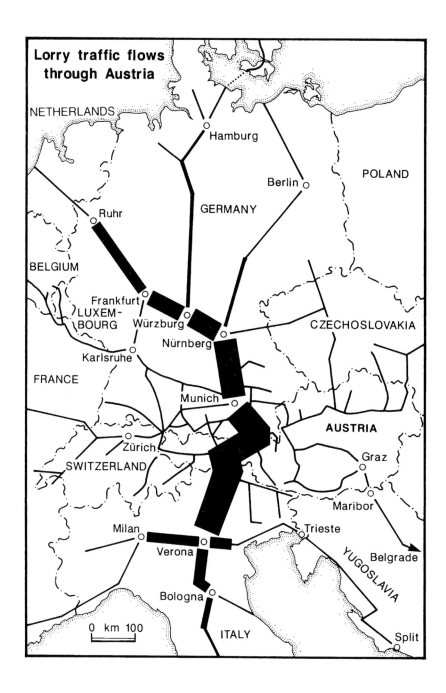

Figure 4.4 Main flows of lorry traffic using the Brenner Pass between Austria and Italy
Source: OECD 1991: 209

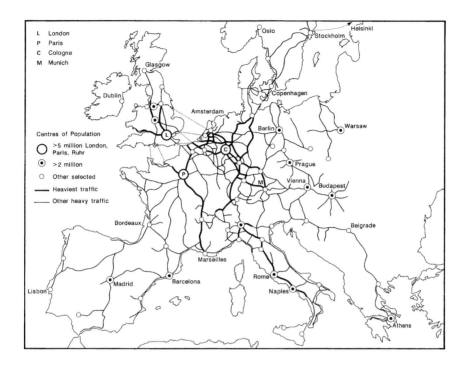

Figure 4.5 Generalised representation of main road traffic flows in Western and Central Europe. The map is based on a more detailed map in which various levels of traffic flow, from less than 1,000 vehicles per day to over 120,000 per day, are shown. The detail has been simplified to distinguish flows of over about 40,000 vehicles per day. Flows of less than about 5,000 vehicles per day have been ignored. Although the original map shows traffic on 'main international traffic arteries', most is either local or internal regional
Source: Economic Commission for Europe 1991, Map 11, Census of Motor Traffic on Main International Traffic Arteries

down to the reality of its geographical position. Improvements are promised on the busy Brenner Pass route (see Figure 4.4), which carries the bulk of the traffic between central Germany and Italy. Slovenia and other parts of the former Yugoslavia have been put in closer contact with Austria thanks to the construction of a road tunnel across the Carinian Alps. In Switzerland it is hoped that the growing flow of heavy goods vehicles across its territory can be handled with minimum environmental damage by dispatching them though long tunnels on vehicle carrying trains between points in northern Switzerland and points near the Italian border. Costs are again astronomical, and unless charges for transit are prohibitively high, would largely be borne by the Swiss taxpayer.

Y Since 1991 transit of traffic between Greece and the rest of the EU across Yugoslavia has been affected by conflicts. It is possible to take a land route avoiding the former Yugoslavia entirely by proceeding from Greece through Sofia (Bulgaria), Romania and Budapest (Hungary) to Vienna (Austria) or Bratislava (Slovakia). There are frequent ferry links between mainland Greece and Italian ports, notably Brindisi and Bari in the 'heel' of Italy, and Ancona further north.

Figure 4.3 gives a general idea of the spatial layout of the main flows of traffic in the EU. A more detailed and precise view of road traffic flows on specific heavily used motorways and other roads in the actual EU highway network is shown in Figure 4.5. The heaviest traffic is in

England, eastern France, the Benelux countries, Germany and Italy. In Central Europe, very few roads outside the immediate spheres of the largest cities carry heavy traffic, while roads in the former Soviet Union, where rail transport still dominates the movement of goods except over short distances, road traffic is very limited.

ACCESSIBILITY IN THE EU

As noted above, most of the traffic carried on the transport networks of the EU consists of short journeys. Many economic activities producing bulky or perishable items such as building materials and food products distribute their products locally or regionally although fish, fruit and vegetables may be taken cooled or frozen over great distances. On the other hand, many of the products of the manufacturing sector of the EU economy and some from the agricultural sector are sold throughout the EU. Various factors determine the choice of location of economic activities, including environmental considerations, and the availability of skilled labour and of capital. For economic activities whose end product is large, bulky or perishable, one consideration in the total cost of production is the cost of reaching the EU market, whatever the mode of transport used. The favourability of a location in relation to the whole EU market can be roughly quantified in terms of aggregate route distance, time or cost of delivery to a representative number of places receiving the products. Such a calculation is described below. Since much of the 'all-EU' flow of goods now goes by road, the example is illustrated by road transport, measured in terms of the estimated time a heavy goods vehicle (HGV) can be expected to take to travel between various pairs of places.

Part A (upper) of Table 4.1 is a matrix of HGV travel time in hours between each possible pair of ten places in the EU (Part B is referred to below). Each of the ten places in Part A, shown

by a black dot in Figure 4.6, is the centre of population (or centroid) of 10 per cent of the total EU population (see also Figure 4.3). The data in Table 4.1 show the travel time in hours, as calculated from the map in Figure 4.6, and is similar to the familiar road distance matrix published in many motoring atlases. For simplicity, there is assumed to be zero distance between each place and itself, although there would be transactions between places within the area represented by each of the centroids.

Column (11) of Table 4.1 shows the sum of all the distances between each place and the other nine, the aggregate travel time of nine separate journeys radiating from each black dot. The mean duration is shown in column (12) (i.e. the score in (11) divided by 10). Since Paris has the lowest index of aggregate travel, it has been chosen as the base against which the scores of the other places can be measured, relatively speaking. Figure 4.7 shows lines of equal accessibility in relation to Paris. From Figure 4.7 and Table 4.1, column (12) it is evident, for example, that the sum of journeys to the rest of the EU from Bari in southern Italy and from Madrid in the centre of Iberia is about twice as great as it is from Paris. A similar exercise was carried out by the authors in Cole and Cole (1993: 197–201) for 20 places in the EU. Results were broadly similar, but the addition of Austria, Finland and Sweden, all located to the east or northeast of the former centre of population of EUR 12, has made some difference in the scores.

In Part B of Table 4.1, 18 other selected cities of importance in the EU have also been assessed with regard to their aggregate travel time to the ten dots. The time distance of each city to each of the ten dots, the centroids of population each representing 10 per cent of the total EU population, has been calculated, and the times summed in Column (11). Column (12) shows the scores of each city compared with the Paris score of 100. Like Madrid and Bari, Dublin and Naples have roughly twice the

Table 4.1 Travel time for HGVs between places in the EU

	(1) Man.	(2) Lon.	(3) Par.	(4) Col.	(5) Cop.	(6) Mun.	(7) Bari.	(8) Milan	(9) Tou.	(10) Madrid	(11) SUM	(12) MEAN	(13) Paris = 100
Part A Centroids of 10% of EU population													
Manchester	—	5	13	17	29	24	41	27	25	36	217	21.7	149
London	5	—	8	12	24	19	36	22	20	31	177	17.7	127
Paris	13	8	—	7	21	13	28	14	12	23	139	13.9	100
Cologne	17	12	7	—	14	10	30	16	19	30	155	15.5	112
Copenhagen	29	24	21	14	—	20	39	29	33	44	253	25.3	182
Munich	24	19	13	10	20	—	19	9	23	36	173	17.3	124
Bari	41	36	28	30	39	19	—	14	27	40	274	27.4	197
Milan	27	22	14	16	29	9	14	—	14	27	172	17.2	124
Toulouse	25	20	12	19	33	23	27	14	—	16	189	18.9	136
Madrid	36	31	23	30	44	36	40	27	16	—	283	28.3	204
Part B Other selected cities													
Dublin	8	13	21	25	37	32	49	35	21	46	287	28.7	206
Glasgow	5	10	18	22	34	29	46	32	18	43	257	25.7	185
Lille	11	6	3	5	18	13	33	18	15	26	148	14.8	106
Brussels	12	7	4	4	17	12	31	17	16	27	147	14.7	106
Amsterdam	17	11	8	4	15	14	34	20	20	31	174	17.4	125
Luxembourg	17	11	6	4	18	8	27	13	17	29	150	15.0	108
Frankfurt	20	15	9	3	13	7	27	13	20	32	159	15.9	114
Strasbourg	19	14	8	6	16	6	24	10	16	30	149	14.9	107
Stockholm	42	37	33	27	25	33	52	42	46	57	382	38.2	275
Helsinki	54	49	45	39	9	45	64	54	58	69	502	50.2	361
Berlin	26	21	18	10	25*	12	31	21	30	42	220	22.0	158
Vienna	32	27	21	15	35	8	22	14	28	41	233	23.3	168
Rome	37	32	22	26	35	15	7	10	20	33	237	23.7	171
Naples	40	35	25	29	38	18	4	13	23	36	261	26.1	188
Palermo	53	48	38	42	51	31	12	26	36	49	386	38.6	278
Seville	47	42	32	39	55	47	51	38	27	11	389	38.9	280
Lisbon	46	41	31	38	54	46	53	41	29	13	392	39.2	282
Athens	63	58	50	52	61	43	22	36	49	62	496	49.6	357

Note: * Avoiding Czech Republic. Via Prague 21

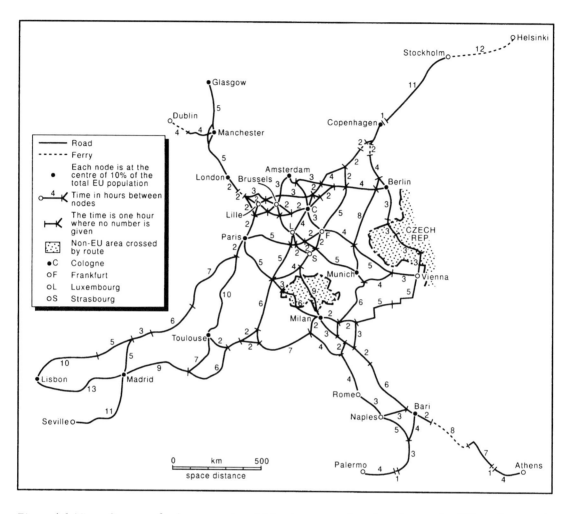

Figure 4.6 Time distances for heavy goods vehicles between selected nodes on the EU road network. Account is taken of the quality of each stretch of road, gradients and curves, and likely delays in and around cities. Ferry crossings allow time for embarkation and disembarkation. No allowance is made for the time taken for stops and rest time of drivers. Calculations of time have been made by the authors

aggregate score of Paris. On the other hand, cities in the most central part of the network, notably Lille, Brussels, Luxembourg and Strasbourg, together with Cologne (in Part A of the table), have scores that are similar enough to that of Paris to make a negligible difference in terms of the cost of delivery to the whole EU market as measured by aggregate time travel. At the other extreme, everywhere in Sweden and Finland except the southern tip of Sweden (around Malmö), in the 'toe' of Italy and Sicilia, in Greece and in the south western half of Iberia has more than twice the score of Paris. Greece and Finland, in particular, are especially locationally disadvantaged. Crete (Greece), the Canarias (Spain), northern Sweden and most of Finland have scores more than four times that of Paris. Some of the more

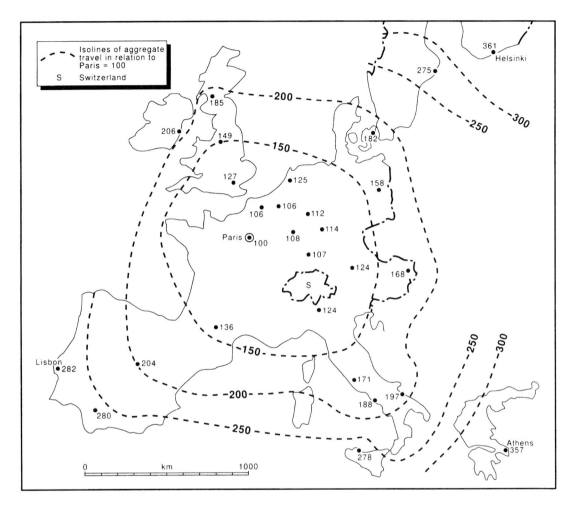

Figure 4.7 Isolines of aggregate travel time from places in the EU to the whole EU, based on data in Table 4.1 and Figure 4.6

general features that are illustrated by this exercise are discussed below.

- Although HGV time 'distances' on roads are used, the principle illustrated applies equally to all modes of transport. Thus, for example, Paris, Frankfurt and London all have much shorter aggregate travel times by air than Helsinki, Athens or Tenerife to any given set of places in the EU, even though air routes tend to be more direct than road or rail routes (e.g. contrast Dublin–Madrid by

air and by land), and the speed of travel (once the plane is in the air) is 10–20 times as fast.

- Major improvements to transport links cut travel distance, time and cost between sets of places (e.g. the Channel Tunnel, the improved roads between Spain and Portugal), often apparently benefiting more peripheral locations, but also making access to these remoter areas more easy from the rest of the EU.

- The impact of the opening of the Channel

Tunnel serves to illustrate the above point. It can if used effectively cut about two hours off the time needed to take an HGV across the Channel compared with the Dover–Calais ferries. Thus, for example, the aggregate score for London of 177 hours (column (11) in Table 4.1) includes eight journeys to places on the mainland of Europe. The score of 177 is therefore reduced by 16 hours to the advantage of London. On the other hand, from Paris, only two journeys are shortened by the opening of the Tunnel (to London and to Manchester), reducing the Paris score of 139 by only four hours.

- Improvements to transport links within the EU produce gradual changes in the relative advantages or disadvantages of places in terms of aggregate travel to the whole of the Union. On the other hand, in the event of the accession of countries of Central Europe, the addition of tens of millions of people would immediately produce a new situation. The EU can only expand eastwards, so the effect would be to pull the most favoured area locationally speaking from northeast France and Belgium into southern Germany. Berlin and Vienna would no longer be capital cities poised on the eastern frontier of the EU but would be well placed to influence Poland, the Czech Republic and Hungary in particular.

THE MODAL SPLIT IN EU TRANSPORT

For about 100 years, roughly from about 1850 to the Second World War, rail transport dominated the medium and long distance movement of passengers and goods in Western Europe. Since the 1920s, other means of transport have, however, taken an increasing share of traffic. The flexibility of road transport in particular has made it attractive compared with rail for two reasons: its capability of moving passengers and goods from 'door to door' and (on appropriate trunk roads) of carrying wide

and high loads precluded by the dimensions of tunnels and other constraints on the railways. Air transport is involved primarily in the movement of passengers, while the inland waterways and special modes, such as pipelines, convey goods.

The relative importance of different modes of transport in the EU can be measured in various ways.

Goods

- Goods carried on the land (ie not by sea or air), regardless of length of haul: road 88.5 per cent, rail 7.5 per cent, inland waterway 4.0 per cent for EUR 12 in 1991 (Eurostat 1995a: 328). Total carried 10,315 million tonnes, an average of almost 3 tonnes per inhabitant of the EU. This method of assessment does not take into account the fact that on average, journeys by rail and inland waterway are longer than journeys by road.

- Tonne-kilometres carried on the land, the length of haul therefore being taken into account: road 73 per cent, rail 18 per cent, inland waterway 9 per cent in the early 1990s. Total carried 1,100 billion tonne-kilometres. In the early 1970s the respective shares were 56 per cent, 30 per cent and 14 per cent (Eurostat 1995a: 327).

- When intra-EU trade is taken separately from all movement of goods, the importance of sea and inland waterway movement is evident: road 38 per cent, sea 26 per cent, inland waterway 18 per cent, fixed installations such as oil pipelines 9 per cent and rail 7 per cent. Air transport accounts for a mere 0.4 per cent.

Passengers

- In 1992, 4,158 million passengers were carried on the railways of the EU. The total passenger-km carried was 272 billion. The apparent average length of journey was therefore about 65 km (the total passenger-km divided by the total number of passengers carried). In contrast, without including the movement of passengers by public transport (buses, trams, metro systems), if the 146 million passenger cars in use in the EU in 1992 on average carried two passengers, and travelled 10,000 km in a year, then the total passenger-km moved on the roads was about 3,000 billion passenger-km, at least 10 times the rail total.

Since the end of the Second World War the movement of goods and passengers in the countries of EUR 15 has grown greatly, but the increase in goods transport has been accounted for mainly by road transport, that of passengers by road transport for shorter journeys and by air transport for longer journeys. On the railways traffic flows of both goods and passengers have not changed to the same extent (see Figure 4.16). Few passengers are carried by inland waterway (too slow) or by sea, except on ferry crossings, but both these modes of transport carry large quantities of bulky goods such as building materials, ores, fertilisers and fuels. Since the Second World War the movement of fuel and power by pipeline and electricity transmission line, often over considerable distances, has increased greatly in the EU, with many trans-border links now in place (e.g. electricity to Belgium and the UK from France).

In spite of the relative decline of rail transport and the closure of many of the less busy routes in the UK in particular, EU transport policy continues to favour a revival of rail transport, either in its own right, or through greater combined rail/road transport. For example, for many decades certain rail tunnels in Switzerland have been used to carry cars and light commercial vehicles on special trains. The use of freightliner terminals with containers transferred where and when appropriate between lorries and trains for long hauls by rail has become widespread.

Outstanding trends in the sphere of transport in Western Europe include the following:

- The closure of routes and stations in less busy parts of the rail network, paralleled, but not offset in terms of route distance, by the construction of some new stretches of route specially designed for high-speed passenger trains.
- The construction of motorways, avoiding city centres and greatly increasing the speed of travel of both commercial vehicles and private cars. Existing roads have also been improved by widening and the provision of better surfaces, but the actual length of road networks has not increased as quickly as the number of vehicles in use.
- A massive increase in air traffic, primarily for carrying passengers, but also to carry mail, and cargoes that are high in value in relation to weight/bulk.
- The completion of a canal link between the Rhine–Main inland waterway system and the Danube basin.
- The construction of pipelines to carry oil and natural gas internally, between EU Member States, and between sources of these fuels outside the EU, mainly in Russia, and the EU market.

RAIL TRANSPORT

Since the Second World War the length of track in the rail network in most countries of Europe has hardly changed, in the face of a general reluctance to close even the least used routes. In contrast, in the UK many routes were closed in the early 1960s, while in the former USSR, both in the Asiatic and European parts, new lines have been built. In spite of the EU policy of attempting to shift traffic from the roads to the railways, the prospect is that many rail routes will soon be closed in parts of the EU, notably in France (see Macintyre 1995).

Since the 1950s there have not been marked changes in most EUR 15 countries in the quantity of passengers and goods carried by rail, as measured in passenger-kilometres and tonne-kilometres respectively. Allowance should be made for an overall increase in population in the EU (not all the 15 were Member States in 1960), and for the fact that the average distance people travel by whatever mode has increased greatly. Thus in 1960s the rail system of what are now the EUR 15 countries handled 202 billion passenger-kilometres, compared with 272 billion in 1992, but there was virtually no difference between 1960 and 1992 in the amount of goods handled, 232 and 234 billion tonne-kilometres respectively.

The movement of passengers by rail roughly doubled in France and Spain between 1960 and 1992, whereas in the UK it hardly changed. The movement of goods by rail increased appreciably in Italy, Sweden and Finland over the same period, but declined sharply after 1960 in the UK. Trends in the last 30 to 40 years will not necessarily continue in the next few decades but, as will be shown in a later section (p.121), for the movement of passengers, rail travel is squeezed between the more attractive possibilities of road travel for shorter distances and air travel for longer distances, especially when time distance is a greater consideration than cost distance. With regard to the movement of goods, the decline of the coal industry in the EU has reduced the flow of one of the main items among the goods carried, while the movement of oil products by rail is very limited.

Table 4.2 shows the length of rail route in EUR 15 in the early 1990s, some 157,000 km, which compares with nearly 39,000 km of

Table 4.2 Railways and waterways of the EU

	Rail system				
	(1) Length in (000s) km	(2) % electrified	(3) bln tonne-km carried	(4) bln pass.-km carried	(5) Waterways in (000s) km
Belgium	3.4	53	8.2	6.8	1.5
Denmark	2.3	8	1.9	4.8	0
Germany	40.8	36	82.1	57.0	4.4
Greece	2.5	0	0.6	2.0	0
Spain	12.6	45	10.8	16.3	0
France	33.6	34	51.5	62.3	5.9
Ireland	1.9	0	0.6	1.1	0
Italy	16.0	53	21.7	46.4	—
Luxembourg	0.3	*	0.7	0.3	—
Netherlands	2.8	65	3.0	15.2	5.0
Austria	5.6	54	9.4	13.2	0.4
Portugal	3.1	13	1.8	5.7	0
Finland	5.9	24	7.6	3.2	—
Sweden	9.8	64	18.8	5.6	—
UK	16.9	25	15.5	31.9	—
EUR 15	157.4		234.2	271.8	16.8

Sources: Eurostat 1995a for (1), (2), (5); *UNSYB* 39th edition for (3), (4)
Notes: * included in Belgium
— very small quantity

motorway, but with over 3 million km of roads of all kinds. Virtually every settlement of consequence in the EU now has access to the road network. In contrast, the former presence of numerous stations on the dense rail network of many parts of Western Europe has been reduced both by the closure of less busy lines and by the closure of less busy stations on lines still in use. Only the most heavily used railway lines in the EU are double track, or in places quadruple track. The percentage of electrified track in each EU country, shown in Table 4.2, varies greatly, one reason for the difference being the early use of hydro-electric power in some countries to run the railways. Electrification is, however, economical only where traffic is above a given level of intensity.

The rail systems of the countries of Europe are now mostly heavily subsidised, and few routes would survive if rail traffic had to make a profit. According to Olins and Lorenz (1992), the level of subsidy varies greatly from one EU country to another, with the following values in millions of pounds sterling per 1,000 km of line in 1989: Britain 38, France 93, the Netherlands 133, FRG 134, Belgium 217, Italy 378. While a rail company has to construct, maintain and operate the track on which it runs its services, road vehicles use public roads, to which the various taxes levied on road users contribute only a limited share of the cost of maintenance and construction. In reality, road users are therefore also subsidised, one reason why long-distance coach services can generally apparently be operated far more cheaply than rail services over similar routes. The privatisation of British Rail will be watched with interest throughout the EU because after more than a century of fierce competition and at times considerable overlap of services, British Rail was finally integrated and rationalised through nationalisation in the late 1940s. In the 1990s it has been broken up again, not only into regional systems, but also between track and rolling stock.

The future of longer-distance rail travel in the EU may be seen in the French model. The lines specially built to carry the TGV are Paris–Lyon, Paris–Tours and Paris–Lille–Channel Tunnel. The Paris–Tours distance is 235 km in length; the 232 km from Paris to St Pierre-des-Corps just outside Tours is covered in 56 minutes, at a speed of 249 km/h (155 mph). The distance from Paris to Lyon (Part-Dieu) is 427 km, covered in 120 minutes, at a speed of 213 km/h. As with the Shinkansen services in Japan, special tracks, freed from the encumbrance of slow passenger or freight trains, allow trains to travel at speeds of about twice those averaged on many conventional main lines in Western Europe.

The proposed high-speed rail network needed to serve Western Europe with a competitive rail system in the twenty-first century, the completion of which is estimated to need about 20 years, is shown in Figure 4.8. In order to get the process moving and achieve rapid results, according to COM 89–564, the routes indicated with a heavy line in Figure 4.8 should be given priority, typically reflecting Community rather than national thinking:

> These projects, which are also interlinked by the French TGV line from Paris to Lyon, represent excellent coverage of Community territory, since nine out of the twelve Member States are involved. They also have the merit of presenting totally different geographical and economic features.
>
> (COM 89–564, 1990: 27)

This situation also exemplifies a problem for those planning developments and investment at the EU level: it may be necessary to involve and 'keep happy' all or most EU Members in a given project even if such a concession is not economically sound.

Given the rapidity of political, economic and technological change in Europe and the eternal financial constraints, the completion by 2010 of the system described above seems very ambitious.

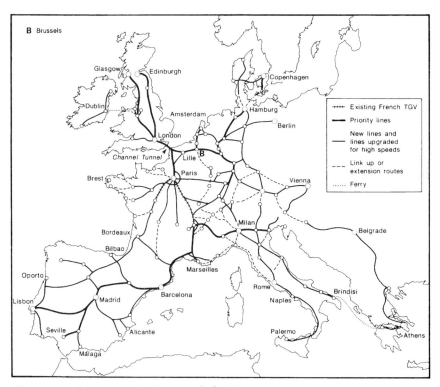

Figure 4.8 The proposed high-speed rail network for Europe
Source: COM 89–564 (1990: 31). This map has appeared in various publications, usually with differences in detail

Dynes (1992) notes ominous disparities at this early stage. French high-speed trains cannot run on German tracks because the power supply is different, while German trains cannot use French tracks because they are too heavy. If the desultory progress of improving the transport infrastructure of the EU in the last 20 years is a guide, some drastic changes in priorities and funding are needed before progress can be made.

The creation of an integrated rail system for the whole of Europe, with uniform conditions throughout, may seem desirable, but even now that barriers to movement within Western Europe and also between Western and Central Europe are expected to be reduced, technical differences make the unimpeded fast movement of trains impossible. Spain and Portugal in the West and the former USSR in the east have

different gauges of track from the rest of Europe, while variations in other dimensions, such as the height of tunnels, cause problems elsewhere. Even so, through passenger services have long been provided between various European countries (see Figure 4.9). For example, many well-known named expresses have for decades carried passengers from the northern cities of the EU to resorts on the French and Italian Rivieras and to cultural centres such as Venice, Florence and Rome. Many of these services cross Switzerland. With the addition of nine Member States of the EU since 1973, new problems of integration of the railway systems have arisen, since all are more peripheral than the original six countries, southern Italy excepted.

Figure 4.9 shows the lines on which international through services run. Examples are the

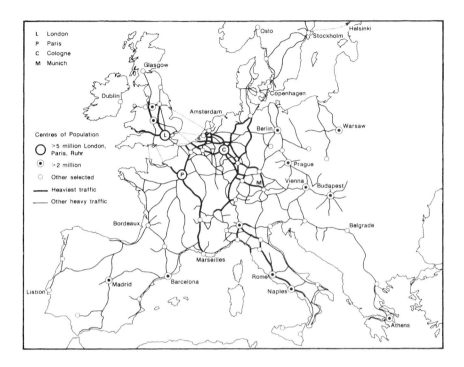

Figure 4.9 Lines used by through international train services in Western and Central Europe, and connections, 1991
Source: Thomas Cook 1991: 50–1

Paris–Madrid Talgo, which takes advantage of some TGV track, making the journey in 12 hours, including an adjustment of wheel sets at the border; the Italia Express, Dortmund to Rome, 18 hours, via Switzerland; and the slower Simplon Express Paris–Belgrade, 25 hours. Such Euro-City and similar expresses linking major urban concentrations at some distance apart usually run overnight, and include sleeping coaches and couchettes. They tend to depart and arrive at convenient times to avoid using working hours for travel, thereby giving them the possibility of competing with journeys on the same routes in a much shorter time by air.

ROAD TRANSPORT

In contrast to rail transport, road transport has expanded spectacularly in Western Europe since the Second World War (see Figure 4.10). The rapid expansion of road transport in the present EU countries can be gauged from the fact that between 1955 and 1992 the number of passenger cars in use increased twelve times, from 11.4 million to 146 million. The rate of growth of car ownership ranged during that period between about six times in Sweden and almost 100 times in Spain. The trend in the growth of commercial vehicles in Western Europe has roughly matched that of passenger cars. In Central Europe the rate of relative growth since the 1970s has been even faster, but the absolute number of private cars in the 1960s was minute.

Columns (1) and (2) of Table 4.3 show the length of motorway and of all roads in the EU in the early 1990s. The data in columns (3) and (4) show that between the early 1970s and the

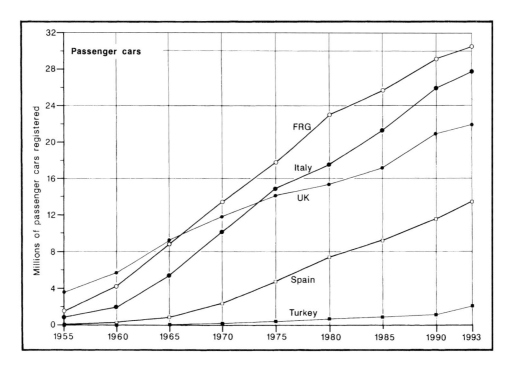

Figure 4.10 Millions of cars registered in selected EU countries and in Turkey, 1955–93. FRG-West Germany only
Source: various numbers of *United Nations Statistical Yearbook*

early 1990s alone the number of passenger cars in the EU almost doubled, with growth in relative terms fastest in Greece, Portugal and Spain during that period, when these countries were catching up. The data in column (5) show that in terms of passenger cars in use per 1,000 people, there is still a big car ownership 'gap', with Greece (178) having less than half of the EU average of almost 400 but with Italy (509) well above it. There remains a large difference between the EU average and the US level, the latter having about 570 cars per 1,000 inhabitants.

Convergence in the level of car ownership among the 15 EU Member States is unlikely for the foreseeable future, given the present great gap, since ownership is expected to increase for some time in all countries. The data in column (7) show the number of cars

in use that could be expected in each EU Member State around 2015 if by then the US level of 570 cars per 1,000 inhabitants was reached; the absolute increase of cars in use would be about 64 million. In practice, the situation varies from country to country in the EU. For example, the natural and economic conditions in parts of Greece may somewhat restrict the usefulness of private motor transport, while there appears to be more resistance in Sweden and especially in Denmark than elsewhere in the EU to the use of road transport, particularly of private cars. There are also still differences between EU countries in the level of purchase tax and fuel tax.

In spite of great improvements in the quality of many trunk roads in the EU, especially with the construction of motorways, the actual length of road in use has not increased greatly

Plate 4.2 Example of a village ill-adapted for heavy motor vehicles. Here, in a village in Lombardia, north Italy, a tanker is emerging with difficulty through the old arch of a farmyard while a lorry manoeuvres along the narrow village street

since the 1950s. The reader is referred to a good road map of Western Europe or the whole of Europe for an appreciation of the extent of Europe's road network and particularly of the motorways. The density of vehicles per unit of road distance has therefore grown almost as quickly as the number of vehicles in use. Saturation is likely to be reached in some regions of the EU long before it is reached in other areas. Typically, perhaps, the Community imagination and vision of road transport is focused above all on upgrading and extending the present motorway system, although many schemes of a more local nature, especially to by-pass larger settlements and to assist in the integration of remote and peripheral areas, are also considered. Some new developments may, how-ever, be prevented through increasing EU con-cern over the negative impact of road traffic on the environment. Environmental impact assess-ments are now required under EU law before major projects are approved, and recent contro-versy has arisen in several Member States due to the failure of authorities to conduct such studies. Examples include new roads in the UK, the tunnel/bridge projects in Denmark and various motorways in Italy and France.

AIR TRANSPORT

Like road traffic, air traffic in Western Europe has grown greatly since the Second World War. Between 1980 and 1992 alone, the passenger

Table 4.3 Roads and passenger cars in the EU

	(1) Length of roads	(2)	(3) Cars in use in millions	(4)	(5) Increase 1973–92* (1973=100)	(6) Cars per 1000 popn	(7) Cars in 2015 in mins
	Motorway	All	1973	1992			
Belgium	1.6	131.8	2.4	4.0	168	402	5.7
Denmark	0.7	70.4	1.2	1.6	129	309	3.0
Germany	11.0	628.8	18.1	32.0	177	397	45.9
Greece	0.3	40.1	0.3	1.8	527	178	5.9
Spain	2.6	156.2	3.8	12.5	330	325	22.0
France	7.4	908.2	14.5	24.0	166	422	32.4
Ireland	—	92.3	0.5	0.9	178	245	2.0
Italy	6.3	297.2	13.4	29.5	220	509	33.1
Luxembourg	—	5.1	0.1	0.2	174	523	0.2
Netherlands	2.1	103.7	3.0	5.7	191	372	8.7
Austria	1.6	30.3	1.5	3.2	211	411	4.5
Portugal	0.5	65.6	0.8	3.1	396	290	6.0
Finland	0.3	76.3	0.9	1.9	217	387	2.9
Sweden	1.0	129.1	2.5	3.6	143	412	5.0
UK	3.3	383.3	13.7	21.9	160	379	32.9
EUR 15	38.8	3,118.9	76.8	146.0	190	396	210.0

Source: (1), (2) Eurostat 1995b: 322; (3)–(6) *UNSYB* various years; (7) estimate by authors

and goods traffic handled by EU airlines have both roughly doubled. Further growth in air traffic in Western Europe is expected to continue well into the twenty-first century in spite of a number of problems that affect air transport in the EU as a whole or in individual Member States in the 1990s.

- According to the Association of European Airlines (AEA) (see Verchère 1994b), in 1992 of Europe's 24 largest scheduled carriers (including some non-EU airlines) only three, British Airways, British Midland and Air Malta made a profit.
- Internal and intra-EU air fares are high compared with fares for journeys over similar distances in the USA (see Table 4.4).
- Traffic is very heavy over many air routes crossing the central part of the EU. In addition to traffic originating in this area, flights

between third countries cross the air space. Congestion is a frequent problem, which greater coordination in air traffic control could reduce if not remedy completely.

- In Eurostat (1995a: 84) the main policy of the EU with regard to air transport is summarised as follows: 'a programme for liberalising air passenger and freight transport, in particular with greater flexibility in operating regulations and rules governing the sharing of capacity'. It is assumed (or hoped) that by allowing greater competition between airlines on routes currently shared bilaterally through agreements between pairs of countries, prices could be lowered on many journeys or at least prevented from rising. On the other hand, if another element of EU transport policy is carried out, then greatly improved, high speed passenger rail services could attract customers from the shorter

Table 4.4 Comparison of air fares in the EU and the USA, 1991

	Distance (miles)	Fare (£)	Pence per mile	Ratio (US = 100)
Birmingham–Paris	309	142	46.0	368
San Francisco–Los Angeles	337	42	12.5	
London–Athens	1,486	254	17.1	214
Dallas–San Francisco	1,465	117	8.0	
London–Nice	645	201	31.2	166
Dallas–Denver	645	121	18.8	
London–Brussels	206	118	57.3	140
New York–Washington	215	88	40.9	

Source: Birrell and Skipworth 1991

intra-EU air journeys. Again, business traffic using air transport could be eroded by the continuing introduction of more sophisticated means of communication such as videoconferencing.

Air transport in the EU or between the EU and the rest of the world falls into three 'geographical' categories: within a Member State, intra-EU (between Member States) and extra-EU. In terms of distance, the extra-EU traffic can be subdivided into intra-European (realistically including the whole of the Mediterranean basin) and inter-continental. EU policy is concerned primarily with the intra-EU traffic.

The growth of the passenger traffic handled by the airlines of the EU Member States is shown clearly in Table 4.5, columns (1) and (2). The quantity of passenger-kilometres carried almost doubled from 183.5 billion in 1980 to 361.5 billion in 1992 (see column (3)). The percentage of passenger-kilometres accounted for by internal flights (column (4)) varies greatly between Member States. It is highest in those that are largest in area or have islands with sizeable populations: France, Spain, Sweden and Italy. Internal air traffic on scheduled flights is negligible or non-existent in territorially small countries: Belgium (and Luxembourg), the Netherlands and Ireland. The

overall importance of air transport in the transportation system of each country and to some degree its importance to the economy as a whole can be judged from the data in columns (6) and (8) in Table 4.5. In terms of passengers and goods handled per inhabitant (of the total population), the Netherlands and the UK stand out, with scores several times higher than those for Italy, Belgium and Austria.

When traffic handled by the airport (or airports) of individual cities is examined (see Table 4.6) it becomes evident that in each EU Member State, one or two cities dominate in the amount of air traffic handled. Ten of the 13 busiest airports (or pairs of airports) listed in Table 4.6 are capitals of EU Member States. In their respective countries, Vienna, Stockholm and Helsinki are likewise the centres of air traffic, while Berlin is likely to increase in importance as it replaces Bonn (served mainly by Düsseldorf airport) as the capital of Germany. Other centres with large amounts of air traffic are listed at the foot of Table 4.6. In addition to national capitals, two other types of centres generate and handle large amounts of air traffic. Frankfurt-am-Main is the hub of Lufthansa Airline, Düsseldorf the gateway to the Ruhr heavy industrial area, Milan to the largest industrial region of Italy. Manchester,

Table 4.5 Air traffic in the countries of the EU

	(1) Passenger-km millions		(3) Change 1980–92 (1980 = 100)	(4) 1992 internal	(5) 1992 internal as %	(6) Passenger-km per capita	(7) Tonne-km 1992 millions	(8) per capita
	1980	1992						
Belgium + Lux	4,907	6,489	132	0	0	624	592	57
Denmark	3,296	4,552	138	793	17	875	542	104
Germany	21,056[1]	52,934	251	4,901	9	657	9,541	118
Greece	5,062	7,262	143	999	14	705	772	75
Spain	15,517	27,400	177	9,202	34	710	3,074	80
France	34,130	56,658	166	19,393	34	996	9,290	163
Ireland	2,049	4,461	218	61	1	1,275	401	143
Italy	14,076	28,491	202	6,783	24	491	3,873	67
Netherlands	14,643	32,611	223	49	0	2,145	5,489	361
Austria	1,120	4,867	435	35	1	616	568	72
Portugal	3,453	7,721	224	1,081	14	735	876	83
Finland	2,139	4,629	216	874	19	926	535	107
Sweden	5,342	8,247	154	2,729	33	948	932	107
UK	56,750	115,199	203	4,737	4	1,993	15,710	272
EUR 15	183,541	361,521	197	44,854	12	982	52,295	142

Note: 1 Former FRG only
Source: UNSYB Vol. 47, (UN 1994, Table 79)

Table 4.6 Commercial traffic in the main international airports of the EU per year

	Passengers in millions	Goods in thousand tonnes		Passengers in millions	Goods in thousand tonnes
London[1]	64.8	965	Palma (Spain)	11.9	15
Paris[2]	49.8	887	Brussels	9.3	314
Frankfurt/Main	30.1	1054	Milan[4]	9.3	66
Rome[3]	18.7	231	Athens	9.0	85
Amsterdam	18.7	695	Dublin	5.8	51
Madrid	18.1	188	Lisbon	5.4	79
Copenhagen	12.1	146			

Sources: Eurostat 1995a: 329 for table; Cole and Cole 1993: 188
Notes:
1   Heathrow and Gatwick
2   Charles de Gaulle and Orly
3   Fiumicino
4   Linate

Additional airports listed as important are Düsseldorf, Manchester, Munich, Barcelona, Las Palmas and Tenerife (Canarias, Spain), Hamburg, Málaga, Nice, Luxembourg, Cologne.

BOX 4.1 AIR SERVICES IN A BYGONE AGE

As the airline services in Europe have become impossibly dense and intricate to map meaningfully on a page size map, Figure 4.11 shows only a selection of the busiest routes of the late 1980s (little change has occurred since then). For comparison, Figure 4.12 shows the 'chief air-routes of Europe' in the late 1920s according to *Modern Teaching in the Senior School* (*c*. 1930: 217), including some of the places outside the present EU. Of the 59 places in the present EU shown on the map, 38 were in Germany or Austria, only 21 in other countries. This disparity may be partly the result of the inclusion of only the central part of Europe in the map, but it does underline the presence of many short distance air links. Since the late 1980s, Germany has been one of the EU countries that has tried to encourage passengers travelling on short air journeys, for example, between such centres as Berlin and Hamburg or Cologne and Frankfurt, to switch from air to rail. Without in any way being critical of the German citizens of the EU, the map is a reminder of the renewed central position of their country after more than four decades of being divided and stacked against the Iron Curtain. The only realistic direction for further EU expansion is eastwards. With EMU in place and Germany powerful enough economically, the next two or three decades could see a re-run of the Second World War, with the Euro replacing the dive bombers and tanks in a bloodless conquest of much of Europe.

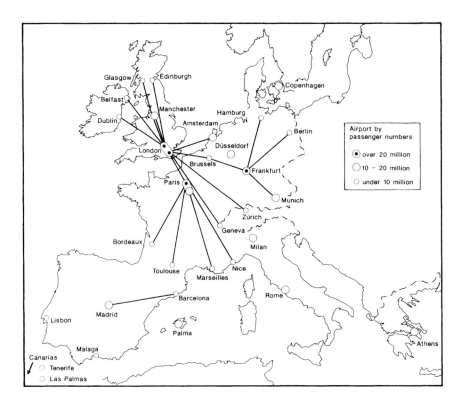

Figure 4.11 Air traffic at major EU airports, and heavily used routes

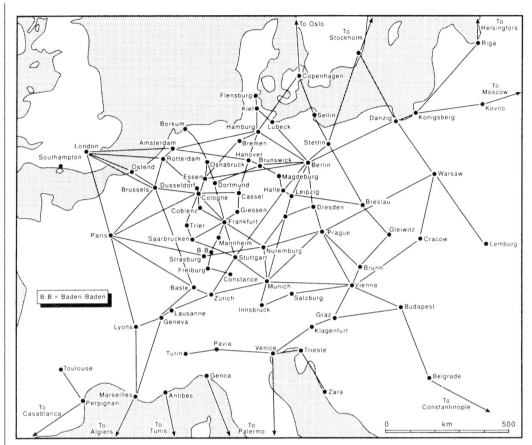

Figure 4.12 The chief air-routes of Europe around 1930 according to *Modern Teaching in the Senior School* (undated), p. 217. Notice how greatly air transport has been developed in Germany (before the Nazis seized power) and Middle Europe generally, and how London is linked with the continent by main air-routes

Munich, Hamburg and Barcelona are industrial centres or ports of EU significance. The second set of major airports consists of a number of tourist centres in the southern part of the EU: Nice, Málaga, Palma de Mallorca (Baleares), Las Palmas and Tenerife (Canarias).

SEA AND INLAND WATERWAY TRANSPORT

According to Eurostat (1995a: 84), the main policies of the EU with regard to sea transport are 'to improve safety at sea, maintain the Community fleet and the level of employment, ensure the freedom to provide services between Member States and with non-Community countries, and to set up a Community shipping register and flag'. The main EU policies with regard to inland waterways are even more modest '[to ensure] technical harmonisation and the recognition of qualifications in the field of inland waterway transport'.

Figure 4.13 shows how close most of Europe is to the coast, with Austria and Luxembourg

Plate 4.3 Reminders of early forms of transport: a canal and a railway viaduct near Llangollen, Wales

the only EU countries that are land-locked. Selected seaports of Europe and the Mediterranean basin are shown, as well as the main inland waterways and most prominent mountain areas; the near circum-Europe shipping route is marked with a thick black line. The northern and southern sea routes of the EU each provide useful, reasonably direct access between pairs of ports (e.g. Hamburg–Bilbao, Barcelona–Athens) but in relation to direct distances and land routes, any journey between northern and southern ports is increased by the doubling-back effect. Thus, for example, internal shipping between Bilbao (1) and Barcelona (2) in Spain has to travel about 2,500 km compared with about one-fifth of that distance for land transport by road or rail. The distance from Le Havre (3) to Marseilles

(4) is 3,350 km against 750, and that from Hamburg (5) to Venice (6) is about 6,100 km against 900.

Rivers and canals penetrate a long way inland into some parts of Europe. The Rhine and its tributaries provide inland navigation in six countries: the three Benelux countries, Germany, France and Switzerland. Navigation on the Danube reaches into Germany and the river serves places in Austria, the Czech Republic, Slovakia, Hungary, Croatia, Serbia, Romania, Bulgaria and Ukraine. Coastal shipping and inland waterway routes are the sector of the total transport system least likely to change greatly in the next two or three decades, apart from the impact of the Rhine–Main–Danube Canal, which links the two largest river basins of Europe outside Russia, an engineering feat

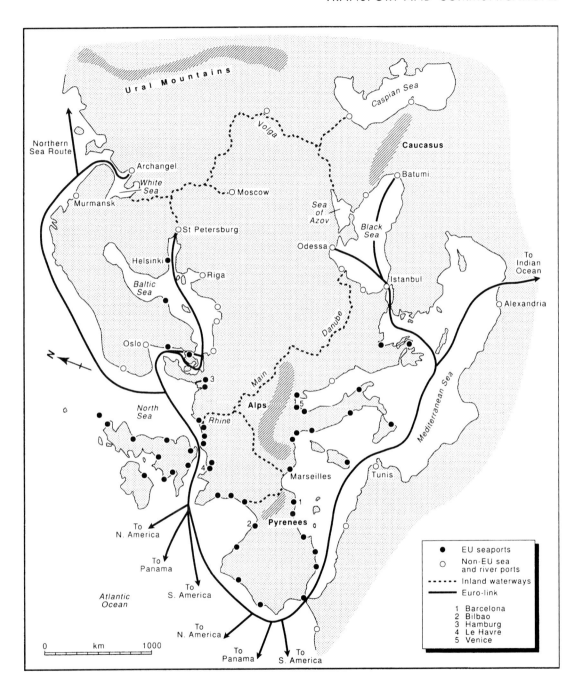

Figure 4.13 Major seaports of Europe. A notional 'round Europe' sea route is marked with a heavy line. It 'starts' at Archangel in Russia, passes Gibraltar, and 'ends' at Batumi in Georgia. No attempt is made to distinguish ports according to capacity or quantity of goods handled. Some ports that are very close together are represented by a single dot. See Eurostat (1995a: 323) for the location of 'major seaports' of the EU, 1991

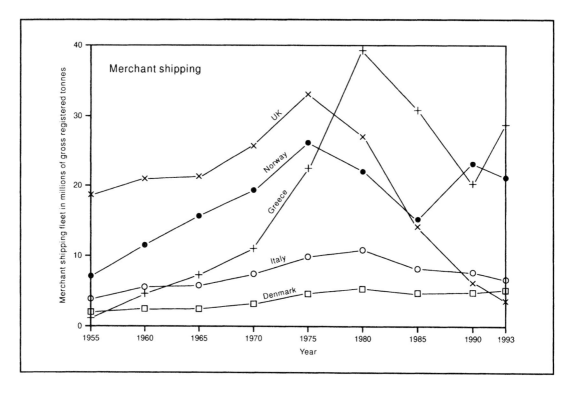

Figure 4.14 The merchant shipping fleet of selected Western European countries
Source: various numbers of *United Nations Statistical Yearbook*

no less impressive than the Channel Tunnel. Russia itself is crossed from north to south by navigable rivers and canal links capable of taking small sea-going vessels to outlets on five seas: White, Baltic, Caspian, Azov and Black Seas.

Since the 1970s there has been an absolute decline in the tonnage of shipping registered in the EUR 15 countries following a relative decline in relation to total world shipping that had already started in the 1950s. In 1955 the present EUR 15 countries had 42.4 million tonnes of shipping (i.e. 42 per cent) out of a world total of 100.6 (the UK alone had 19.3). By 1983 the EUR 15 tonnage had grown to 110.7 million tonnes, which was only 26 per cent of the world total of 422.6 million, while by 1991 the EUR 15 fleet had dropped to 66.7

million tonnes, only 15 per cent of the world total. Figure 4.14 compares the volatile tonnage of shipping registered in four selected EUR 15 countries and in Norway. For some time now, many of the ships of the fleets of EU countries have largely been manned by non-EU employees. At the same time, the trend to convert vessels from EU national flags to flags of convenience has continued, with, for example, Malta's fleet growing from almost nothing in the early 1980s to 11 million tonnes in 1991.

Great changes have taken place in the types of vessel registered in or serving EU countries and their ports. Intercontinental passenger liner services are now rare and the long distance transport of passengers is accounted for mainly by cruise ships. The capacity of vessels to carry both vehicles and passengers on the numerous ferry services has grown enormously over the same period. Ships carrying mixed or general cargoes have largely been replaced by more specialised ships carrying mineral cargoes (oil, LNG, ores) or containers. Ports have changed correspondingly, with ferry ports accounting for the bulk of the movement of passengers, specialised ports, with nearby oil refineries, receiving crude oil, and new dock facilities created since the 1960s to handle containers.

Many of the regions of the EU at NUTS level 2 or below have been affected adversely (see Chapter 12), or in a few cases favourably, by the decline of shipbuilding and changes in the favourability of the location of different ports. Thus, for example, the port of Liverpool in Merseyside in the northwest region of the UK has declined in importance while the port of Manchester has ceased to exist. In contrast, in the Netherlands, the development of Europort in Zuid-Holland (West Nederland) has greatly boosted the economic life of the region. The unification of Germany has brought much of East Germany and regions beyond back into the hinterland of Hamburg. At a more local level, the opening of new fixed links under or across the sea has already affected the fortunes of places in east Kent, especially Dover and, with the completion of the Øresund link, will affect ferry ports in Sjaelland and the extreme south of Sweden. At a still more 'micro' level, the opening of a bridge to link the Isle of Skye to the mainland of Scotland, a project supported with EU funding, has resulted in the closure of a ferry service, with the consequent loss of jobs in an area with high unemployment.

COMPETITION BETWEEN ROAD, RAIL AND AIR FOR PASSENGER TRAFFIC

As a result of the very dense networks of road, rail and air routes, travellers in the EU often have several possible modes of transport to choose from when planning to make a journey. Since the majority of families in the Union own at least one car, the choice for many people is between private car, bus, rail and air. The choice of mode, therefore, depends to a large extent on the speed of travel and the distance to be covered. Distance can be measured in a straight line; according to the road or rail distance; by time; by cost; or even by convenience. In the following study made by the authors, time is used to represent distance, although the relative cost of different modes might outweigh an advantage in time among less affluent travellers.

The time taken to make most journeys by public transport (bus, train, plane) can be divided into discrete 'legs'. For example, time on a rail journey would include travel from the place of origin of the journey to a station, waiting for a train to leave, the train journey itself, waiting for transport at the destination station, and journey to destination. To illustrate the data used in this study, the calculations of travel times for two journeys, Paris–Lyon and Paris–Milan, one internal, one international, are shown in Table 4.7. Even though the components of each journey have been simplified, it is evident that a large number of influences and variables must be considered. Some time additional to the main journey will be spent in almost all travel. While very short journeys will be made on foot or by bicycle, most comparatively short journeys will be made by car or bus, except those by rail on a metro or suburban train service in larger cities. In contrast, on most journeys over a considerable distance, air will be preferred.

In Table 4.8 ten selected journeys are compared and the information needed to construct the graph in Figure 4.15 is shown. The relationship is shown on the graph between the three modes of travel: road, rail and air. In Figure 4.15 the angles of the three lines reflect the average speed of travel on a journey and the points at which the lines intersect the vertical

Table 4.7 Calculation of travel time on two journeys: Paris–Lyons and Paris–Milan (travel times are in minutes)

	(1) Distance km	(2) Built-up area initial time	(3) Await departure	(4) Travel	(5) Leave conveyance	(6) Built-up area	(7) Total time	(8) Speed km/h	(9) Time air = 100
Paris to Lyon									
Road	400	30	—	310	—	15	355	68	182
Rail (TGV)	461	30	20	120	—	15	185	130	95
Air	427	60	30	60	15	30	195	123	100
	(400)								
Paris to Milan									
Road	628	30	—	570	—	15	615	61	246
Rail	838	30	20	515	—	15	580	65	232
Air	821	60	60	85	15	30	250	151	100
	(628)								

Table 4.8 Travel between ten pairs of places by three modes

	Direct distance km	Total time in minutes			Average speed km/h		
		Road	Rail	Air	Road	Rail	Air
1 London–Birmingham	160	145	155	150	66	62	64
2 Milan–Zurich[1]	216	310	290	190	42	45	68
3 London–Manchester	260	255	215	190	61	73	82
4 London–Paris*[2]	340	430	465	260	47	43	78
5 Paris–Lyon**	400	355	185	195	68	130	123
6 Milan–Rome	472	420	295	170	67	96	167
7 London–Edinburgh	518	465	308	210	67	101	148
8 Paris–Milan	628	615	580	250	61	65	151
9 London–Milan	948	985	1,185	280	58	48	203
10 London–Rome	1,420	1,375	1,430	310	62	60	275

Notes:
* before opening of the Channel Tunnel
** using TGV
1 Journey 2, Milan–Zurich, shows the effect of the Alps on road and rail times
2 Journey 4, London–Paris, shows the effect of crossing the Channel on road and rail times, also influencing to a smaller degree London–Milan and London–Rome

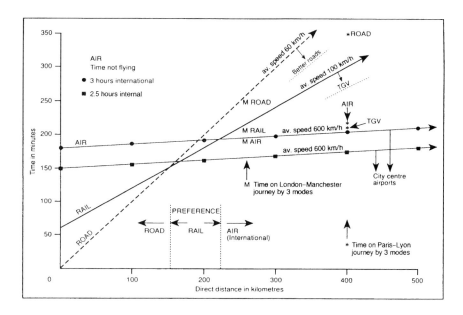

Figure 4.15 The choice of mode of travel for passengers in Western Europe. Calculations by the authors (see text). For simplicity, precise distances and speeds are shown on the diagram but, in reality, each line should be a band of considerable width to include a variety of actual journeys

axis are related to the combined time taken before and after the actual main journey is made. For simplicity, no such delay is counted for road, since it is assumed to be door-to-door, but for rail 60 minutes are allowed, while for air two variants are used, internal flights 150 minutes, international 180 minutes. The average speed of road travel is 60 km/h, that of rail travel 100 km/h and that of air travel 600 km/h. In all these modes, time is measured against *direct* distances, whereas in reality there is always some additional distance to be covered through curves, detours and delays (e.g. waiting at ferry crossing, circling above airports), which make the actual main journey longer in distance and time than the ideal.

Some implications of the situation illustrated in Figure 4.15 will now be discussed, with reference mainly to passenger rather than goods traffic. Road and rail are in competition for passengers making relatively short journeys, rail and air for passengers making longer jour-

neys. Competition between road and air is limited because the railways occupy the 'middle distances'. Road and rail compete for goods traffic over medium and long distance journeys and air is not seriously in contention here.

The trend in the EU in the last forty years has been to extend greatly roads that carry fast traffic and to upgrade many others, thereby producing a 'motorway effect' that reduces time travel, but not necessarily distance or cost. So long as road improvements continue, road traffic will in principle be able to compete with rail traffic over increasingly large distances. On the other hand, the increasing number of vehicles in use will cause greater congestion, while the imposition of lower speed limits for goods vehicles and coaches would reduce the present advantage commercial road vehicles have over rail.

More drastic in its effect both on road traffic and on the whole motor vehicles industry would be an increase in road tax for cars, for

[Plate 4.4] London's orbital ring road, the M25, was already carrying as much traffic a few weeks after it was opened as it had been expected to carry eight years after completion. With the continuing growth in car ownership throughout the EU, increasing congestion may be expected on many roads in the next century. Jam today and jam tomorrow?

Plates 4.4 and 4.5 The motorway age
Source: European Commission, Direction Générale X, Audiovisuel

[Plate 4.5] Construction work on a new motorway in Portugal, supported by the ERDF. One of the least motorised countries of the EU, Portugal is catching up quickly

example from around the present £130 a year in the UK to £630, according to Nuttall (1994a) the estimated real cost in terms of maintaining the road system, pollution, accidents, traffic congestion and other aspects. Such a measure might reduce the average mileage covered by private cars and also the number of cars actually registered, but it is no guarantee that people would automatically switch to rail travel. A further example of the complex relationship between road and rail travel is the increasing use of taxis to carry passengers on 'cross-country' journeys between smaller places not linked by direct, fast or frequent rail services, reported by Ramesh (1996).

Just as motorists are heavily subsidised

because the road tax they pay does not match up to the real cost of road travel, so the rail users of Western Europe have been heavily subsidised throughout the last few decades. Some loss-making lines have been closed, but many others are still in use, the view of governments throughout Western Europe being that in the last resort all routes, loss-making or not, have either a strategic importance or help to keep remote communities in touch with national core areas.

A more positive approach to the modernisation of the European rail network is included in the trans-European networks (TENS) programme. According to Bounadonna (1996):

> The TENS programme is aimed at rationalising and harmonising transport facilities in the 15 EU countries by building or revamping road and rail links, with a special focus on the need to connect more peripheral Member States. The Horizon 2000 network, with a high speed, harmonised cross-border rail network, is central to the plan. The nature of the exercise is exquisitely transnational.

Since 14 major projects in Horizon 2000 would alone cost over 90 billion ECU, far more than the amount available from the EU budget, national governments and private sources would have to contribute. The plan described would increase the ability of the railways to attract some of the passenger traffic at present carried by air, but it is difficult to see a revival of medium-distance rail travel on a scale large enough to justify the huge investment; another Channel Tunnel experience is more likely. The problem of keeping the transport of goods on the rail network is illustrated in Figure 4.16.

As noted earlier in this chapter, only three of the main airlines of the EU were making a profit in the mid-1990s. In various ways, national governments are therefore subsidising travel on their carrier airlines. However, Gibson (1996) cites the establishment by smaller airlines of many new air services between pairs of medium-sized places, avoiding the need to change flights at the main hubs. In this area,

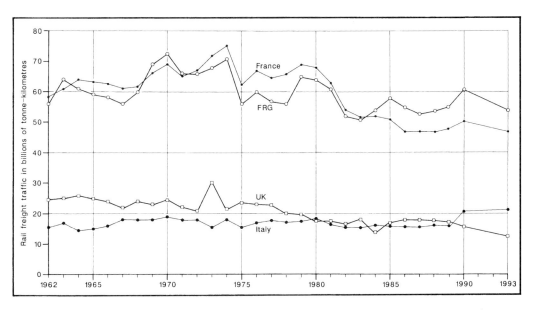

Figure 4.16 The movement of goods by rail in West Germany, France, the UK and Italy, 1962–93
Source: various numbers of *United Nations Statistical Yearbook* and Eurostat 1991, Table 7.4

air transport threatens to take passengers not only from the railways but also from the roads. The airports of medium-sized cities tend to be closer to city centres than the airports of large cities. The case of London City airport shows how the delays common in travelling between out of town airports and city centres can be overcome. Similar in concept to Santos Dumont airport on the edge of the central business area of Rio de Janeiro, Brazil, London City airport is a matter of minutes away from London's commercial heart. Planes with a range of 1,600 km can reach most of Western and much of Central Europe from it. Here is a challenge to the attempt to attract passengers to new (costly) high speed rail links.

Before central planning collapsed in the Soviet Union a basic principle was to ensure that different modes of transport complemented one another and did not compete. Since the whole economy was 'owned' by the state it was only an academic matter as to whether or not the whole system of Soviet transport was subsidised. Such is not the case in Western Europe where, as already shown above, road, rail and air travel are all subsidised to some extent, whether by EU funds, or by national or local governments, yet at the same time are supposed to be in competition. Although it is not fashionable to contemplate any kind of central planning in the EU, it might be a useful exercise to calculate the amount of waste through overlap of services that allow the traveller to choose between travel by private car, bus, taxi, train or aircraft between innumerable pairs of places in the Union.

TELECOMMUNICATIONS AND THE TRANS-EUROPEAN NETWORKS

The increasing importance of telecommunications as the medium for the transmission of data or information has led to the need for a policy at EU level on the development of a modern and efficient infrastructure. Based on the principles of the Single Market, the EU has concentrated its efforts on the harmonisation of technical standards in the sector to ensure there are no obstacles to the free movement or compatibility of equipment nor to the freedom to provide services in this sector. Traditionally in most Member States the control of the telecommunications sector has been either in state monopolies or in firms with exclusive rights in most Member States and it has been difficult for the EU institutions to free up the sector. Full liberalisation will not be achieved before 1998, when telecommunications networks should be fully open, with common rules established for access and use. Exemptions have already been agreed for Luxembourg, which has until the year 2000, and Spain, Portugal, Ireland and Greece, which have until 2003 to free up their infrastructures completely.

The telecommunications sector in Europe will account for 7 per cent of EU GDP by the year 2000 and much investment is required not only to modernise systems and equipment but also to make European telecommunications companies more competitive against US and Japanese firms, which are technologically more advanced and better able to compete on the open market. Europe as a whole lags behind the USA and Japan, not only in terms of technology but also in terms of use of telecoms infrastructure. The traffic by telephone transmission, both voice telephony and data transmission, is far more dense in Japan and the USA, and traffic within Europe is also shown to be highly uneven (see Figure 4.17), centred around countries such as Germany, and very limited within the peripheral regions and between them and the central countries.

Globally, the telecommunications sector accounts for 400 billion ECU, is growing at 7 per cent per year and therefore merits special attention in the World Trade Organisation (WTO), in which negotiations are under way between the USA, Japan and the EU to improve

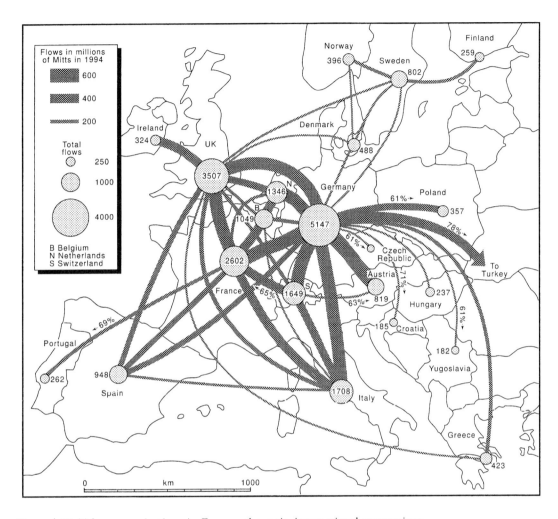

Figure 4.17 Telecommunications in Europe: the main international transactions
Source: *Geo*, November 1995: 128

access for each other's firms to their expanding markets. The EU remains reluctant over foreign or non-EU ownership of phone companies in Europe, in particular in France, Spain, Portugal and Belgium, where opposition is the strongest.

The EU is also investing 2.6 billion ECU in the sector over the period 1994–8 as part of the framework programme for R&TD. Funding is provided for three main areas of activity: the development of technologies and standards required for a future broad band integrated network, further development of information technologies linked to telecommunications, and the establishment of telematic systems for general use, such as data exchange between administrations, and distance education.

The European telecommunications network is regarded as one of the trans-European networks that the Treaty of Maastricht identified as crucial to the further economic and social development of Europe. It states that 'the Community shall contribute to the establishment and development of trans-European networks in the area of transport, telecommunications and energy infrastructures' (EC Art. 129b), focusing on the interconnection and interoperability of national networks as well as access to them. Particular emphasis is placed on the regional policy importance of the TENs, when reference is made to 'the need to link island, landlocked and peripheral regions with the central regions of the Community' (ibid).

The TENs, which are mainly the improvement of transport links for road and rail, are concentrated on 14 priority projects, worth a total of 91 billion ECU up to the year 2000. They will benefit from funding from the EU budget, in particular from the financing of studies and the establishment of guidelines, costing 2.3 billion ECU, but the bulk of funds will come from Member States and the European Investment Bank.

FURTHER READING

Button, K. (1990) 'The Channel Tunnel – the economic implications for the South East of England', *The Geographical Journal*, 156 (2), July: 187–99.

COM 89–564 (1990) *Communication on a Community Railway Policy*, Brussels, 25 Jan.

Giannopoulos, G. and Gillespie, A. (1993) *Transport and Communitcations Innovation in Europe*, London: Belhaven Press.

Gibb, R. A., Knowles, R. D., and J. H. Farrington (1992) 'The Channel Tunnel Rail Link and regional development: an evaluation of British Rail's procedures and policies', *The Geographical Journal*, 158 (3), Nov.: 273–85.

Mellor, R. (1992) 'Railways and German Unification', *Geography*, 77/3, July: 261–4.

Nicholson, D. (1992) 'Unblocking an inland artery', *Geographical*, LXIV (9): 30–3. Completion of the trans-Europe North Sea–Black Sea Canal, a dubious prospect?

Plon, U. (1996) 'Dryshod across the Baltic', *The European Magazine*, 4–10 July, pp. 10–11.

Romera, I. Alcazar (1991) *Draft Report on Community Policy on Transport Infrastructure*, European Parliament. Committee on Transport and Tourism, 22 Feb., DOC EN/PR/103678 PE 148.168.

Tickell, O. (1993) 'Driven by dogma', *Geographical*, LXV, (10), October: 20–4. Does road building bring prosperity to backward areas?

Tully, C. (1992) 'High-speed trains in Europe', *Geographical*, LXIV (3), March: 10–14. Debates the advantages of a high-speed rail network for Europe at the expense of more appropriate forms of public transport.

Vickerman, R. W. (1994) 'Transport infrastructure and region building in the European Community', *Journal of Common Market Studies*, 32 (1), March: 1–24.

Whitelegg, J. (1993) *Transport for a Sustainable Future*, London: Belhaven Press.

5

ENERGY AND WATER SUPPLY

GREECE IS GOING DRY
PLEASE SAVE WATER
Take a shower, not a bath.
Report any leaks.
Don't let water run non-stop.
Saving energy also saves water.

Bathroom, Athens Center Hotel, March 1996

- The energy and water supply industries of the EU have a high level of investment per worker and a relatively small labour force.

- The energy industry is difficult to plan at EU level because each country is concerned with its own particular problems and because half of the energy consumed is imported and is therefore subject to conditions and changes in the world beyond the EU.

- The fossil fuels coal, oil and natural gas provide about 85 per cent of EU energy.

- Since the 1960s coal production in the EU has declined dramatically.

- Since the 1940s the use of oil and natural gas has increased sharply.

- The generation of electricity from nuclear power has increased greatly since the early 1960s.

- Alternative energy sources that do not threaten the environment account for only a few per cent of the total energy consumption of the EU.

- Concern over water supply has grown in various regions of the EU in the 1980s and 1990s. There is abundant water in Western Europe but not in some of the areas of greatest consumption.

Until the Treaty on European Union, EU energy policy was characterised by sectoral policies for coal through the ECSC and for nuclear energy through Euratom. No corresponding policy body exists for oil and natural gas. The ECSC Treaty created a common market for coal with common objectives, including equal access for customers to sources of production, measures to increase productivity and the promotion of the growth of international trade. State aids have been authorised if they promote economic viability, solve social and regional problems due to closures, or assist the coal industry in adapting to new environmental standards. The EAEC (Euratom) Treaty has been aimed at promoting the development of the nuclear industry by encouraging investment, providing for research, disseminating knowledge for health protection, setting safeguards and controlling safety and supply through the Supply Agency.

ENERGY POLICY

The Treaty on European Union did not lay down an energy policy as such, but has considerable impact on the energy sector through provisions for the internal market, rules on competition, the development of trans-European networks, commercial policy, cooperation with third countries, environmental protection and policies for consumers and R&TD. Apart from the nuclear and coal sectors, however, there is no clear delineation of responsibility for energy policy at Community level.

With a view to establishing a more coherent energy policy, the EU Commission published a Green Paper (COM (95) 659) in 1995 entitled 'For a European Union Energy Policy' which sets out energy policy objectives and argues for greater EU responsibility in the field due to changes that have taken place in the legal, institutional and economic environment of the EU. These changes include the Single Market, in force since 1 January 1993, but still incomplete for the energy industry and energy consumers, and the increasing concern about energy in the EU due to constraints of environmental protection arising from growing energy consumption, and to geopolitical changes affecting supplies to the EU from third countries.

The Green Paper sets a general context for a policy framework for energy, focusing on those objectives of the TEU that apply to energy production and supply, including economic and social progress, environmental protection, a common foreign and security policy and economic and social cohesion, including regional policy. Priorities and objectives are also identified for the energy policy itself. The main objectives are competitiveness, security of supply and respect for the environment, to be achieved through internal harmonisation and cooperation, international cooperation, energy efficiency and R&TD.

The Commission also proposes that more specific provisions for an EU energy policy be written into the Treaty during its revision in the 1996 Inter-Governmental Conference. The energy sector is of crucial importance to the future development of the EU and its dependence on third countries for supply gives it an added international dimension. Nevertheless, cooperation in the sector is handicapped by various obstacles, in particular the differing traditions that govern national energy policies, such as the availability of energy sources in certain Member States, the use of nuclear energy and the high degree of state control over energy supplies, in particular of electricity and gas. Some Member States argue that these differences, together with the principle of subsidiarity, mean that a true common energy policy at EU level is unnecessary.

GENERAL FEATURES OF THE ENERGY SECTOR

In the general industrial classification of economic activities within the European Union (NACE), energy and water form a major subdivision (1). In 1992 (Eurostat 1994b, Table 3.19) there were 2,061,000 employees in energy and water in the EUR 12 countries. Over half (1,156,000) were in category 16,

Table 5.1 Selected large companies

(1) Company name	(2) Nationality	(3) Sector	(4) Capital employed £ million	(5) Number of employees 000s	(6) Capital per employee £ 000s
1 Royal Dutch/Shell	UK/Nld	Fuels	50,207	106	474
2 Electricité de France	France	Electricity	45,016	118	381
3 British Gas	UK	Fuels	24,639	70	352
4 British Petroleum	UK	Fuels	20,777	67	310
5 Ste Elf Aquitaine	France	Fuels	22,252	94	237
6 RWE	Germany	Electricity	20,837	118	177
7 SNCF	France	Railways	22,416	221	101
8 Volkswagen	Germany	Motor vehicles	21,922	244	90
9 Daimler-Benz	Germany	Motor vehicles	27,503	342	80
10 Fiat	Italy	Motor vehicles	20,041	257	78
11 Siemens	Germany	Electricals	25,450	394	65
12 Unilever	UK/Nld	Foods	10,719	304	35

Source: Times Books 1995: 90–1

the production and distribution of electricity, gas, steam and hot water. The energy and water industries of the EU thus accounted for only 5.1 per cent of total industrial employment (40,501,000 persons) and a mere 1.8 per cent of all employment (115,342,000) yet this sector accounts for about 6 per cent of total GDP of the EU. By comparison, in the agricultural sector the contribution per person employed to GDP is only half the EU average.

In view of the relatively small labour force in the energy and water supply sector, this branch of industry makes only a small direct contribution to employment. The decline of coal mining and the privatisation of some utilities have resulted in job losses in the 1980s and early 1990s in various regions of the EU. The industry is dominated by very large companies, many still in the public sector, each following its own agenda. The data in Table 5.1 for 12 of the largest companies in the EU show that those connected with the production of fuel and power have a much higher amount of capital per person employed than those in manufacturing and transport. Thus, for example, the capital employed per employee in Shell is more than 13 times that in Unilever (Foods). In general,

the energy companies employ considerably more capital per person employed even than companies in the heavy industrial sector and much more than companies in light industry.

The average pay for workers in the energy industry is considerably higher than in agriculture and in most branches of manufacturing and services. In view, however, of the small number of jobs in most sections of the energy industry and the great financial resources of the large companies that dominate it, already referred to above, little can be done at EU level to attract new employment in the energy sector to backward areas or areas of high unemployment. Indeed, one of the causes of economic decline in many regions of the EU has been the closure of numerous smaller, high-cost coal mines. The location of oil refineries in backward regions, for example in southern Italy, in Puglia (at Brindisi) and Sicilia (Augusta), has made only a limited economic impact. At the same time, the presence of the energy industry has made a large statistical impact on the Gross Regional Product of certain regions with a small population, notably Groningen in the Netherlands, where much of the country's natural gas is extracted or 'landed'

Table 5.2 Consumption of the main sources of energy in Western Europe (OECD definition)

	Consumption in millions of tonnes of oil equivalent			Percentage of total consumption		
	Total	Home	Imported	Total	Home	Imported
Coal	261.8	153.2	108.6	18.3	10.7	7.6
Oil	652.5	286.7	365.8	45.7	20.1	25.6
Natural gas	263.2	188.8	74.4	18.4	13.2	5.2
Nuclear	209.3	209.3	—	14.6	14.6	—
Hydro	42.4	42.4	—	3.0	3.0	—
	1,429.3	880.4	548.8	100.0	61.6	38.4

Source: BP 1995, various tables

from the North Sea, and Grampian in the UK (Scotland), serving the North Sea oil industry in a similar fashion.

In spite of the limited influence of EU policymakers on the energy industry at the EU level, the industry is of such importance to the economy of every EU Member State that its main features, problems and prospects will be discussed at some length in the next two sections. There follows an account of the main types of energy used in the EU and the main consumers.

Almost all of the commercial primary energy consumed in the EU at present is provided by five types. The three fossil fuels, coal, oil and natural gas account for about 85 per cent of the total, nuclear power and hydro-electricity for most of the remainder. Table 5.2 shows the consumption, home production and imports of primary sources of energy in OECD Europe, which in addition to EUR 15 includes Norway, Switzerland and Turkey. In 1994 Western Europe as defined above satisfied 61.6 per cent of its energy needs. Without Norway, Switzerland, Turkey and some smaller countries, the home production of all types of primary energy in EUR 15 in 1994 was 674 million tonnes of oil equivalent (t.o.e.) compared with a total consumption of 1,325 million tonnes, requiring a net import of 651 million t.o.e. The EUR 15 therefore produces only about 50 per cent of the energy it consumes, a proportion that has changed little in the 1980s and early 1990s.

The outstanding feature of the energy supply

of the EU is the fact that each year about 450 million tonnes of oil are imported, about two-fifths from the Middle East and about one-fifth each from North Africa and the former USSR. Norway and the UK export some oil, both to other West European countries and, outside Europe, mainly to North America. Thus the EU depends on imported oil for about one-third of all its energy needs. Several of the main suppliers, notably Algeria, Libya, Iraq (until 1990), Iran and some of the former Soviet Republics, are unstable politically or openly hostile to the West. There is speculation that a change of government in Saudi Arabia, which has a quarter of the world's oil reserves, could be followed by oil price rises in OPEC countries comparable with those of the 1970s. Nevertheless, British Petroleum has started to develop gas deposits in a 25,000 sq km concession in central Algeria to supply gas to Spain (Mortished 1995).

There are three main users of energy in the EU. Final energy consumption by sector for the EUR 12 in 1991 was accounted for as follows (percentages of total): industry 29.1, transport 30.5, residential, administrative and commercial 40.4. During 1983–91 total demand rose by 16 per cent. The most marked increase (40 per cent) was in transport, partly the result of the great growth of private motoring during that period. Slower growth occurred in the amount taken by commercial, administrative and residential users (22 per cent) and by industry (11 per cent).

During the 1950s and 1960s the energy base of Western Europe was transformed by the decline of dependence on coal and the great increase in the use of imported oil. After the swingeing rises in oil prices in 1973–4 and again in 1979 the policy of national governments both individually, and collectively through the EU, has been to encourage the more efficient application of energy and the use of indigenous sources. To some extent this has prolonged the life of production of high cost and heavily polluting coal and lignite.

For some purposes, as in the generation of thermal electricity, fossil fuels are interchangeable, but for other purposes one source of energy has advantages over others. Thus oil, in particular, as well as being cheaper than coal to move, is the cheapest source of fuel for the transport sector. On the other hand, high-grade coking coal has been essential in some processes in the iron and steel industry. Whether produced directly as primary energy from hydro- or nuclear generators, or as secondary energy from the burning of coal, oil or natural gas, electricity is the form in which many modern industrial and non-industrial users need or prefer their energy. The principal users of electricity in the EU are industry, which in 1991 used 43.3 per cent, and households, which took 29.3 per cent. In contrast, transport consumed only 2.6 per cent (Commission of the EC 1994b: 1–35).

The EUR 15 has about 6.5 per cent of the total population of the world, produces between 6 and 7 per cent of all primary energy, but consumes about 15 per cent. In 1994 Western Europe imported 486 million tonnes of crude oil and oil products. Western Europe, the USA and Japan together have less than 15 per cent of the total population of the world but account for almost 70 per cent of all oil imports in world trade. At the end of 1994 the reserves to production ratio for the world's oil gave a 'life' of 43 years, assuming no change in the level of consumption, and no major new discoveries. In view of the expected rapid growth of industry and of motor transport in the developing world in the next few decades there is no place for complacency over oil supplies in the EU, either over the scale of current reserves or over the reliability of supplies from a political point of view.

SOURCES OF ENERGY

In terms of energy consumption per inhabitant, the EU is second only to North America among the major regions of the world. In contrast, its shares of the world total of reserves of oil (about 0.5 per cent) and of natural gas (about 2.0 per cent) are very modest. Norway alone has bigger oil reserves than the whole of the EU, as well as substantial natural gas reserves. On the other hand, the EU has about 5 per cent of the world's higher grade coal reserves and over 10 per cent of the lignite.

Although various types of coal are found in small quantities in many parts of the EU, almost all the reserves are located in a broad zone extending from the coalfields of the UK in the west through northern France, southern Belgium and central Germany to the lignite deposits of the former East Germany in the east (see Figure 5.1). The coal deposits of northern Spain and the lignite deposits of Greece are the main areas of production elsewhere in the EU. There are small deposits of oil and natural gas in many places in the EU, but by far the largest deposits so far discovered are under the North Sea. Norway and the UK (Scotland) share most of the oil deposits, which are located mainly in the northern part of the North Sea. The Netherlands and the UK share most of the natural gas deposits, located mainly in the southern part of the North Sea and onshore in the Netherlands. In this general area of the EU, Denmark (offshore) and northern Germany

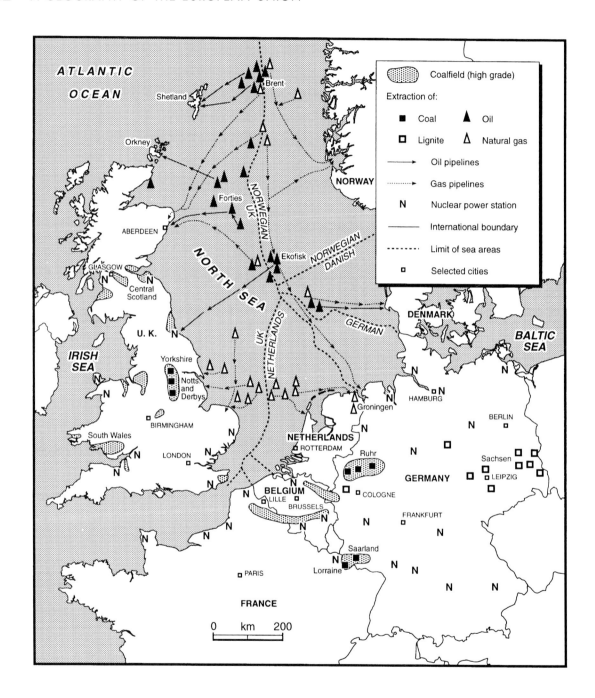

Figure 5.1 The main sources of fossil fuels in the northwestern part of the EU and oil and gas pipelines in the North Sea
Source: based on Spiessa 1993: 45

Plate 5.1 The first hydro-electric power station in Italy: Vizzola Ticino, near Milan. This represented the beginnings of modern, clean energy. Most hydro-electric power in the EU has now been harnessed

(Niedersachsen) have smaller oil deposits. In two other regions of the EU, natural gas deposits have been of importance for considerable periods since the Second World War, those of southwest France (Aquitaine) and Italy (particularly Emilia-Romagna).

The hydro-electric potential of the EU is located in areas in which, geologically, fossil fuels are rare or non-existent. France, Italy, Germany and Austria (together with Switzerland) share in the Alps the largest hydro-electric power (HEP) capacity in the EU. Stations in the Pyrenees supply France and Spain. Sweden and Finland, together with Norway, also have a considerable hydro-electric potential. While the location of the extraction and supply of fossil fuels and HEP is determined by the location of commercially suitable deposits and sites, the distribution and location of thermal electric power stations, whether run on fossil fuels or with nuclear energy, are independent of geological and topographical features, and are related particularly to the availability of water for cooling purposes and to markets.

Although coal and falling water were used for domestic and industrial purposes, wood was the main source of energy in Western Europe until the eighteenth century. With the development of steam power to operate appropriate pumps, coal could be mined at considerable depths below the surface in areas with coal deposits. Coke from coal gradually replaced charcoal from wood in the smelting of iron ore, while machinery in factories and, from the 1820s, railway locomotives and ships, could also be driven by steam power. The development of industry in Western Europe was most rapid in locations on or near coalfields, although by the second half of the nineteenth century coal could be transported commercially by rail or sea over considerable distances from places of extraction. National governments and private entrepreneurs alike influenced the pace of development in different areas.

The commercial use of HEP since the 1880s provided a new energy base of regional importance, as in Lombardia in north Italy and in Catalunya in Spain. Since the Second World War, oil, natural gas and nuclear power have gained relatively as sources of energy in Western Europe and the extraction of coal has

BOX 5.1 ALTERNATIVE SOURCES OF ENERGY

Ultimately both fossil fuels and nuclear fuels (principally uranium) are exhaustible, although strictly speaking fossil fuels are renewable, but at a very slow rate. The growing interest in the EU and elsewhere in the world in alternative energy sources arises from at least three different developments in the late 1960s and the 1970s: awareness that the principal energy sources of the twentieth century are non-renewable, the added political interference with the economic cost of extracting oil in the oil price rises of the 1970s, and the growing concern over the negative environmental effects of the burning of fossil fuels and of nuclear accidents. In effect, most of the so-called minor sources of energy have been used for thousands of years, although overshadowed in the last two hundred years first by coal, then by oil, natural gas and nuclear power. In the early 1990s, 3–4 per cent of energy consumption in the EU was provided by alternative sources, a proportion that it is hoped can be raised to about 8 per cent by the year 2005. Several of the alternative sources are mainly used to generate electricity, a feature also of nuclear power.

In the early 1990s about 30 per cent of renewable energy in the EUR 12 countries came from hydro-electric power and 60 per cent from biomass. For simplicity the sources are listed below.

- Peat is a non-renewable source of energy and should be classified in this respect with the solid fuels coal and lignite. It is used both for domestic heating and to generate electricity. In the EU, Ireland is the only country in which much peat is now used (about 5 million tonnes a year) but in Russia and several other former Soviet Republics it is also widely cut. As well as serving as a fuel, peat is increasingly sold in garden centres. As a result, the unique ecosystems of the few remaining peat bogs in the EU are threatened.

- Biomass fuels include ethanol from agricultural products, vegetable oils (as diesel fuel), wood and wood wastes. Fuelwood can be considered renewable so long as the replanting of trees makes up for cutting. Out of a world annual consumption of fuelwood of some 1,800 million cubic metres, Europe consumes only about 130 million, of which the EUR 15 countries use some 40 million (about 2 per cent of the world total). In relation to population, Sweden and Finland are the largest consumers (United Nations 1994)).

- Biogas, recovered from sewage, landfill sites and agricultural wastes consists mostly of methane; it is of local use only at present.

- Large-scale hydro-electricity generation has already been discussed in the text of this chapter. Its contribution to overall energy consumption appears much larger if it is calculated in terms of fossil fuel saved during generation than if its final thermal value is calculated. By the latter criterion the 27 million tonnes of oil equivalent attributed to hydro-electricity in EUR 15 out of 1,287 million consumed altogether was a mere 2.1 per cent. Moreover, there is little scope for increasing the hydro-electric capacity.

(continues)

- Wind power requires considerable space to set up the generating capacity and since the best locations are generally on higher open ground, is aesthetically a conspicuous negative feature in the landscape. At present Denmark obtains 2 per cent of its electricity from wind power but very little is generated elsewhere (e.g. in northern Germany, Wales). Larger amounts of electricity could be obtained in the EU, but one problem is the sheer amount of space needed. For example, to construct a 1,000 MW wind farm near Pori on the coast of Finland it is estimated that 2,500 generators occupying an area of 100 sq km would be needed.
- Tidal power has been used to generate electricity in the Rance estuary, Bretagne, France for some time and there is a potential elsewhere in the northern EU countries with suitable estuaries and a large tidal range.
- Wave power experiments continue with technology suitable for generating electricity from this source.

- Solar energy produces about 200,000 tonnes of oil equivalent, a minute share of the total energy budget at present, most in Greece, France, Spain and Portugal.
- Geothermal electricity is produced in Italy and Iceland, but the potential in the EU is small and eventually the pressure of steam from beneath the surface diminishes, making it in reality a non-renewable source.

Secondary and alternative sources of energy in the EU have been discussed at some length in this Box because both economically and environmentally there is an increasing need to reduce dependence on fossil fuels, while the future for nuclear power is far from clear. It seems only a matter of time before much more funding and effort are provided by the EU to encourage research and pool national experience and findings.

Sources: Commission of the EC 1994b: 1–46 – 1–52; Stanners and Bourdeau 1995: 406–7.

---

declined. Nevertheless, about a quarter of the 170 NUTS level 2 units of the EUR 12 countries have coalfields, although production in most has ceased or has been greatly reduced since the 1960s. Most of these coalfield areas come within one of the categories of problem region of the EU, Objective 2 (see Chapter 12).

THE PRODUCTION AND CONSUMPTION OF ENERGY (See Table 5.3)

In the early 1990s three NUTS level 1 regions of the EU accounted for about 80 per cent of

the high grade coal produced: the Ruhr area in Nordrhein Westfalen (Düsseldorf, Cologne and Munster) in Germany, and Yorkshire and Humberside, and East Midlands (Nottinghamshire and Derbyshire) in the UK (Eurostat 1994a: 140–7). On the Franco-German border, Lorraine accounts for almost all of French coal production, while Saarland is Germany's second supplier. Northern Spain (Asturias, Aragón and Galicia) is the only other area in the EU producing hard coal in substantial quantities. Production in such formerly distinguished coalfields as those in Scotland, South Wales and northern France-southern Belgium, has virtually ceased. Most of the lignite extracted in

Table 5.3 Production and consumption of primary sources of energy in the EU and EFTA by country in millions of tonnes of oil equivalent, columns (1)–(10)

	Production				Consumption						(11) Per cent home produced[2]	(12) Consumption t.o.e. per capita
	(1) Coal	(2) Oil	(3) Gas	(4) Total[1]	(5) Coal	(6) Oil	(7) Gas	(8) Nuclear	(9) HEP	(10) Total		
Belgium/Lux	—	—	—	10	8	27	10	10	—	55	18	5.2
Denmark	—	9	2	11	8	10	3	—	—	21	52	4.0
Germany	77	4	14	136	96	135	61	39	2	333	41	4.1
Greece	8	1	—	9	8	17	—	—	—	26[3]	35	2.5
Spain	14	1	1	32	18	54	7	14	2	95	34	2.4
France	6	3	3	112	14	90	28	93	7	232	48	4.0
Ireland	—	—	1	1	2	5	2	—	—	9	11	2.5
Italy	—	4	18	26	13	92	41	—	4	150	17	2.6
Netherlands	—	3	59	63	9	36	34	1	—	80	79	5.2
Austria	1	—	—	4	3	11	6	—	3	23	17	2.8
Portugal	—	—	—	1	3	12	—	—	1	16	6	1.6
Finland	—	—	—	6	4	10	3	5	1	23	26	4.5
Sweden	—	—	—	24	2	17	1	19	5	44	55	4.9
UK	29	127	59	239	50	83	61	23	1	218	110	3.7
EU	135	152	157	674	238	600	257	204	26	1,325	51	3.6
Norway	—	129	28	167	1	10	—	—	10	21	795	4.9
Switzerland	—	—	—	9	—	13	2	6	3	24	38	3.4

Source: Main source BP (1995); also consulted Eurostat 1994a. The calculations of oil equivalents used are those made by BP

Notes: — indicates production of less than 0.5 million t.o.e. or no production
1 Column (4) includes consumption of nuclear and hydro-electricity from columns (8) and (9), assumed for simplicity to be equivalent to production
2 Column (11) production to consumption ratio, consumption = 100
3 Differs from sum of amounts in columns (5) and (6) due to rounding and inclusion of small amounts in columns (7) and (8)

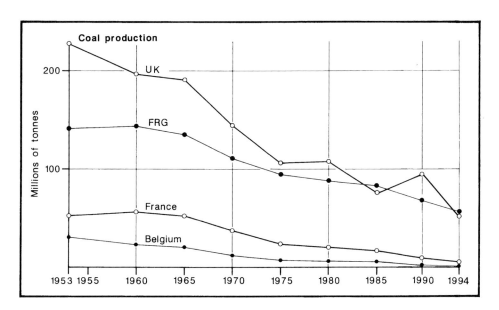

Figure 5.2 Coal production in selected EU countries 1953–94. Note that the German production refers only to hard coal in West Germany and excludes lignite in both Germanies. In 1994 production was 52.0 million tonnes in Germany, 48.0 in the UK, 7.5 in France and zero in Belgium
Source: various numbers of *United Nations Statistical Yearbook* and *BP Statistical Review of World Energy*

the EU comes from the southern part of the former GDR (especially the 'new' Land of Sachsen) and from Nordrhein Westfalen. Secondary areas of production are found in Greece in the regions of Voreia Ellada (Kentriki Makedonia and Ditiki Makedonia) and Kentriki Ellada (Peloponnisos). Greece is being encouraged by the EU in the interests of the environment to reduce lignite production and to change to imported natural gas.

The rises in oil prices in 1973–4 and 1979 arguably slowed down the decline in coal production in the EU for a time. Between 1988 and 1994 alone, however, total coal production in the EU dropped from 238 million tonnes of oil equivalent to 134 million. The 1988 total consisted of 216 million tonnes of hard coal and 486 million tonnes of lignite, while the 1994 total was made up of 123 million tonnes of hard coal and 284 million tonnes of lignite. It should be appreciated that one tonne of oil is equivalent to about 1.5 tonnes of hard coal and to 3

tonnes of lignite. Output in the two major producers of hard coal (see Figure 5.2), the UK and Germany, roughly halved during 1988–1994 (104 million tonnes to 48 million in the UK, 79 million to 52 million in Germany). Further job losses are expected in the UK, where production is mainly confined now to a few mines with high output per worker. Productivity is lower in Germany, but government support is still applied to keep the industry operating. If trends in the last decade or so continue, then the EU could end up in a decade or two with a few token mines producing hard coal and a greatly reduced output of lignite, the latter trend affecting some of the poorer regions of the EU in Sachsen, northern Spain and mainland Greece.

Virtually no oil was produced in the present countries of the European Union until the development of North Sea oil deposits in the 1970s. In 1994 the UK accounted for about 90 per cent of the total EU production. Although

Plate 5.2 Natural gas centre at Groningen, north Netherlands: a cleaner fuel than coal

the extraction of oil in the EU gives direct employment to a very small number of workers, its impact is considerable in terms of supplying equipment for a highly capitalised activity, while oil refining in the EU, supplied mainly from imported oil, is an associated source of employment. Almost all of the extraction of oil in the EU and Norway takes place offshore. Small quantities of oil are extracted in many other localities in most Member States. The prospect for the next decade seems to be more of the same, provided new reserves are found. If not, then the proved reserves of the UK as of the end of 1994 would last only until the year 2000 at the 1994 rate of output, those of Norway only until 2005. Exploration is proceeding in the Atlantic Ocean to the north and west of Scotland as well as around Ireland. New fields have been located to the west of the Shetland Isles. Brierley (1995) describes the activities of various transnational oil companies in the exploration of the ocean floor at depths far below the level of the continental shelf. Floating production, storage and offloading systems (FPSO) give much greater flexibility than the conventional fixed oil platforms used in the North Sea. By comparison, the Middle East has almost 100 years of oil reserves at current rates of production.

The extraction of natural gas, like that of oil, requires a very small labour force in relation to the value of the product and is actually cheaper to transport by pipeline than oil. Most of the natural gas produced in the EU is extracted offshore in the North Sea and moved by pipeline to convenient localities on the coast of the UK and the Netherlands for processing. The gas reserves of the EU have a longer life than the oil reserves. At the 1994 rate of extraction the present reserves of the UK would last about 10 years, those of the Netherlands 25 years. Norway consumes only a small amount of the natural gas it produces. A deep trench separates the coast of southwest Norway from the generally shallow North Sea where its oil and gas deposits are located. Most of Norway's natural gas output is therefore taken directly by pipeline to the UK and Germany. The prospects are that the present natural gas industry in and around the North Sea will not change much. Other deposits of natural gas in the EU are mainly of local importance. About half of the natural gas now consumed in the EU is imported, arriving through various pipelines from the former USSR into and through Ger-

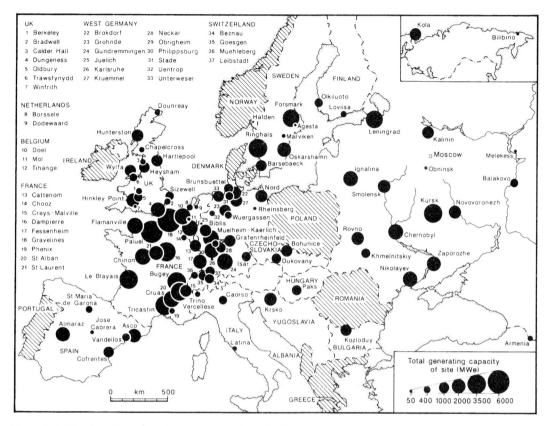

Figure 5.3 The location of nuclear power stations in Europe
Source: Mounfield 1991, maps on pp. 101 and 133 combined with the permission of the author. *Note*: countries with no nuclear power stations are shaded

many and Austria, from Tunisia into Italy also by pipeline, and by liquefied natural gas carriers mainly from Algeria and Libya.

The cost of moving nuclear fuel is minute compared with that of transporting fossil fuels, and has no influence on the location of nuclear power stations. Like thermal power stations using fossil fuels, the generating process requires large quantities of water for cooling the steam driving the turbines. Location by the sea or by a major river is an important factor, as are also the contradictory considerations of avoiding large concentrations of population yet not being in areas that are so remote from consumers that transmission costs to mar-

kets are excessively high. The location of major nuclear power stations is shown in Figure 5.3. Among the EUR 15 countries, Denmark, Ireland, Portugal and Austria do not produce electricity from nuclear energy. In Italy the modest nuclear industry has been abandoned since 1986, while Sweden is to phase out its 12 nuclear reactors. Although the UK was a pioneer in nuclear power, with accompanying technology and know-how, no expansion of capacity is contemplated in the near future. In Europe, France, Russia and Ukraine continue constructing new nuclear power stations, while Japan, South Korea and China are among some 15 countries outside Europe constructing new

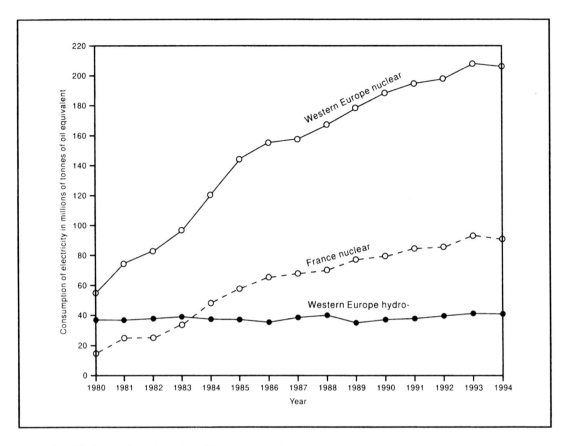

Figure 5.4 Hydro- and nuclear electricity consumption
Note: 1 Western Europe consists of the EU, EFTA and Turkey
2 Hydro-power is converted to oil equivalent on the basis of the energy content of the electricity generated
3 Nuclear is converted on the basis of the average thermal efficiency of a modern nuclear power plant (i.e. 33 per cent efficiency)
Source: BP *Statistical Review of World Energy* 1991 and 1995

plant (Wavell 1995). The data in Figure 5.4 show that nuclear energy consumption in Europe is increasing while hydro-electric energy consumption is stagnating.

As noted earlier in this chapter, the EU produces only about half of the energy it consumes. The direct impact of imported energy on employment in the EU is very small, although affecting a large number of seaports. Most oil refineries are in coastal locations, although some are situated inland, receiving crude oil either by waterway (e.g. the Rhine) or by pipeline. Much

of the hard coal and lignite extracted in the EU is fed directly to nearby thermal electric power stations and no longer forms such a large element in the rail transport system as it did some decades ago. Refined oil products are distributed mainly by coastal shipping, inland waterway or road. Natural gas, on the other hand, is distributed almost exclusively by pipeline. It would be illuminating if rather spurious to have data for energy consumption levels in the NUTS level 2 or even NUTS level 3 regions of the EU. The broad contrasts can be worked

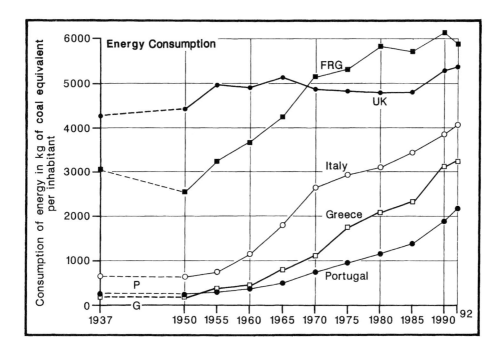

Figure 5.5 The consumption of energy in kg of coal equivalent per inhabitant in selected EU countries 1937–92
Source: various numbers of *United Nations Statistical Yearbook*

out easily. Regions with heavy industry or with large urban centres would tend to have much higher levels of consumption than predominantly rural, agricultural regions. The northern countries could be expected to use more energy also for heating, at least until, if ever, air conditioning becomes widespread in the southern countries.

Even at national level, there are great variations among the countries of the EU in both the degree to which they depend on imported energy and the level of consumption per capita (see columns (11) and (12) of Table 5.3). The UK was the only country in the EU in 1994 in which the production of energy exceeded consumption. In contrast, Ireland satisfied only 11 per cent of its needs and Portugal only 6 per cent. Consumption of energy per inhabitant in the EU also varies greatly among the countries

(see Figure 5.5), ranging from 5.2 tonnes of oil equivalent per capita in the Netherlands to 1.6 tonnes per capita in Portugal. Energy consumption per capita is not an area in which the EU could expect to make much impact in its policy of convergence.

The processing and distribution of energy in the EU forms an extremely complex and sophisticated system, mainly organised by large state or private companies, with an increasing proportion in the private sector since the 1970s. It is doubtful if the organisation would be very different whether the EU existed or not. The fundamental problem remains: the EU cannot plan its energy industry since so much of the fuel it uses is imported. In the early years of the EEC the availability of food preoccupied if not obsessed policy-makers. In the future, if coal production is run down to almost nothing,

Plate 5.3 Power on the River Loire, Middle Ages style: the château at Saumur, north central France

Plate 5.4 Power on the River Loire, twentieth-century style: nuclear power station at Chinon, France

Although successful in reducing French imports of fossil fuel, the nuclear power programme is not without controversy, including its high cost and the ultimate problems of safety and de-commissioning power stations. In the visitor centre at the French nuclear power station at Chooz by the River Meuse close to the Belgian border a prominent poster is entitled 'Centrales Nucleaires Françaises: pas un seule victime'. It is pointed out that except in the Soviet Union there have been no deaths in the nuclear energy sector in 40 years, in contrast to other branches of the energy industry. French families living near commercial nuclear power plants are 'compensated' in a number of ways, for example with local tax concessions and contributions to community funds

BOX 5.2 THE RAIN IN SPAIN FALLS MAINLY ON THE MOUNTAINS

While the population of northern Europe was enjoying or suffering one of its hottest and driest summers on record, in late summer of 1995 many parts of Spain experienced exceptionally heavy rain, which caused considerable damage in flash floods and interfered with tourist activities. The wet weather followed several years of below average precipitation in most of Spain, which left water reserves in the southern half of the country as low as 10–20 per cent of normal and some places with acute shortages. In 1995 losses from desiccated crops were estimated to be about 5 billion dollars, while in resorts on the main tourist centres of the Mediterranean coast water supply was restricted to a few hours a day. Just when the housewives of northern EU countries were becoming brainwashed into believing in the superiority in health terms of olive oil over other cooking oils and fats, the EU 'olive oil lake' vanished because in the autumn of 1995 it was too wet in many parts of Spain to harvest the olive crop; the price of olive oil more than doubled.

In the years immediately preceding the summer of 1995 the problem of water supply in Spain was becoming acute as the *guerra de agua* (water war) between different regions grew in intensity. Meanwhile the water question took on an international dimension as water was shipped from Glasgow in Scotland to Málaga and Mallorca, and serious proposals were made to build a pipeline to transfer water from the lower Rhône in southern France into Catalunya.

In terms of water resources per inhabitant Spain is more generously endowed than several other EU countries. Precipi-tation is heavy in the Cantabrian and Pyrenees mountains of the north. On the other hand the lowest annual precipitation in the EU (apart from the cold extreme north of Sweden and Finland), with under 40 cm a year, is experienced in much of the plateau of Castilla to the north and south of Madrid as well as along the Ebro valley, and in the Mediterranean coastal lowlands of the southeast of the country.

There is enough water in Spain for the reasonable needs of its population, but it falls in the wrong areas. One of the main means of transferring water from wetter to drier areas is the river Ebro and its tributaries, which are fed by water from Spain's northern mountains. However the Aragonese, whose region, Aragon, occupies much of the upper Ebro basin, are reluctant to allow the building of numerous reservoirs on their territory and to lose water to Catalunya and even Baleares (by sea). Similarly, citizens of the region of Castilla la Mancha, one of the poorest (and driest) in the EU, are not happy that their water should help the fruit farmers of Murcia.

The proposal to transfer water by pipeline across the Franco-Spanish border is politically even more complex. A pipeline almost 300 km in length, costing about 1.7 billion US dollars, would be needed to transfer some 15 cubic metres of water per second from near Montpellier to a location north of Barcelona. Such a project is just what the Committee of Regions of the EU encourages, and the idea is indeed attributed to Jacques Blanc, its former President, and Jordi Pujol of Catalunya. The French farmers of the region of Languedoc-Roussillon could have the final say.

They would stipulate that French water should be used exclusively for non-agricultural purposes in Catalunya to prevent rival fruit and vegetable growers of Spain from benefiting. Such is the complicated and paradoxical nature of collaboration among EU Member States.

Sources: Smith 1994; Bowditch 1995; Webster 1995b

further expansion of nuclear power is curtailed for safety or other reasons, and North Sea oil and natural gas output declines, the EU could become even more vulnerable than it is at present to crises elsewhere in the world, with energy replacing agriculture at the centre of EU policy preoccupations.

WATER RESOURCES AND SUPPLY

Although in recent decades population has not grown as quickly in Western Europe as in many other parts of the world, the consumption of water has increased greatly. Since only negligible quantities of water are transported from one country to another, issues and problems of water supply are essentially matters of national or regional significance, and up to the present EU policy on the subject has been limited to directives on the quality of drinking and bathing water. Concern over the supply of water has, however, been expressed in some EU countries as also in many others around the world.

Unlike fossil fuels and non-fuel minerals, fresh water is a renewable natural resource, although its usefulness in various places and contexts may be reduced by man-made pollution. In any given country or region the availability of water may come from two sources, precipitation over its territory (see Figure 6.4) or water from rivers from elsewhere entering and passing through (eg the supply from the Rhine to the Netherlands). Although precipitation is considerable in most of Western Europe compared with precipitation at comparable latitudes elsewhere in the world, it is not entirely reliable. In the southern parts of the EU, precipitation is concentrated mainly in the winter months.

The water resources of a country can be calculated approximately by multiplying the average annual precipitation by the area of the country. The resulting volume of water is usually measured in cubic kilometres, but when it is expressed as the average amount available per inhabitant, it is more convenient to use cubic metres (one cubic kilometre contains one billion cubic metres). In Table 5.4, columns (1)–(3) show the estimated water resources of EU countries (except Luxembourg) and of Switzerland and Norway, measured in thousands of cubic metres per inhabitant. Barney (1982) has one of the first comprehensive sets of data for water availability by country. Column (1) in Table 5.4 shows data for 1971, column (2) estimates for the year 2000. The estimates in column (3) from *HDR 1995* (UNDP 1995) for internal renewable water resources broadly agree with Barney's figures.

Of the EU countries, Sweden and Finland have the most abundant water resources per inhabitant, while among developed countries of the world, Norway is surpassed only by Iceland, New Zealand and Canada. On the other hand, the Netherlands, Belgium and Germany are very poorly endowed, as are Hungary, Poland, the Czech Republic and Slovakia in Central Europe. Columns (4) and (5) in Table 5.4 show the level of consumption of water in EU countries. The contrast in the volume of annual fresh water withdrawals in relation to water resources (1980–9) is striking. Ireland, Austria, Finland and Sweden use a minute

Table 5.4 Water resources and supply

	(1) (2) Water availability in cubic metres per head		(3) Water resources 1992	(4) Fresh water withdrawals as %	(5) per capita	(6) Irrigated land %	(7) Domestic bill, ECU
	1971	2000					
Belgium	0.9	0.8	0.8	72	917	—	330
Denmark	3.0	2.7	2.1	9	228	17.1	680
Germany	1.4	1.3	1.2	55	1,274	4.0	610
Greece	7.6	6.4	4.4	12	720	31.5	210
Spain	3.9	2.8	2.8	41	1,184	17.1	250
France	4.6	3.8	3.0	24	783	6.2	410
Ireland	13.7	11.1	14.3	2	235	—	n.a.
Italy	3.0	2.4	3.1	30	984	26.3	80
Netherlands	0.8	0.6	0.7	16	993	60.7	300
Austria	7.7	7.6	7.2	2	279	0.3	500
Portugal	2.8	2.5	3.5	16	1,075	—	120
Finland	22.5	19.9	22.0	3	605	2.5	330
Sweden	24.1	21.3	20.3	2	356	4.3	360
UK	2.7	2.0	2.1	12	253	1.6	210
Norway	96.9	81.4	94.5	—	490	11.0	190
Switzerland	7.3	6.5	6.2	2	170	5.4	600

Sources: For (1), (2) Barney 1982: 156
(3)–(6) HDR 1995 (UNDP 1995), Table 434, p. 209
(7) Born and Paterson 1995

Notes: — negligible quantity, column (4)
Columns: (3) Internal renewable water resources per capita, 1,000 cubic metres per year, 1992
(4), (5) Annual fresh water withdrawals, (4) as % of water resources 1980–9, and (5) per capita, cubic metres, 1980–9
(6) Irrigated land as a percentage of total arable land area, 1992
(7) Total household bill for water sewerage services 1993 in ECU

Table 5.5 Main uses of water supply in selected countries of the EU, 1990 or nearest year

	Total withdrawal billion cubic metres	Public water supply %	Irrigation %	Industry %	Electrical cooling %
West Germany	44.4	11.1	0.5	5.0	67.6
France	43.3	13.7	9.7	10.4	51.9
Italy	56.2	14.2	57.3	14.2	12.5
UK	13.2	48.6	0.3	10.8	18.8
Spain	45.8	11.6	65.5	22.9	—
Finland	4.0	10.6	0.5	37.5	3.5
Sweden	3.0	32.4	3.1	40.2	0.3

Source: OECD 1991: 54, 66
Note: — negligible

fraction of the water available, whereas Belgium and Germany use over half. The volume of water withdrawals per capita varies less markedly among the countries of the EU than water resources themselves.

Each country has its own profile of water uses. The data in Table 5.5. are from a reputable source, but some of the uses appear to take excessively large percentages of the total withdrawal. In the UK public water supply dominates the uses. In Italy and Spain, irrigation takes up most of the amount consumed (see column (3) of Table 5.5). In Finland and Sweden the processing of raw materials accounts for a large share, while in Germany (using fossil fuels) and France (with nuclear power) the process of cooling in thermal electric power stations takes much of the supply. Not only do different users of the water supply predominate in different countries, but the charge to domestic households varies greatly, reflecting the complexity of systems of supply needed in some countries (e.g. Denmark) and the presence of many dwellings, especially in rural settlements, using local water sources (e.g. Italy, Portugal) at little or no charge.

The per capita consumption of water has risen throughout the EU with the proliferation of domestic appliances, car washes, garden hoses and swimming pools, and the increasing use of sprinklers for irrigation in the northern EU countries. Water is already moved considerable distances both by 'naturally' conveniently oriented rivers and by pipelines (e.g. mid-Wales to Birmingham, the Apulian (Puglia) aqueduct in south Italy). Drinking water has occasionally been transported by rail (e.g. Puglia) and by ship (e.g. from Lake Katrine in Scotland via Glasgow to Spain). Indeed, in the future the international trade in water may grow. Nuttall (1996) reports plans by the small Folkestone and Dover water company (Kent), located conveniently near the entrance to the Channel Tunnel, to pipe water from France through the tunnel, while also obtaining supplies by tanker from Norway. In relation to its value, however, water is very bulky, and the establishment of an EU water grid is at present only a remote prospect.

In both North America and the former USSR the management of water resources has been regarded as a high priority. Various users compete: agriculture, industry, cooling water for thermal electric power stations, domestic supply, fisheries, navigation and leisure uses. Unlike electricity (except in hydro-electric stations), water can be stored for use as needed, but the prospect that large quantities could be moved by long-distance water transfers, as, for example, from Norway or Sweden to Germany

Plate 5.5 Dam and reservoir, Embalse Ricobayo, Northwest Spain, Spring 1992. In the early 1990s many parts of Spain experienced serious water shortages. The water level in this dam was well below full capacity. See Box 5.2

or the Netherlands, seems far off. Ambitious proposals aired earlier this century by Soviet planners to transfer water from Siberian rivers into Central Asia and by US water resource managers from Alaska and Canada to California (NAWAPA scheme) have been dropped. EU transport planners are concerned about moving passengers, energy supplies and containers at speed. With the privatisation of the water industry in various parts of the EU a grand design seems far off.

FURTHER READING

Band, G. C. (1991) 'Fifty years of UK offshore oil and gas', *The Geographical Journal*, 157 (2), July: 179–89.

British Petroleum (BP) (1995) *Statistical Review of World Energy*, yearly, from BP Corporate Communications service.

Eurostat (1995) *Europe in Figures*, 4th edition, 'Energy', pp. 276–85.

Mounfield, P. R. (1991) *World Nuclear Power*, London: Routledge.

Shcherbak, Y. M. (1996) 'Ten years of the Chernobyl era', *Scientific American*, 274 (4), April: 32–7. A clear account of the incident itself, its aftermath and the technological and political measures needed to contain the lasting danger. An issue of great relevance to the EU.

6

AGRICULTURE, FORESTRY AND FISHERIES

'The challenge therefore is for agricultural policy to recognise that there are three different basic needs to reconcile: the production of food and agricultural products, the protection of the environment, and the maintenance of the socio-economic fabric of rural areas.'

(Stanners and Bourdeau 1995: 460)

- More than any other major sector of the economy in the EU, agricultural policy and practice is determined at EU level, through the Common Agricultural Policy.

- Western Europe has extensive areas of good quality farmland, but the area under cultivation has hardly changed in the last fifty years.

- The labour force in agriculture has been declining sharply for many decades, accelerated by widespread mechanisation since the Second World War.

- Through the intensive use of chemical fertilisers, yields have increased massively in EU countries in recent decades and surpluses of agricultural products have been a problem.

- The fishing industry has been a cause of controversy and conflict in the EU whereas the forestry industry has few problems.

- The greatest uncertainties for EU agriculture in the next century include the possible negative effect of global warming and the implications of the prospect of food shortages elsewhere in the world.

When the EC founding treaties were adopted in the 1950s, Western Europe was emerging from a period of war, food shortages and rationing. Agriculture was a key sector of the economy and it provided for the livelihood of a large proportion of the population in many regions. It is not surprising, therefore, that a comprehensive agricultural policy was included in the EEC Treaty, laying down the following objectives:

a) To increase agricultural productivity by promoting technical progress and by ensuring the rational development of agricultural production and the optimum utilisation of the factors of production, in particular labour;

b) thus to ensure a fair standard of living for the agricultural community, in particular by

Plate 6.1 Fortress Europe, illustrating the effect of the Common Agricultural Policy of the EU
Source: 'Le douanier se fait la malle', cartoon by Plantu in *Le Monde Éditions*, Paris, 1992, pp. 88-9

increasing the individual earnings of persons engaged in agriculture;

c) to stabilise markets;
d) to assure the availability of supplies;
e) to ensure that supplies reach consumers at reasonable prices.

(Art. 39 EC)

THE COMMON AGRICULTURAL POLICY

The Common Agricultural Policy is based on the principles of a Single Market, Community preference and financial solidarity, effectively ending all national pricing and markets for agricultural production. In order to attain the objectives set out above, a highly efficient common market organisation system was set up, based on intervention (minimum guaranteed price to producer) and external protection against cheaper imports. A wide range of specific price mechanisms, import levies and export refunds have been introduced for different types of agricultural production ranging from wheat to wine. The levy and refund

mechanism is illustrated in Figure 6.1 with the example of wheat.

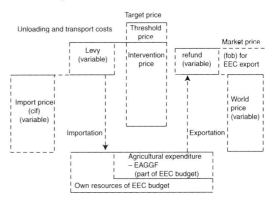

Figure 6.1 The levy and refund mechanisms of the Common Agricultural Policy, illustrated with the case of common wheat
Source: European Parliament 1994b: 391

The CAP has unfortunately been too successful in attaining certain objectives, and too rigid to adapt to changing market realities. Agricultural yields have increased substantially through advances in mechanisation, animal husbandry, plant health and particularly the use of fertilisers. Farmers have been encouraged to produce as much as possible in the knowledge that a guaranteed price would be paid regardless of market conditions. This policy led to overproduction in a variety of products, with infamous 'mountains' of butter, milk powder and beef, all piled up in intervention storage, as well as wine 'lakes' and the destruction of huge amounts of fruit and vegetables. This situation has not only upset public opinion, with consumers forced to pay prices higher than world prices in spite of surpluses in many product areas, but has also been a costly burden on the EU budget, consuming more than two-thirds of total expenditure through the 1980s and still accounting for well over half in 1996 (see Chapter 2, p. 52). In addition, the CAP has distorted world markets and soured relations with major trading partners including the USA and the Cairns Group of agricultural exporters (including Australia, New Zealand, Brazil and Argentina).

During the 1990s, Member States became increasingly aware of these difficulties and overcame strong opposition from the powerful farming lobby to introduce major reforms in the sector. The reforms, adopted in May 1992, were long term in nature and effect, with the budgetary impact only beginning to be felt in 1996. They focused on the sectors with the greatest problems, cereals, beefmeat, sheepmeat, tobacco, dairy produce and oil-bearing seeds, and consisted in substantially reducing levels of intervention and guaranteed prices, while extending instead the system of direct aids paid to farmers. This income support system, as opposed to one based on price support, is used by the other major agricultural producers in the world and is less prone to distort world market prices. The EU aids are also aimed at promoting agriculture that is less intensive and more environmentally friendly through encouraging the 'set-aside' of agricultural land, as well as supporting smaller-scale agricultural activity.

The reforms also entail increased funding made available for so-called socio-structural measures, some of which are horizontal, throughout the EU, others of which are more targeted. The former include aid for modernisation of farms to improve working conditions, safety and hygiene and energy saving, and aid paid for improving processing installations and product quality. The more specific measures include aid for less-favoured and mountainous agricultural zones and aid and interest-free loans to assist young farmers in setting up for the first time. At the same time, early retirement schemes are promoted to encourage farmers to cease farming activity and to pass their holdings on to young farmers, or to diversify. There are also region-specific measures, notably Objectives 1 and 5b of the Structural Funds aimed at developing rural and declining agricultural areas (see Chapter 12). Finally there are special aids paid to farmers to promote environmentally-friendly agriculture, such as the reduction of fertiliser use, the breeding of endangered species, the establishment of fallow land or pastures, and the promotion of set-aside for rural tourism or leisure activities.

PHYSICAL CONDITIONS AFFECTING AGRICULTURE

Only slightly more than 10 per cent of the total land area of the world is cultivated, whereas in the EU the figure is 28 per cent, a proportion that has hardly changed in the last 50 years. The EU has comparatively benign environmental conditions for agricultural activities. These conditions, together with man-made improvements such as drainage, irrigation and the widespread use of fertilisers, have made possible the achievement of some of the highest yields of crops and livestock products in the world. Nevertheless, physical constraints set ultimate limits to the growth of agricultural production. The enabling factors of the physical environ-

[Plate 6.2] Glasshouses on terraces by the Mediterranean coast, Imperia province, Italy

[Plate 6.3] Vineyards in the Moselle valley, Germany

Plates 6.2 and 6.3 Making the most, agriculturally, of sloping ground

Plate 6.4 Landscape in Alicante, southern Spain. This lends credence to the old saying that 'Africa begins at the Pyrenees'

Plate 6.5 Maximising a poor soil and a dry climate: olive trees as far as the eye can see in Andalucía, southern Spain

ment are summarised below under four headings: thermal resources, slope, soil and moisture resources.

Thermal Resources

Each agricultural crop has different requirements, with barley or potatoes, for example, requiring more limited thermal resources than grain maize, vines or the sunflower. Accumulated temperatures decrease broadly northwards with latitude, and also with altitude, and the threshold temperature at which growth starts in each type of plant also varies greatly. Thermal resources can be enhanced in a limited way both by providing artificial conditions such as hothouses, common for example in the Netherlands, and by developing strains of plant that mature quickly, shortening the length of the growing season needed. Thermal resources are adequate almost everywhere in the EU for some kind of arable farming. Cultivation in the Nordic countries extends nearer to the Arctic Ocean than anywhere else in the northern hemisphere, thanks to the warming effect of the Gulf Stream and the moderating effect of the North Sea and Baltic Sea. On the other hand, the cultivation of warm temperate crops such as rice, the olive and citrus fruits is restricted to the southern countries of the EU, while subtropical crops such as sugar cane and cotton can be grown commercially in only a few favoured localities in Spain and Greece. Figure 6.2 shows July temperature conditions in the EU.

Slope

There is a slope limit beyond which commercial cultivation is impracticable. The terracing of steep slopes has been practised in southern Europe at least since Roman times, but for economic reasons the cultivation of field crops has now largely ceased in such areas, although bush and tree crops are still grown. In general, terraces have the drawback of precluding the efficient use of agricultural machinery. Figure 6.3 distinguishes areas defined as mountainous (Commission of the EC 1993a: 32). Here steep slopes predominate, while temperatures increase with altitude, bringing cold conditions more characteristic of very high latitudes.

Soil

Some local areas in the EU have soils that are inherently very fertile. Examples include the volcanic soils in Italy around Vesuvius and Etna, the drained marshlands in the Netherlands, the Fens in England and the North Italian Lowland. An extensive area of reasonably good soils, formed on sedimentary rocks or on glacial deposits, extends across northern France, southern UK, northern Belgium, the Nether-

lands and parts of Germany into Denmark (see Figure 6.3). The extent and continuity of areas of good soil is more restricted in the southern part of the EU and in all but the most southerly parts of Sweden and Finland.

Moisture Resources

There are few areas in the EU where conditions are too dry for cultivation. Central and southern Spain is the area most frequently affected by dry conditions, the long, hot, but dry summer preventing the full use of the advantageous thermal conditions unless irrigation is provided. Italy and Spain have the most extensive gravity systems of irrigation fed by rivers. Increasing use has been made in recent years throughout the EU of sprinkler and drip irrigation, using local water supplies to distribute water, putting additional pressure on limited water resources. It was shown in Chapter 5 that by world standards the water resources of most of the countries of Western and Central Europe are small, the result of a combination of the high density of population and the heavy use per capita of water for various purposes.

The effectiveness of precipitation (rain, snow, hail) is only broadly related to the quantity falling (see Figure 6.4). Since Europe extends over 35° of latitude, from 70°N in the north to 35°N in the south, the average temperature varies greatly, increasing broadly in a north–south direction. Evaporation increases with temperature, so that a given quantity of precipitation in the cooler north is more effective than the same amount in the warmer south. The season in which the precipitation comes also affects its usefulness for some purposes. In general, the period May to October is the wettest part of the year in Central and Northern Europe, whereas it is the drier half of the year in Mediterranean Europe.

Farmers in some parts of southern Europe have used water for irrigation for many centuries, but in other areas the practice is very recent.

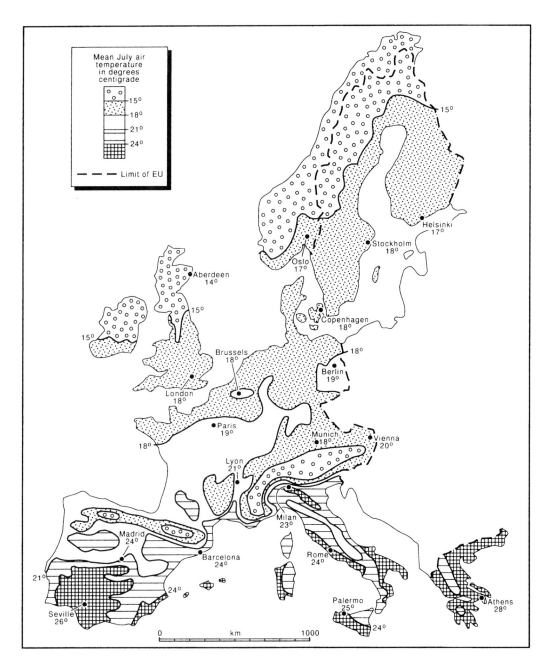

Figure 6.2 Mean July air temperature in the countries of the EU, Norway and Switzerland. To highlight the coolest and warmest parts of the region, different types of shading are used for temperatures under 18° and above 21°, while the intermediate area with between 18° and 21° is left unshaded. Temperatures are also given for selected cities. Note that temperature in the rest of Europe is not indicated. For simplicity the boundary of Switzerland is omitted
Source: based on Spiessa 1993: 82

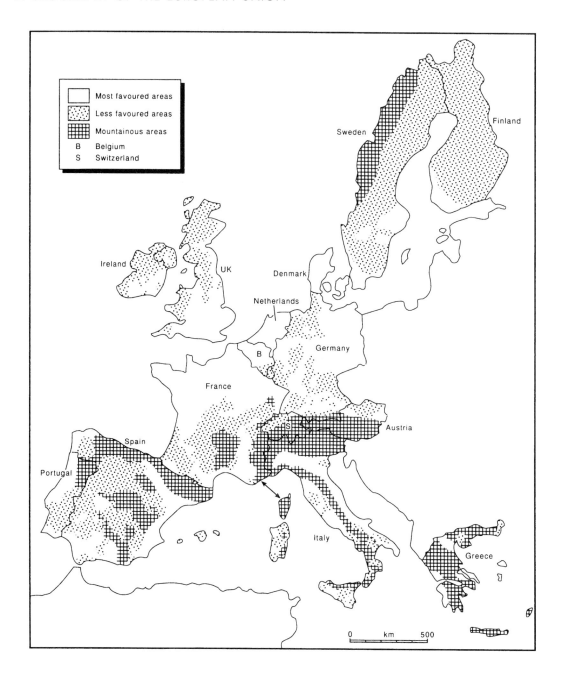

Figure 6.3 Rural development in the European Union. Note: the farmland located in less favoured and mountainous areas makes up over 50 per cent of the total area farmed, including arable and permanent crops and permanent pasture. Austria, Finland and Sweden have been added by the authors to the original map. Switzerland is included but is not a member of the EU
Source: based on Commission of the EC 1993a: 32

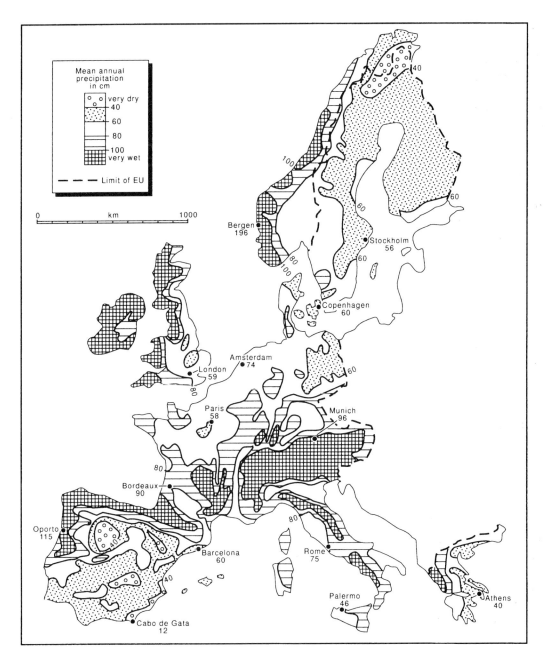

Figure 6.4 Mean annual precipitation in the countries of the EU, Norway and Switzerland. To highlight the wettest and driest parts of the region, different types of shading are used for precipitation levels under 60 cm and over 80 cm, while the intermediate area with between 60 and 80 cm is left unshaded. Precipitation is also given for selected localities. Note that precipitation in the rest of Europe is not indicated. For simplicity the boundary on Switzerland is omitted
Source: based on Spiessa 1993: 84

In column (11) of Table 6.1 it can be seen that in the four southern countries of the EU, Greece, Italy, Portugal and Spain, a considerable part of the land used for cultivation is irrigated. For example, in the North Italian Lowland very high yields of rice and fodder crops are obtained thanks to the use of water carried by rivers flowing from the Alps, while along the Mediterranean coastlands of Spain there are many small irrigated lowlands on which vegetables and fruit are widely grown. Irrigation from sprinklers is now extensively used in the Netherlands and Denmark, especially in periods of dry weather.

LAND USE IN THE EU

In the publications of the Food and Agriculture Organisation (FAO) four broad classes of land use are recognised: arable land and permanent crops; permanent pasture; forest; and other uses, including waste such as desert, and built-up land. In Table 6.1 (columns (2)–(4)) the area occupied by the first three of the above categories is shown for the countries of the EU in 1994. Before arable land is examined more closely, some aspects of the other two uses will be noted.

Permanent Pasture

Permanent pasture varies greatly in quality both between and within countries. For example, in the UK and Ireland much of the pasture in mountainous areas supports only a very low density of livestock, whereas water meadows provide good supplies of fodder. There is scope for converting some of Europe's permanent pasture into arable, should the need arise, but much of the fodder produced to raise Europe's livestock already comes from arable land. Column (3) in Table 6.1 shows the area of permanent pasture in each EU country, while column (6) shows the proportion of the total EU land area in this category, 18 per cent, compared with a world figure of 25 per cent. The contrast between EU countries is very great. Thus, for example, over two-thirds of the total land area of Ireland is classed as permanent pasture compared with a mere 1 per cent in Sweden.

Forest and Woodland

About 33 per cent of the world's land area is classified as forest and woodland, while the proportion in the EU is 36 per cent (see columns (4) and (7) in Table 6.1). Almost half of the area of forest and woodland in the EU is in the three newest Member States, Austria, Finland and Sweden. In contrast, Russia has almost seven times as much forest as the whole of the EU, and Canada has more than four times as much.

Other Uses

In spite of the publicity given to the growth of built-up areas (residential, industrial, transport uses), in most of the countries of the EU only a very small fraction of the residue 'other uses' land is so used. Most is waste, whether rocks, bare sand, swamp or Arctic waste. Beyond the limits of the EU, 70 per cent of the land area of Norway and 80 per cent of the land area of Algeria have no bioclimatic use, but the percentage in the EU as a whole is only 18 per cent.

Arable Land

Arable land and permanent crops include land growing field crops (mostly annuals such as wheat, sugar beet), fallow land, and bush and tree crops of a more permanent nature (orchards, vineyards, olive groves). The EU has about 88 million hectares of arable land and permanent crops, 28 per cent of its total land area of 313 million hectares. In comparison with the USA and Russia, and in particular with Canada

Table 6.1 Land use in the countries of the EU

		(1) Millions of hectares Total land area	(2) Crops	(3) Permanent pasture	(4) Forest and woodland	(5) Percentage of total area % Crops	(6) % Pasture	(7) % Forest	(8) Other	(9) Persons per ha of cropland	(10) Permanent crops % of all cropland	(11) Irrigated land % of all cropland
1,9	Belgium-Lux.	3.3	0.8	0.7	0.7	24	21	21	34	13	2.1	—
2	Denmark	4.2	2.5	0.2	0.4	60	5	10	25	2	—	17.1
3	Germany	34.9	12.1	5.3	10.7	35	15	31	19	7	3.6	3.9
4	Greece	12.9	3.5	5.3	2.6	27	41	20	12	3	30.8	37.6
5	Spain	49.9	19.7	10.3	16.1	39	21	32	8	2	23.8	17.6
6	France	55.0	19.4	10.8	14.9	35	20	27	18	3	6.1	7.6
7	Ireland	6.9	0.9	4.7	0.3	13	68	5	14	4	—	—
8	Italy	29.4	11.9	4.3	6.8	40	15	23	12	5	23.9	22.8
10	Netherlands	3.4	0.9	1.1	0.4	28	31	10	31	16	3.0	60.0
11	Austria	8.3	1.5	2.0	3.2	18	24	39	19	5	5.3	—
12	Portugal	9.2	3.2	0.8	3.3	34	9	36	21	3	25.3	19.9
13	Finland	30.5	2.6	0.1	23.2	8	3	78	13	2	—	2.5
14	Sweden	41.2	2.8	0.6	28.0	7	1	68	24	3	—	4.1
15	UK	26.2	6.3	11.0	2.4	25	46	10	19	10	0.8	1.8
	EU	315.3	88.1	57.2	113.0	28	18	36	18	4	12.7	12.9

Source: *FAOPY 1994 (FAO 1995)*

and Australia, the EU is poorly endowed with arable land in relation to its population size, but it is much better off than China, Japan and South Asia. The EU has much higher yields for comparable crops than most other parts of the world, reflecting both the presence of much good quality land and the heavy use of fertilisers. From the data in column (9) of Table 6.1 it is evident that the Netherlands and Belgium are the EU countries with the smallest amount of arable land per inhabitant; they are followed by the UK. Among the larger EU countries in area, France and Spain are more generously endowed.

Since the Second World War, the arable area of Europe has not increased greatly anywhere except in the Volga and Ural regions of the former USSR, as also in West Siberia and Kazakhstan in Asiatic USSR, during the new lands campaign of the 1950s. The population of Europe though has grown considerably. In the EU itself, the total area of arable land and permanent crops in most countries has actually diminished since the 1950s, but changes in the definition of arable make precise comparisons between different countries through time impossible. Even so, the evidence from FAO Production Yearbooks points to a peak period in the EUR 12 countries roughly between 1955 and 1965, when the area under arable and permanent crops was 93–5 million hectares. There was a sharp decline between the mid-1960s and early 1970s to around 84 million. The decline continued more slowly, to reach about 81 million hectares in 1994. The addition of some 7 million hectares in the three newest Member States raised the total for EUR 15 to 88 million in 1995.

In the 1980s it became clear that the EU could satisfy many of its needs from the agriculture sector very comfortably, although it does not have high enough temperatures to produce commercially such crops as tea, coffee or tropical fruits. On the other hand, it can grow oilseeds and sugar beet, substituting imports of oilseeds from areas such as West Africa (oil palm) and of sugar cane from the Caribbean and elsewhere in the tropics.

In spite of the current policy to reduce land under cultivation in the EU through set-aside measures, it must be remembered that the world agricultural scene is changing rapidly. Conceivable futures include the possibility that significant areas of good farmland in the EU could be flooded through a rise in sea level, or that the EU will have to help to feed the growing population in the less developed countries. Under these circumstances it could be advantageous both environmentally and politically to use existing land less intensively in terms of the application of fertilisers, while maintaining or actually increasing the area cultivated, rather than allowing existing arable land to disappear irrevocably from use. Under current EU guidelines for set-aside procedures, if the choice arises, individual farmers are more likely to take the less productive land out of use in their farms rather than the more productive land, assuming they are paid according to the extent of land taken out of use rather than according to the potential value of production from the area.

In the category of arable and permanent crops, the latter accounted for over 11 million hectares, 12.7 per cent of the total. Tree and bush crops are cultivated most widely in the four southern countries of the EU, Greece, Italy, Spain and Portugal, where they occupy roughly a quarter of the total crop area (column (10) in Table 6.1). In the Nordic countries virtually no tree crops are grown commercially other than for timber. In planning agriculture at the EU level it is more easy to switch annual field crops from one crop to another or from cultivation to fallow than to remove plants such as fruit and olive trees, since their subsequent replacement may take many years before new trees mature.

THE ECONOMICALLY ACTIVE POPULATION IN AGRICULTURE

The agricultural sector conventionally includes not only farming but also forestry and fishing. The processing of the products of these activities is regarded as part of the industrial sector. In 1994 somewhere between 5 per cent (according to the Food and Agriculture Organisation) and 5.6 per cent (*Regional Profiles* 1995) of the economically active population of the EU was engaged in the agricultural sector. In this section attention will focus on two aspects of employment in agriculture – the present uneven distribution in relation to total employment at national level, and at regional level, and efficiency and productivity.

In 1994 agriculture in EUR 15 employed about 8.5 million people, compared with 24.6 million only three decades previously. The absolute numbers are shown for 1965, 1980 and 1994 in columns (1)–(3) of Table 6.2. In 1994 there was a great difference at national level between at one extreme Belgium and the UK, with less than 2 per cent of their economically active population in agriculture, and at the other, Greece with 22 per cent. At NUTS level 1 the extremes are even more marked. Among other factors the population size and the level of urbanisation of each region affect the results. Several regional divisions covering a small area and containing a large city have less than 1 per cent of their economically active population in agriculture: Brussels, the Ile de France (Paris), Berlin and Madrid. At the other extreme, three of the four NUTS level 1 regions of Greece had 39, 31 and 28 per cent of their economically active population in agriculture. At NUTS level 2 the contrasts are even more marked than at NUTS level 1. Figure 6.5 shows the situation at NUTS level 1.

The differences in the proportion of economically active population in agriculture among the regions of the EU are explained not only by the area and population size of the regions and the presence of other activities but also by the productivity of workers in the agricultural sector itself. One way of comparing the productivity of labour in agriculture with that in other sectors is to compare the share of GDP accounted for by agriculture with its share of economically active population. Such an index is calculated by dividing the percentage of GDP accounted for by agriculture by the percentage of the economically active population employed in agriculture. The result is multiplied by 100 to remove decimal places. Table 6.3 shows the result of such a calculation for the ten EUR 15 countries for which recent data were available. GDP percentages are rounded to a whole number and the indices are therefore very approximate. Even so, they do show marked contrasts, with the UK, Denmark and Sweden having the most efficient agricultural sectors as measured in this way, Italy, Austria and Germany the least efficient. The reasons for such disparities may relate to some extent to policies on food supply prevalent in the period during and between the First and Second World Wars.

It is possible also to obtain the above measure of efficiency for NUTS level 1 regions of the EU/EC before the entry of the three newest members. Appropriate data are found in COM 87–230 (1987) and Eurostat (1989: 56–64). Figure 6.6 shows the efficiency rating of the regions. At NUTS level 1, parity is achieved in the region of Zuid-Nederland, giving it an index of 100. This is the only EU region in which agriculture can hold its own against industry and services. Of the eight most efficient regions, scoring 80 or above, apart from Denmark, all are in the Netherlands or the UK (East Anglia 83, Wales 82, Northern Ireland 80) except Ile de France (the Paris area, 80). The lowest scoring regions are all in Spain or Germany, but the reasons for such low scores clearly differ. For example, Baden-Württemberg is not particularly fertile agriculturally, whereas it has a very large and

Table 6.2 Economically active population in agriculture

		(1)	(2)	(3)	(4)	(5)	(6)	(7) per 100 ha of arable + pasture	(8) Tractors thousands 1993	(9) Tractors per 100 agric. workers	(10) Tractors per 100 ha of arable
		Total in thousands			As percentage						
		1965	1980	1994	1965	1980	1994				
1,9	Belgium - Lux.	233	120	66	6	2.9	1.5	4	113	171	14
2	Denmark	330	190	114	15	7.3	3.9	4	156	137	6
3	Germany	4,442	2,601	1,625	17	6.9	4.0	9	1,300	80	11
4	Greece	1,935	1,134	906	53	30.9	22.0	10	216	24	6
5	Spain	4,143	2,215	1,340	34	17.1	8.9	4	775	58	4
6	France	3,600	2,024	1,133	18	8.6	4.3	4	1,460	129	8
7	Ireland	360	234	168	32	18.6	11.9	3	168	100	18
8	Italy	5,005	2,600	1,343	25	12.0	5.7	8	1,430	106	12
10	Netherlands	405	302	199	9	5.5	3.1	10	182	91	20
11	Austria	690	306	186	20	9.0	4.8	5	343	184	23
12	Portugal	1,410	1,109	620	40	25.6	13.6	16	131	21	4
13	Finland	670	286	179	32	12.0	6.9	7	232	130	9
14	Sweden	400	236	148	12	5.7	3.3	4	165	111	6
15	UK	961	706	511	4	2.6	1.8	3	500	98	8
	EU	24,584	14,063	8,568	17.6	9.1	5.0	6	7,171	84	8

Sources: FAOPY 1994 (FAO 1995, Table 109 for tractors); FAOPY 1967 (FAO 1968, Table 3) for 1965 data

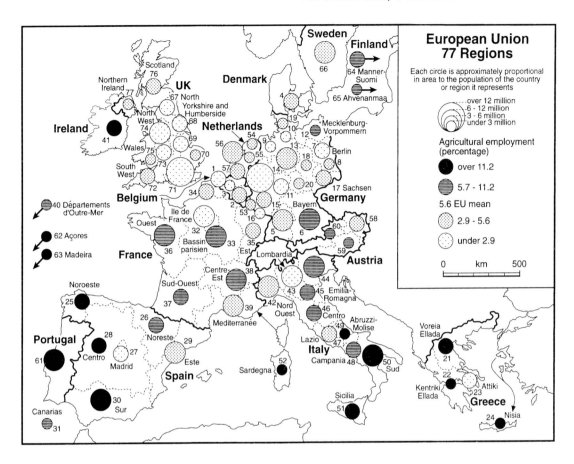

Figure 6.5 Employment in agriculture, forestry and fishing as a percentage of total employment. Each NUTS level 1 region is represented by a circle roughly proportional in area to the size of its population. See Table A1 (Appendix) for key to the numbering of unnamed regions
Source: Regional Profiles 1995

Table 6.3 Efficiency of agriculture ratio

	Per cent				Per cent		
	GDP	Emplt	Ratio		GDP	Emplt	Ratio
UK	2	1.8	111	France	3	4.3	70
Denmark	4	3.9	103	Ireland	8	11.9	67
Sweden	2	2.0	100	Italy	3	5.7	53
Greece	18	22.0	82	Austria	2	4.8	42
Finland	5	6.9	72	Germany	1	4.0	25

Source: World Bank 1995, Table 3

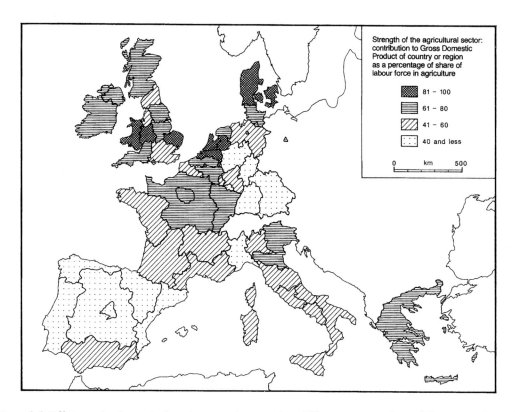

Figure 6.6 Efficiency in the agricultural sector. A somewhat different measure is used for comparison with that described in Figure 6.7. The index is measured by dividing the share of total GDP of each region contributed by the agricultural sector, by the share of total employment in the region accounted for by agriculture; the result is multiplied by 100. The index is 100 if the two shares are equal. Comparable data for the former GDR are not available and only EUR 12 countries are shown on the map

successful industrial sector. In Bremen, West Berlin and Madrid, agriculture is overshadowed by a successful service sector. In the Noroeste of Spain, agriculture is very backward, while the industrial and service sectors, although not outstanding in comparison with those in the EU as a whole, are more successful.

In the view of the authors, the index explained above is a useful general guide to the agricultural health of different parts of the EU. Whether assistance to the agricultural sector should go to raising efficiency in the regions with a low index (so long as changes could be guaranteed to improve standards) or to phasing out agriculture where further investment would

be wasted, must be a consideration in applying the Common Agricultural Policy. The NUTS level 1 or even the NUTS level 2 regions would not, however, form a detailed enough base for the application of some policies without further refinements, given the great variations in practices and performance among quite small areas in Western Europe.

A different measure of productivity in the agricultural sector is the agricultural value added per work unit, showing labour productivity. Commission of the EC (1992: 38) includes a map of labour productivity at NUTS level 2. The distribution of high and low productivity shown in Figure 6.7 is broadly

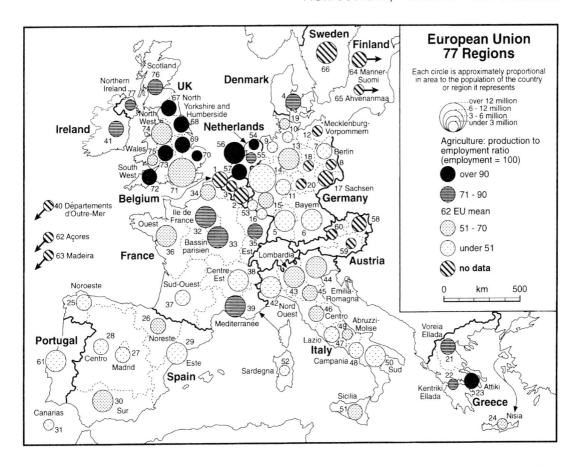

Figure 6.7 The ratio of the share of employment in the agricultural sector in each region to its share of the total ouput of all sectors. A few regions are close to parity, but in most the agricultural sector's contribution per person employed is well below average

Source: *Regional Profiles* 1995 for employment data; Eurostat 1994b, Table 2.25, for output data

similar to that in Figure 6.6. The highest labour productivity in agriculture in relation to other sectors of the economy is found in Denmark, the Netherlands, much of the UK, northeast France and Greece, the lowest in Portugal, Spain and much of Italy and Germany. Some of the former areas have a level of labour productivity at least four times as high (at over twice the EU average) as the latter areas (with under half), thus providing further evidence of the great gap in economic conditions between various parts of the EU.

The relative success or lack of success of the agricultural sector in different parts of the EU can be gauged by other measures in addition to those discussed above. Yields per area cultivated or per unit of livestock (covered later in this chapter) can be taken as a broad measure of success. Alternatively the level of mechanisation can be used, a feature referred to in the section below. It may not be long before 'points' are awarded or deducted according to appropriate indicators of damage to the land and soil.

MECHANISATION IN AGRICULTURE

Almost all the countries of Europe are still experiencing a substantial decline in employment in the agricultural sector, and it can be expected that the trend will continue for several decades in many regions of the EU as well as in EFTA, Central Europe, the former USSR, Turkey and the Maghreb. The single most influential factor in reducing the labour force in agriculture in Western Europe has been the diffusion of agricultural machinery of various kinds. In the late nineteenth century such machinery was driven by steam powered engines, to be superseded in the twentieth century by petrol and diesel fuelled tractors. The widespread availability of electricity in the second half of the twentieth century has also saved labour, mainly on the farms themselves.

The use of tractors is a useful broad measure of mechanisation in agriculture. In Table 6.2, column (8) shows the number of tractors in use in the countries of EUR 15 in 1993, almost 7.2 million. The number was 3 million in 1960 in the same fifteen countries (not all in the EEC then). In the early 1960s, agriculture in the UK, Sweden and Denmark was already highly mechanised, whereas in Spain, Portugal and Greece the number of tractors in use was still very small.

In the last three decades, the 'tractor gap' has been narrowing. In Table 6.2, column (9) shows that there is still a great disparity in the number of tractors in use in relation to the number of people employed in the agricultural sector. It must be noted, however, that the average size of tractor differs from country to country and that in general the cultivation and harvesting of field crops is more easy to mechanise than the cultivation and harvesting of tree crops. In the EU as a whole there were 84 tractors per 100 people employed in agriculture in 1993, but with an enormous range, extending from 21 in Portugal to 184 in Austria. In so far as greater mechanisation is the

cause of a decline in employment in agriculture, Portugal, Greece and Spain are the EU countries most likely to experience a further increase in mechanisation and the greatest decline in the labour force.

The final column (10) in Table 6.2 shows the 'intensity' of tractor use in relation to land rather than to agricultural workers (column (9)). The differences in level are explained by a combination of factors. First, the level of mechanisation must be taken into account. Then the intensity of land use (high in the Netherlands), generally difficult terrain (mountains in Austria) or flat and gently undulating terrain (northern France, Denmark), and the presence of high quality natural pastures (Ireland, UK) not counted under the cultivated area all contribute to the diversity of levels and conditions of mechanisation.

Agriculture in the northern countries of the EU is more highly mechanised than elsewhere in Europe. Figure 6.8 shows the large increase in the number of tractors in use since the Second World War. The mechanisation of agriculture has, however, proceeded quickly in Central Europe in the last three decades, but in most countries from very modest beginnings in the 1950s. The level of mechanisation in Central Europe and the former USSR in the mid-1990s was roughly comparable with that in Portugal and Greece, while in Turkey and the Maghreb countries it was lower still. In all these regions further mechanisation can be expected in the next few decades. For example, in Poland, the Czech Republic, Slovakia and Hungary the labour force in agriculture declined by 2.5 million during 1980–94 from 7.3 to 4.8 million. In Turkey agriculture still accounts for 45 per cent of the economically active population, compared with 59 per cent in 1980, but as a result of the increase of total population, the actual number remained between 11.0 and 11.7 million, more than the total of 8.5 million in the whole of the EU.

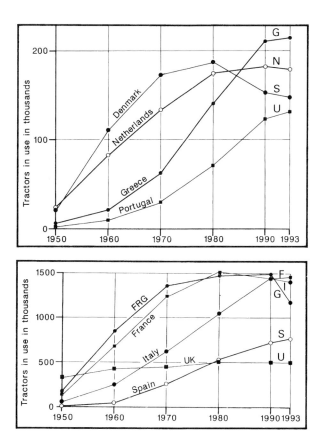

Figure 6.8 Tractors in use in selected EU countries 1950–93. Data for the early 1990s are only rough estimates
Source: various years of *FAO Yearbook, Production*

AGRICULTURAL PRODUCTION

Although products from bioclimatic resources, before processing and preparation for sale, account for only 3–4 per cent of GDP in the EU as a whole, the agricultural sector has a prominent place in EU policy. In this section reference will be made to only a few of the most important crops and types of livestock. The Production Yearbook of the Food and Agriculture Organisation can be consulted for further information. After some introductory points, attention will focus on increases in yields of certain products. Because yields have increased

much more quickly than land has been taken out of agricultural use, overall agricultural production in the EU has increased continuously since the 1950s, when the Common Agricultural Policy was started, a situation that gives rise to one of the principal problems of the EU budget. During 1992–4 the area under field and permanent crops in the EU averaged almost 88 million hectares, of which almost 77 million were under field crops, just over 11 million under tree and bush crops.

1 Permanent crops are relatively far more prominent in Greece, Portugal, Italy and Spain than in the remaining EU countries. The

[Plate 6.6] Old vines grubbed up in foreground, new vines planted in the centre of picture and bare limestone hills in the distance. Field crops are difficult to grow in such areas of poor soil as this, located on the southern margin of the Massif Central

Plates 6.6 and 6.7 Contrasts in French agriculture

[Plate 6.7] High technology in the agriculture of Dreux, northern France, where slope, soil and climatic conditions are good, yields high, fields large and mechanisation widely applied

cultivation of vines accounts for about one-third of the area under permanent crops in the EU while that of olives also occupies a large area. The cultivation of nuts is more limited in extent. The vines, olive trees and nut trees are widely grown in areas where either the slope or the quality of the soil is not suitable for the cultivation of high yielding field crops. Soft fruits such as apples, plums and pears are more widely grown in

the EU. Almost all the production of permanent crops grown in the EU is for human consumption as food or beverages.

2 Field crops grown mainly or partly for human consumption, usually after appropriate processing, include wheat, other cereals, sugar beet and vegetables, including potatoes. Cereals occupy almost half of the area under field crops in the EU.

3 Some crops are grown mainly or exclusively

as feed for livestock. These include some cereals, such as green maize, as well as lucerne and rapeseed, turnips and wurzels. Some fodder for livestock is also imported from outside the EU.

4 Permanent pastures provide the remainder of the fodder for livestock. Animals are increasingly kept part or all of their lives in confined spaces in order to reduce their intake of fodder. The relative contributions of crop and livestock products to total production in the countries of the EU vary markedly, with the livestock share generally highest in the regions of the northern countries.

5 A third set of field crops provides raw materials for manufacturing industry, as opposed to processing for human food. Most of the agricultural raw materials used in the EU are imported, but cotton (Greece) and flax are among those produced 'at home', together with livestock products such as wool and hides.

6 Forestry will be covered more fully later in this chapter. Although the use of fuelwood is widespread in the EU, the quantity is small in relation to total energy consumption. The use of forest products in construction or as raw materials is generally regarded as more desirable, especially as the EU is a net importer of forest products. For some decades now it has been a practice locally, as for example in the North Italian Lowland, to plant 'crops' of fast growing trees such as poplars, in place of field crops.

Although cereals do not produce such a high value of output per unit of area as some other crops, for example vegetables, sugar beet and cotton, they account for almost half of all the arable area in the EU and have therefore been chosen to illustrate the changes in yield in

Table 6.4 Cereal cultivation in the EU (excluding Luxembourg)

| | 1979–81 | | | 1992–4 | | | |
	(1) Area million ha	(2) Yield tonnes per ha	(3) Production million tonnes	(4) Area million ha	(5) Yield tonnes per ha	(6) Production million tonnes	(7) Fertiliser kg/ha
Belgium	0.4	4.9	2.1	0.3	6.5	2.3	238
Denmark	1.8	4.0	7.3	1.5	5.2	7.7	209
Germany	7.7	4.2	32.0	6.4	5.6	35.6	248
Greece	1.6	3.1	5.0	1.4	3.6	5.0	181
Spain	7.4	2.0	14.7	6.8	2.3	15.8	101
France	9.8	4.7	46.1	8.7	6.5	56.6	251
Ireland	0.4	4.7	2.0	0.3	6.5	1.8	704*
Italy	5.1	3.5	18.0	4.1	4.7	19.5	207
Netherlands	0.2	5.7	1.3	0.2	7.5	1.4	607
Austria	1.1	4.1	4.4	0.8	5.3	4.4	186
Portugal	1.1	1.1	1.2	0.7	2.0	1.4	109
Finland	1.2	2.5	3.0	0.9	3.4	3.1	136
Sweden	1.5	3.6	5.4	1.1	4.1	4.5	108
UK	3.9	4.8	18.8	3.1	6.6	20.4	323
EUR 15	43.2	3.7	161.3	36.3	4.9	179.5	

Source: *FAOPY 1994* (FAO 1995), Table 15; Column (7) *UNSYB* 40th edition (UN 1995: 356–77)
Note: *Ireland has only a small area of arable land and much of the fertiliser applied is actually used in areas of pasture

recent decades. Wheat and barley are the most widely grown cereals in the EU, the former occupying an average of about 16 million hectares in 1992–4, barley about 8 million. They are grown in all EU Member States, wheat primarily as a bread and pasta grain, barley partly for malting, partly as feed for livestock.

Barley tends to be grown in areas with lower accumulated temperatures than wheat. For example, barley covers a larger area than wheat in Finland, Sweden, Denmark and Ireland, while wheat predominates in the Mediterranean region of the EU. Two salient features of the cultivation of both wheat and barley are their

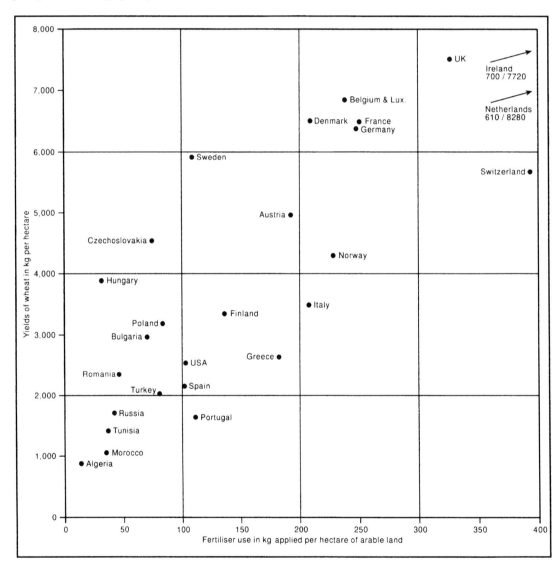

Figure 6.9 The relationship between the use of fertilisers and yields of wheat in EU and other selected countries in 1992

Table 6.5 Wheat yields in the EU and in other selected countries

| | (1) | (2) | (3) | | (4) |
| | Tonnes per hectare (five or three year average) | | | | Tonnes per hectare |
	1921–5	1961–5	1992–4		1992–4
Netherlands	2.8	4.4	8.3	Mexico	4.2
Ireland	2.3	3.3	7.7	Hungary	3.9
UK	2.3	4.0	7.5	China	3.4
Belgium-Lux	2.6	3.9	6.9	Poland	3.2
Denmark	3.0	4.1	6.5	USA	2.6
France	1.5	2.9	6.5	India	2.4
Germany	1.8	3.3	6.4	Romania	2.3
Sweden	2.0	3.4	5.9	Canada	2.2
Austria	1.3	2.6	5.0	Argentina	2.2
Italy	1.2	2.0	3.5	Turkey	2.0
Finland	1.4	1.7	3.3	Russia	1.7
Greece	0.6	1.5	2.5	Australia	1.6
Spain	0.9	1.1	2.2	Kazakhstan	1.0
Portugal	0.7	0.8	1.6	Algeria	0.9

Sources: League of Nations (1933), *Statistical Yearbook of the League of Nations 1932/33*, Geneva: Economic Intelligence Service; *FAOPY 1970* (FAO 1971, Table 14); *FAOPY 1994* (FAO 1995, Table 16)

differing relative importance in different countries and regions of the EU, and the almost universal increase in yields achieved in the last few decades.

Table 6.4 shows the change in the area, yield and production of all cereals in the countries of the EU over a comparatively short period of 13 years. Three-year averages have been calculated to reduce the effect of year to year fluctuations. The area sown has diminished by 14 per cent, the yield has increased by 32 per cent and production has risen by 11 per cent. Even ignoring the special cases of Ireland and the Netherlands, column (7) in Table 6.4 shows marked differences in the use of fertilisers per hectare of arable land. Figure 6.9 shows the relationship between the application of nitrate, phosphate and potash fertilisers and yields of cereals. Not all the fertiliser is actually applied to cereal crops so the relationship is only a broad indicator.

Wheat yields in the countries of the EU over a much longer period are shown in Table 6.5. They show very marked contrasts between the 15 present Member States during all three periods for which the average yields have been calculated. They also show an increase of between two and three times almost throughout the EU. There is no reason why wheat yields could not be increased further even in the countries achieving the highest yields in the 1990s, although there are both commercial and environmental reasons why the application of very large amounts of fertiliser per unit of area is undesirable. Diminishing returns set in, while excessive chemicals, including nitrates, seep through the underlying subsoil and rock, and pollute underground water, a problem widespread in northern France and eastern England.

Figures 6.10 and 6.11 show the growth of wheat yields in selected countries of the EU and in non-EU countries in the 'near abroad' of the EU. A selection of non-EU countries has been included in Table 6.5 to allow a comparison of wheat yields on a global scale. Wheat is one of the major cereal crops in all the countries listed. It is thought-provoking if somewhat spurious to argue that if all the EU countries could raise their wheat yields to the level of 7–8 tonnes per

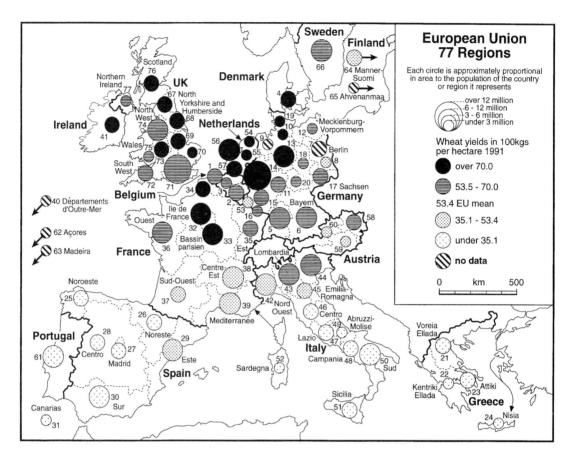

Figure 6.10 Yields of wheat in kg per hectare in NUTS level 1 regions of the EU in 1991 (Austria, Sweden and Finland were not members then)

hectare already achieved in parts of northwest Europe, wheat production could be increased greatly. A similar argument could be applied to the rest of the world. A more detailed examination of conditions and practices would show that such a view is simplistic. In Italy, for example, much of the best agricultural land in the North Italian Lowland is used to cultivate rice (irrigated) or grain maize, both used for human consumption, rather than wheat, because they give higher yields of grain. In contrast, in Mexico wheat is grown mainly on fertile northern oases and gives high yields, whereas maize is grown widely in areas with

difficult environmental conditions, and gives low yields. In Central Europe, the former USSR and Turkey, all of which are of increasing interest to the EU on account of growing trade and possible EU membership, cereals in general and wheat in particular at present give yields well below those achieved in regions with broadly similar environmental conditions in the EU. In non-EU Europe there is indeed the prospect of a grain mountain of enormous proportions if the path taken by farmers in the UK, Denmark and France is followed.

In the livestock sector of the EU a large increase has been achieved since the 1950s in

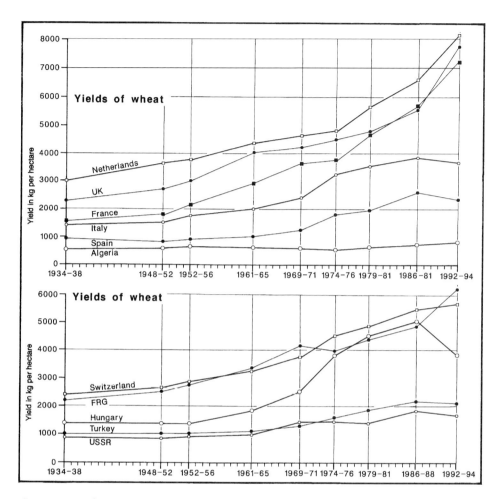

Figure 6.11 Yields of wheat in selected EU and non-EU countries from 1934–8 to 1992–4. To reduce the effect of year to year variations, 5-year or 3-year averages have been used
Source: various years of *FAO Yearbook, Production*

the amount of meat and other products obtained from each animal raised. The total number of livestock units, like the arable area, has changed little in the last decade. Indeed, the number of cattle kept has declined slightly, but the number of pigs and particularly sheep has risen. The weight of mutton, lamb and pig-meat per sheep and pig has changed little in the 1980s, but that of beef and veal per animal has continued to rise, while in four countries milk yields per animal have more than doubled

between 1960 and the early 1990s, as can be seen in Table 6.6. Variations among the countries of the EU in the meat yields per animal are considerable, but are not nearly so marked as the extremes in cereal yields. Milk yields, on the other hand, do vary greatly, with each dairy cow kept in the Netherlands giving more than twice as much milk as each one in Greece.

The level of provision of fodder crops and the quality of pastures result in a marked difference between the northern and southern regions of

BOX 6.1 MAD COW DISEASE

In the spring of 1996 new evidence was publicised in the UK that linked cases of bovine spongiform encephalopathy (BSE) in cattle with Creutzfeldt-Jakob disease (CJD) in human beings. Fears about the spread of BSE through contaminated animal feed containing offal from cattle had already led to a ban on such animal feed being used, but cases of BSE continued to occur, mainly in the UK but also in France and Switzerland. The new evidence in 1996 led to emergency measures being adopted by the European Commission not only to limit the circulation of British beef for health grounds but also to attempt to restore confidence in beef products among consumers, since there was a fall in consumption of up to 50 per cent in some EU Member States as a result of health concerns.

The ban on worldwide exports imposed by the Commission and endorsed by Agricultural Ministers in the Council was a drastic measure aimed at restoring public confidence, intended to force the UK to adopt a comprehensive slaughtering policy to limit the spread of BSE. Unlike in France, where the entire herd is destroyed when one case is discovered, the UK still preferred a policy of selective slaughtering, questioning the validity of the scientific evidence. Further evidence of infection in sheep combined with the disclosure that BSE can be transmitted from cow to calf led to calls for even stricter controls to be applied in the summer of 1996.

The BSE crisis and the measures adopted are a clear indication of the way in which the Common Agricultural Policy (CAP) operates, replacing national policies for agriculture and effectively removing decision-making on farming policy and trade in farm products from the Member States. The worldwide ban on British beef was successfully enforced by the EU until the UK was forced to adopt stricter measures. The financing sought by the UK to assist in the slaughtering programme also shows the extent to which the CAP is based on the principle of solidarity, with EU funds paying a large share of national costs in each case, just as the outbreak of swine fever in Germany was supported in the 1980s. In the preliminary draft budget for 1997, 368 million ECU was allocated for eradication measures and a reserve of 505 million 'to cover the other costs linked to the epidemic which will no doubt arise but for which the underlying parameters are not yet known' (COM (96) 300).

The UK Government was in the impossible position of objecting to a decision adopted under the CAP but at the same time demanding funding from the CAP to cover the cost of measures to eradicate the problem. The BSE crisis led to further calls from the Euro-sceptic camp in the UK for withdrawal from the EU and its centralist policies, ironically, in some cases, from those who benefited most from funding of agricultural surpluses during the fat years of the 1970s and 1980s. It will take the beef producing sector at least several years to recover from the impact of the BSE crisis on consumption throughout the EU.

Table 6.6 Annual average milk yield per milking cow in the countries of the EU, tonnes per animal

	1959–61	1979–81	1992–4		1959–61	1979–81	1992–4
Netherlands	4.2	5.0	6.3	Belgium-Lux.	3.8[1]	3.9	4.7
Sweden	3.0	5.3	6.3	Spain	1.4	3.3	4.2
Denmark	3.7	4.9	6.2	Austria	2.5	3.5	3.9
Finland	3.0	4.6	5.9	Italy	2.0	3.5	3.9
UK	2.9	4.8	5.5	Ireland	2.2	3.2	3.7
France	2.2	3.7	5.2	Greece	0.8	1.9	3.1
Germany	3.2	4.2	5.2	Portugal	2.3	2.1	1.9

Sources: FAOPY 1962 (FAO 1963, Table 84); *FAOPY 1994* (FAO 1995, Table 99)
Note: 1 Belgium only in 1959–61

the EU in the dairying sector. It can be assumed that the amount of fodder provided for each animal corresponds roughly with what it yields, so the saving in higher yields is more in the provision of facilities for handling a reduced number of animals than in the fodder required. As with cereals, there seems to be some slack to be taken up in the livestock sector of the EU if standards in lagging regions are brought up to those in the regions that achieve the highest yields.

TRADE IN AGRICULTURAL PRODUCTS

Since the sixteenth century, Western Europe has traditionally been an importer of agricultural products from other continents, for example sugar and beverages from Latin America, cereals, cotton and tobacco from North America, livestock products from Australia and New Zealand, and various tropical fruits. In the 1930s, British policy was to encourage trade within the Empire, but Germany and Italy, anticipating war and a possible blockade, were even then concerned with self-sufficiency in food. Kuczynski cites the Germany Institute for Business Research (23 May 1939):

> The struggle for an expansion of the Germany foodstuff productive capacity has entered a new and decisive stage since the assumption of power by the present German government. All efforts are being made to make a greater acreage available and to secure from the present acreage the greatest possible yield.
>
> (Kuczynski 1939: 28)

According to Kuczynski, Germany was 83 per cent self-sufficient in foodstuffs, Great Britain only 25 per cent, while Italy managed 95 per cent, the latter thanks to a 'battle' to achieve self-sufficiency in grain. Experiences in the 1930s must have helped to account for concern over food supply in the young EEC of the 1950s.

Whereas Western Europe was a net importer of most agricultural products before the Second World War, it has largely become self-sufficient in those items that can be produced in cool and warm temperate climatic conditions. However, it still has to import tropical beverages (coffee, tea, cocoa), tropical fruits (bananas), subtropical and tropical raw materials such as cotton, and more esoteric items such as spices. The data in Table 6.7 show the position of each EUR 15 country with regard to trade in agricultural products. Four countries are net exporters of agricultural products, while 11 are net importers. Germany, Italy and the UK are the largest net importers. Most EU countries are importers of forest products, but Sweden and Finland are large exporters. Their exports of forest products averaged about 9 and 8 billion US dollars respectively in 1991–3, far exceeding their trade in agricultural products.

Most of the foreign trade of the EU countries in food, beverages and tobacco is actually

Table 6.7 Trade in agricultural products in billions of US dollars, average 1991–3

	Imports	Exports	Balance		Imports	Exports	Balance
Belgium-Lux	13.5	13.4	−0.1	Austria	3.0	1.4	−1.6
Denmark	3.4	8.3	+4.9	Portugal	3.1	1.0	−2.1
Germany	40.0	22.2	−17.8	Finland	1.3	0.7	−0.6
Greece	3.1	2.8	−0.3	Sweden	3.4	1.1	−2.3
Spain	9.5	9.4	−0.1	UK	22.6	13.8	−8.8
France	23.3	34.0	+10.7	EU all	169.8	157.8	−12.0
Ireland	2.4	6.0	+3.6	EU - non-EU	42.5	30.5	−12.0
Italy	23.2	12.3	−10.9	USA	28.0	46.9	+18.9
Netherlands	18.0	31.4	+13.4	Japan	30.9	1.4	−29.5

Source: FAO 1994, Table 6

between Member States. In 1990, 72 per cent of EUR 12 trade in such products was accounted for by 'internal' transactions, about another 5 per cent by trade with the EFTA countries of the time. A comparison with the USA and Japan shows that the former has a considerable surplus of agricultural products in its foreign trade while the latter has a massive deficit. The EU has an intermediate position.

The EU cannot isolate itself entirely from the rest of the world with regard to the production of and trade in agricultural products. The population of the present developing countries of the world is expected to increase from about 4.5 billion in 1995 to about 7.0 billion in 2025, an addition of 80–85 million people per year. The area under cultivation in the world cannot grow that fast, while increases in yields will depend on great improvements in the quantity and quality of means of production. Indeed, from being currently a net importer of food and beverages from many less developed countries, in the future the EU may find itself politically persuaded or forced to sell or even give food to various parts of the developing world.

FORESTRY IN THE EU

Very few people are directly employed in forestry in the EU, although forests cover about 36 per cent of the total area of the Union and are valued for both their products and their recreational and aesthetic functions and properties. Table 6.8 shows that since 1960 the forested area in most of the EUR 15 countries increased, in some cases considerably. The notable exception is Spain, although the reduction of its forest area appears to result partly from the reclassification of some forest as pasture. With 113 million hectares of forest and woodland altogether, the EU has only 2.7 per cent of the world total. Almost 20 per cent of all the EU wooded area, most of it in Greece, Spain and France, consists of brushland, scrubland and garrigue, rather than true forest.

It can be seen in Table 6.8 that the area of forest to population varies markedly among the EUR 15 countries. Spain has about 10 times as much forest per 1,000 inhabitants as the UK, while Finland has over 100 times more than the UK. EU policy favours the maintenance of the forests of the Union and even some extension of their area. There could be opposition to afforestation projects in some areas, however, both from farmers wishing to retain arable and pasture land, and from conservationists wishing to protect natural habitats from forestry development.

At present the forests in most regions of the EU supply only negligible quantities of fuelwood, and the EUR 12 countries produced only half of their needs of wood for construction

Table 6.8 The forest and woodland of the EU, 1960 and 1994

	Area of forest (million hectares)		Hectares of forest per 1,000 people in 1994		Area of forest (million hectares)		Hectares of forest per 1,000 people in 1994
	1960	1994			1960	1994	
Belgium-Lux.	0.7	0.7	66	Italy	5.8	6.8	118
Denmark	0.4	0.4	77	Netherlands	0.3	0.4	26
Germany	10.1	10.7	131	Austria	3.1	3.2	395
Greece	2.5	2.6	248	Portugal	2.5	3.3	333
Spain	24.3[1]	16.1	412	Finland	21.8	23.2	4,549
France	11.6	14.9	256	Sweden	22.5	28.0	3,146
Ireland	0.2	0.3	83	UK	1.7	2.4	41
					107.5	113.0	305

Sources: FAOPY 1962 (FAO 1963, Table 1); FAOPY 1994 (FAO 1995)
Note: [1] The reduction between 1960 and 1994 is mainly the result of a redefinition of forest and woodland

purposes and as a raw material. The UK, Germany and Italy accounted for over two-thirds of EU imports of roundwood, while only Portugal had a surplus. The UK produces only about one-tenth of its total needs. The entry of Finland and Sweden has changed the situation radically, while closer association with Russia in the future could benefit that country, especially if and when new areas in Siberia are exploited.

The map of EU forests (Commission 1987b) shows in great detail the distribution of trees of all kinds. The proportions of coniferous and non-coniferous species vary greatly between countries and regions, with coniferous species generally in the more northerly parts of the Union, at higher altitudes, and often on areas of sandy soil. Broadleaf deciduous species are common in central regions and in higher areas in southern Europe, while evergreen species, often of limited size due to dry summer conditions and poor soils, are common in the Mediterranean coastlands. Each type has its own advantages and disadvantages, particular products and scenic values.

The future of the forests is one of the less sensitive and urgent issues facing EU policy-makers, but there is no reason for complacency. According to Born (1995b) forestry experts argue that Europe's forests are facing an unprecedented crisis. A quarter of Europe's trees suffer advanced defoliation with the loss of at least a quarter of their leaves or needles. The main causes appear to be attacks by insects and fungi, forest fires, air pollution and unfavourable weather conditions. In particular, the dispersal of pollutants from thermal power stations, falling from the atmosphere as acid rain at great distances from sources, is regarded as a transnational pollution problem (see Chapter 10).

FISHERIES

The Common Fisheries Policy (CFP) began in the 1980s, growing in importance with the accession of Spain and Portugal in 1986, in order to provide common management of fisheries resources and markets in the EU, in particular due to increased demand for fisheries products and concerns over the overfishing of stocks. Just as with agriculture, technological advances in fishing tackle and navigation equipment have increased productivity considerably, with concerns over whether certain species can replenish their stocks. The fishing industry employs nearly 300,000 people directly

Plate 6.8 One problem in the fishing industry of the EU has been the use of nets with too fine a mesh, not allowing younger fish to escape. The breeding process is thus interfered with
Source: European Commission, Direction Générale X, Audiovisuel

and several hundred thousand additional jobs depend directly on the fisheries sector for their livelihood, in particular in the processing and canning industry. The amount of fish caught by each EU Member State is shown in Table 6.9.

The CFP is based on certain principles whereby EU waters are open to vessels from all Member States except within a 12 mile exclusion zone in which a Member State may limit access to vessels that have 'historical' fishing rights. The policy is also based on Total Admissible Catches (TACs) and quotas that are agreed upon between Member States for specific species of fish and for specific zones in EU waters. Quotas are negotiated on the basis of traditional fishing activity, and the possibilities

that fleets have of fishing in third country waters in the light of increases in the application of 200 mile exclusive zones by certain third countries. Provision is also made for the management of resources through the rigorous enforcement of quotas by means of inspections of Community and third country vessels, including their tackle and catches, and through the introduction of a Community fishing licence.

Scientific evidence has shown over the years that certain species of fish, including plaice, whiting and mackerel, are threatened through overfishing, and special measures are enforced through the CFP to limit fishing of such species to allow for stocks to renew themselves, in spite of reluctance on the part of the major fishing

Table 6.9 Average yearly fish catch 1990–2 in EU and other selected countries

		Thousands of tonnes	Kg per person			Thousands of tonnes	Kg per person
1	Faeroe Isles	270	5,740	11	Netherlands	450	29
2	Iceland	1,380	4,600	12	Finland	90	18
3	Greenland	120	2,400	13	*WORLD*	97,570	17
4	Norway	2,120	482	14	Greece	160	15
5	Denmark	1,770	340	15	France	830	14
6	Japan	9,370	74	16	UK	820	14
7	Ireland	260	72	17	Italy	350	6
8	Spain	1,350	34	18	Belgium	40	4
9	Portugal	310	31	19	Germany	330	4
10	Sweden	270	31				

Source: UNSYB 40th edition (UN 1995, Table 39)
Note: The fish catch of Luxembourg, Austria and Switzerland is negligible

nations to make concessions. As far as agreements with third countries are concerned, the EU has exclusive competence to negotiate and conclude international fisheries agreements, such as reciprocal agreements with Norway, Iceland and the Baltic States, or agreements for conditional access with countries such as Canada and Argentina. As in the case of agriculture, the CFP also finances restructuring and modernisation activities through the Financial Instrument for Fisheries Guidance (FIFG), which provides funding for research, aquaculture activities, training or retraining of fishermen and marketing and quality enhancement of fisheries products, with a budget of 450 million ECU per year.

CLIMATIC CHANGE AND EUROPEAN AGRICULTURE

While future trends in agriculture in the EU will be affected by as yet unknown policy and practical decisions at all levels from the EU Commission to the individual farmer, there is also growing uncertainty over possible changes in climate related to the rapidly increasing emissions and accumulation of greenhouse gases in the atmosphere. It is by no means unanimously accepted by researchers on the subject whether or not there has been an increase in the temperature of the earth's atmosphere over the last 50–100 years. For example Olstead (1993) (see reading list for this chapter) urges caution in accepting evidence presented from various quarters. If and when it *is* established that the temperature is rising, then again it is not clear what the effects might be. While common sense would indicate, for example, that some of the ice in the world's glaciers and ice caps would melt, thereby causing world sea level to rise, some argue that more ice could accumulate because the amount of moisture carried in the atmosphere would be greater; sea level could then fall. A number of papers on the subject of atmospheric change are included in the reading list for this chapter.

Parry (1990) discusses various models and forecasts related particularly to the general circulation of the atmosphere. The effects of various possible future rises in the level of temperature are considered in relation to agriculture. Increases in temperature would probably change precipitation in different parts of the world but in ways that are difficult to anticipate and if some of the ice at present in glaciers and ice caps were to melt, a small rise in sea level could be expected, bringing serious consequences in some low-lying areas.

Parry (1990: 65) shows, for example, that various models agree that a doubling of carbon dioxide in the atmosphere could shift the thermal limit of the successful commercial cultivation of grain maize in Europe some 200–350 km north. The limit at present runs roughly eastwards from the northern extremity of France across central Germany and Poland to Ukraine. Much of the UK, southern Scandinavia and the rest of the former European USSR could become suitable for maize cultivation depending, however, on adequate precipitation in the right season.

Parry's broad conclusion is that increased global temperature could substantially benefit northern Europe, whereas southern Europe could suffer through decreases in soil moisture in the summer due to a reduction of summer rainfall. Such a change could decrease the biomass potential in Italy by 5 per cent and in Greece by 35 per cent. According to Parry (1990: 83) 'This implies an important northward shift of the balance of agricultural resources in the EC.'

Outside Europe, some expected changes could also affect the EU. Grain production is expected to be reduced by 10–20 per cent in the USA and to fall also in Canada, as well as in Ukraine and southern Russia. The current surplus of grain from North America could be reduced or eliminated. Of most immediate concern to the EU is the prospect that the Maghreb countries will become even hotter and drier than they are now, with even greater pressure on the growing population to seek employment in Western Europe.

FURTHER READING

Eurostat (1995) *Europe in Figures*, 4th edition, 'Agriculture', pp. 238–71.

Hendriks, G. (1991) *Germany and European Integration: The Common Agricultural Policy: An Area of Conflict*, Oxford: Berg.

Houghton, R. A. and Woodwell, G. M. (1989) 'Global climatic change', *Scientific American*, (4), April: 18–27.

Jones, P. D. and Wigley, T. M. L. (1990) 'Global warming trends', *Scientific American*, 263 (2), August: 66–73.

Mander, N. (1992) 'The bottle stopper tree', *Geographical*, LXIV (11), November: 26–9. The traditional harvesting of bark from the cork oak in Iberia is threatened by the introduction of lucrative eucalyptus plantations.

Olstead, J. (1993) 'Global warming in the dock', *Geographical*, LXV (9), Sept.: 12–16. Science in the dock over global warming. Few scientists are sure that man-caused warming of the earth's atmosphere is occurring but politicians and the public take global warming as a fact.

Ritson, C. and Harvey, D. (eds) (1991) *The Common Agricultural Policy and the World Economy*, Wallingford: CAB International.

Safina, C. (1995) 'The World's Imperiled Fish', *Scientific American*, 273 (5), November: 30–7. Global view of the causes of stagnation or decline in the size of fishing catches. Catches peaked in the northeast Atlantic in 1976 and in the Mediterranean and Black Seas in 1988.

White, R. M. (1990) 'The great climate debate', *Scientific American*, 263 (1), July 18–25.

7

INDUSTRY

'During the eighties the EU moved rapidly towards the post-industrial phase of economic development . . . The majority of manufacturing sectors stagnated, while metal production, mining, transport equipment industries, and textiles and clothing shrank'.

(Commission of the EC 1995a: 82)

- The EU has no common industrial policy as such, rather a sectoral approach through separate programmes for funding and legislation to promote competitiveness and fair competition.

- The EU has experienced much decline in traditional industrial sectors, with high unemployment and relocation becoming commonplace in specific EU regions.

- In most industrial sectors the existence of the Single Market has not led to the complete removal of protectionism and other obstacles to the free movement of goods within the EU.

- In the high-technology sectors of the future, Europe remains characterised by strong competition from the USA, as well as Japan and other eastern nations, and is handicapped by a lack of innovation and investment.

Industry accounts for the largest share of EU exports to third countries and still provides jobs for over 30 per cent of all persons employed in its Member States. Of the three sectors of economic activity it is the most open to competition from the world outside, both in the form of more technologically advanced products from the USA and the Far East and, increasingly, from goods produced in the developing countries and emerging economies with far lower labour costs. Nevertheless, industrial policy in the EU remains one of the least coherent policy areas of EU activity, due principally to the different industrial traditions that have governed the nations that make up the EU and that, in many cases, continue to do so.

The ECSC treaty laid down a number of specific sectoral provisions for the coal and steel industries which still remain in force in spite of the major upheavals in the sector over the intervening decades. It can be argued that cooperation through the ECSC has enabled both industries to survive recession and painful restructuring and remain intact, albeit on a small scale. The majority of other industrial sectors have not, however, benefited from such detailed policy and have endured similar difficulties without the same degree of assistance.

INDUSTRIAL POLICY

EU industrial policy in general consists of policy formulation in a number of areas that have a direct or indirect impact on EU industry, together with specific programmes that are aimed at providing funding to assist in restructuring of capacity through R&TD or training. The former areas include legislation under the Single Market, rules governing competition and state aid to industry, and international trade agreements. Under the Structural Funds there are specific programmes funded to assist particular industrial regions in decline as well as the various R&TD programmes aimed at improving the competitiveness of EU industry. The basis for the global strategy for industry is to be found in the EC Treaty, Article 130, which sets the objective of enhanced competitiveness through the following:

- Speeding up the adjustment of industry to structural changes.
- Encouraging an environment favourable to initiative and to the development of undertakings throughout the Community, particularly small and medium-sized undertakings.
- Encouraging an environment favourable to cooperation between undertakings.
- Fostering better exploitation of the industrial potential of policies of innovation, research and technological development.

Although incomplete for certain categories of goods and services, Single Market legislation, has changed the EU market for industry, both domestic and foreign. Harmonisation of technical standards and common rules governing marketing of goods as well as health and safety for consumers have enabled EU industry to produce for a Single Market of 370 million users rather than the protected and compartmenta-lised markets that existed before. The market has, of course, also become easier for imports from third countries, and many US, Japanese and Korean firms have established a plant in a single Member State in order to be able to produce for the entire market.

As far as rules on competition and state aid are concerned, the EU Commission has a strong record in stamping out abuses of dominant position, cartels and other illegal agreements that lead to price fixing or market distortions at the expense of the consumer. The Commission has found it more difficult to apply its powers to certain public sector industries, such as energy supply, defence, telecommunications and transport, where strong national opposition still prevents rules on competition and state aid from being applied.

The EU has great powers and responsibilities under the common commercial policy through which it signs and administers international trade agreements on behalf of Member States. International agreements under the WTO have a clear impact on customs duties on trade in industrial and manufactured goods which, in turn, determine the fate of many smaller scale industries throughout the EU, whose competitive position is often precarious. International conventions on trademarks, industrial and intellectual property and product safety also have far reaching consequences for EU industry.

Through specific programmes under the Structural Funds, the EU institutions attempt to assist particular industrial sectors or regions through direct financing, a policy also applied in the area of R&TD. The emphasis in EU policy remains, however, on establishing an environment within which coordination can function and competitiveness can improve, but with Member States retaining overall control for their various industrial sectors. It would be inconceivable for a common policy, such as the CAP, to be developed for the EU automobile or textile industry and for production and markets to be controlled as is the case for steel products.

This chapter analyses different industrial sectors in the EU, beginning with traditional types of production which are now in general decline due to changes in demand and competition from other parts of the world. The biggest industrial sectors are then examined before the new industries of the future are presented, and throughout there is a focus on those activities undertaken by the EU in order to improve competitiveness and future prospects. The problem of competition from third country imports is of particular significance given the worldwide trend towards more open markets through the GATT agreements establishing the World Trade Organisation (WTO) and the explosive development of advanced production of many goods in the Far East and the Third World.

In order to compete with advanced technologies from outside, the EU can do much to promote R&TD and improve the workforce through training programmes. The answer to cheaper labour costs is less easy to find as there is clear opposition, in particular on the part of unions in the EU, to removing any of the social protection and generous working conditions that have been achieved over the decades and which weaken the competitive power of many sectors of EU industry. Many in the EU argue that protectionism against imports is the only way in which jobs in EU industry can be guaranteed and that competing with foreign imports is an impossible objective. Before individual sectors are presented in more detail, the next sections analyse the situation regarding employment in industry and the availability of raw materials in the EU.

EMPLOYMENT IN INDUSTRY (See Figure 7.1)

The importance of the industrial sector to the economic life of the EU can be measured by the fact (*Regional Profiles* 1995) that in 1993 in the EUR 12 countries it accounted for 31.6 per cent of all employment in the Union, while agriculture accounted for 5.6 per cent, services for 62.8 per cent. The contribution of industry to the total GDP of the EU is greater than its share of employment, and the sector accounts for a very large part of the exports of EU countries both to one another and to countries outside the EU. It is ironical, therefore, that agriculture receives about half of the EU budget, 'industry' about a tenth whereas industry employs about five times as many people and produces about ten times as much in value.

Industry itself is a term that covers extractive activities, processing and construction as well as the more central area of manufacturing. According to Eurostat (1994a: 140), industry employed 40.5 million people out of 115.3 million, about 35 per cent of the total labour force. Energy and water (2.1 million), together with building and civil engineering (8.2 million), accounted for about 25 per cent of industrial employment, manufacturing for 75 per cent (non-fuel minerals and chemicals 4.9 million, metals and engineering 13.4 million and other (mainly light) 11.9 million). There has been little change overall in the number of people employed in industry in the EU since the mid-1980s but in France, Italy and Spain the workforce has grown while in Germany, the UK and the Netherlands it has declined.

The proportion of the total labour force engaged in industry varies considerably among EU Member States, the extremes being Germany with 38.3 per cent and the Netherlands with 23.9 per cent (see Table 7.1). The disparities at NUTS levels 1 and 2 are much greater. Table 7.1 shows the most and least highly industrialised NUTS level 2 regions in 1993 of eight of the EUR 12 countries and for NUTS level 1, for the UK. Appropriate data were not available for Denmark, Ireland and Luxembourg. In almost all of the countries there is roughly a 2:1 disparity between the most

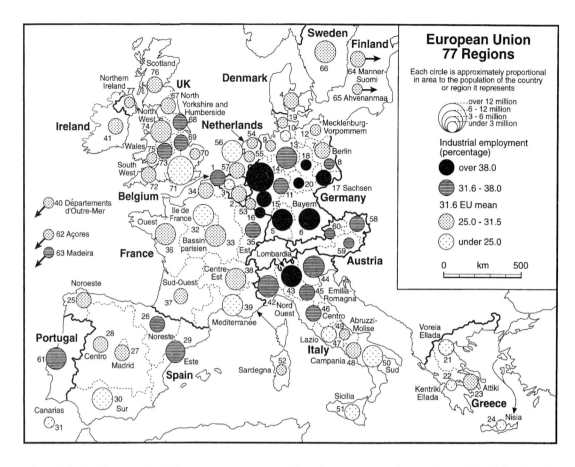

Figure 7.1 Employment in industry as a percentage of total employment in the EU at NUTS level 1. See Table A1 (Appendix) for key to the numbering of unnamed regions
Source: *Regional Profiles* (1995)

highly industrialised region in employment and the least highly industrialised region. In Denmark and Ireland much of the industry is concentrated in and around the national capital. The lack of industrial employment in regions at the lower end of the scale does not necessarily mean that they are backward or lagging in development. In the cases of Hamburg, Algarve and the southeast of the UK, industry is overshadowed by services, while in Kriti and Calabria agriculture is still a large employer of labour.

ENERGY AND RAW MATERIALS IN THE EU (See Figure 7.2)

In the late eighteenth and early nineteenth centuries, before the era of railways and steam navigation, modern industry in Western Europe was, with some exceptions such as cotton, based on local resources. These resources included falling water and coal for providing power to drive machinery, metallic minerals, and products from agriculture and forestry. As coking coal replaced charcoal for the smelting of iron ore, and as coal also replaced water

Table 7.1 Employment in industry in EUR 12 countries and regional extremes

	(1) Per cent in industry 1993	(2) Annual per cent change in industrial employment 1983–93	(3) (4) Per cent in industry at NUTS level 2			
			Highest		Lowest	
Belgium	29.5	0.4	Namur	42.4	Luxembourg (B)	21.0
Denmark	26.0	−0.1	n.a.			
Germany	38.3	−2.5	Stuttgart	47.5	Hamburg	24.6
Greece	24.2	0.6	Dytiki Makedonia	33.8	Kriti	14.6
Spain	30.8	1.1	Catalunya	41.0	Canarias	16.7
France	27.1	2.9	Alsace	36.3	Corse	15.2
Ireland	27.9	0.6	n.a.			
Italy	32.4	1.2	Lombardia	42.9	Calabria	19.6
Luxembourg	26.4	0.7	n.r.			
Netherlands	23.9	−1.0	Limburg	32.0	Utrecht	16.0
Portugal	32.9	n.a.	Norte	42.2	Algarve	18.4
UK[1]	27.5	−1.8	West Midlands	36.8	Southeast	23.6

Source: Regional Profiles (1995)
Notes: 1 NUTS level 1
n.a. Not available
n.r. Not relevant

power for driving machinery, locations in coal-field areas became particularly attractive to industrialists. Rivers and canals were widely used, but compared with the railways that followed, had very limited capacity to move raw materials and finished products,

As population grew and industrial development in Western Europe expanded rapidly, many parts of the region could no longer produce enough food or raw materials for their needs. The development of the steamship made it economically realistic to transport increasing quantities of food and raw materials from other continents over great distances. In the later nineteenth century, for example, guano, nitrates and copper were exported from Peru and Chile to Europe. Iron ore, on the other hand, was still mainly obtained from local ores of comparatively low grade, or from northern Sweden. The movement of high-grade iron ore from other continents to Western Europe developed on a large scale only after the Second

World War. Some non-fuel minerals, including bauxite, have been used commercially only in the twentieth century. Coal, on the other hand, was widely available, and production adequate to supply almost all of the energy needs of Western Europe well into the twentieth century. Railways and shipping services could move coal easily from coalfields such as those of South Wales and the Ruhr to other parts of Europe, such as Switzerland and Italy, with no coalfields. Since the Second World War, the EU has become largely self-sufficient in foodstuffs, but it imports about half of its energy needs, and about 90 per cent of its non-fuel mineral needs, the most important of which are shown in Table 7.2. It should be appreciated that Japan is even more dependent on the rest of the world for its food, fuel and raw materials.

The EU has about 6 per cent of the total population of the world but it consumes about 20 per cent of the non-fuel minerals used. In relation to both consumption and population,

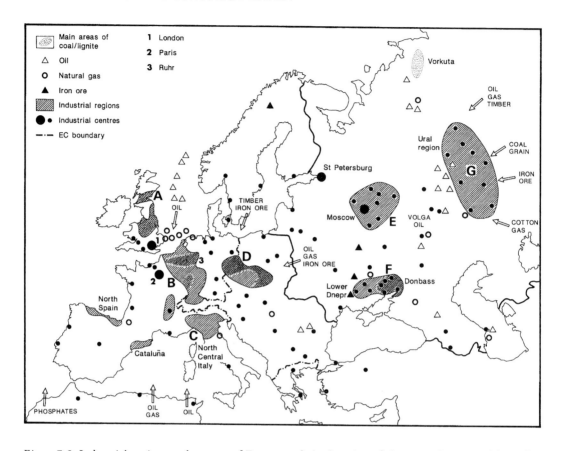

Figure 7.2 Industrial regions and centres of Europe and the location of the main deposits of fossil fuels

its share of the reserves of most of the world's non-fuel minerals is very limited. The countries with the largest reserves of 11 of the most widely used non-fuel minerals are shown in Table 7.2, together with the reserves, if more than negligible, located in Western and Central Europe. While there has been concern in the Western industrial countries over the location of some mineral reserves of world significance in regions regarded as unstable (e.g. southern Africa) or unfriendly (e.g. the former USSR), as can be seen in Table 7.2, enough of the reserves are in regions that are unlikely suddenly to cut off exports of primary products to Western Europe, the USA or Japan in the event of a conflict or of political discord. The

break-up of the USSR means that its large reserves of eight out of the 11 minerals in Table 7.2 could be of increased importance to the EU in the future.

Raw materials are being used increasingly efficiently and sparingly in the developed industrial countries as time passes (Larson *et al.* 1986), and substitutes and synthetic materials are giving increasingly flexibility and choice (Clark and Flemings 1986) in the use of various non-fuel minerals. The supply situation for non-fuel minerals should therefore be of less concern in the EU than energy, but there is no reason for complacency. Another trend is the increasing proportion of the processing of non-fuel minerals carried out in the countries in

Table 7.2 The location in the world in 1985 of the estimated reserves of 11 major non-fuel minerals (metallic content of 1–9), 1985

	Largest reserves (% of world total)	*Europe, excl. USSR*
1 Iron ore	USSR 35, Brazil 15, Australia 14, India 7	Sweden 2 France 1
2 Chromite	South Africa 78, USSR 12	—
3 Manganese ore	South Africa 41, USSR 37, Australia 8, India, Brazil, China, 2 each	—
4 Nickel ore	Cuba 34, Canada 14, USSR 13 Indonesia 7	Greece 4
5 Bauxite	Guinea 27, Australia 21, Brazil 11, Jamaica 10	Greece 3 Yugoslavia 2 Hungary 2
6 Copper	Chile 23, USA 17, Zambia 9, Zaire 8	
7 Lead	USA 22, USSR 13, Canada 13	Yugoslavia 4 Bulgaria 3 Spain 2
8 Tin	Malaysia 36, Indonesia 22, Thailand 9, Bolivia 5	UK 3
9 Zinc	Canada 15, USA 13, USSR 6, South Africa 6	Spain 4 Ireland 3 Poland 2
10 Phosphates	Morocco 49, South Africa 19, USA 10, USSR 9	—
11 Potash	Canada 48, USSR 33	Germany 14

Source: Bureau of Mines 1985

which they are extracted. Such a trend benefits developing countries in a limited way by adding value to the products they export, but it also means that pollution is 'exported' from Europe to them.

TRADITIONAL INDUSTRIES (See Figure 7.3)

Like the rest of the industrialised world, the European Union has seen substantial changes in demand for industrial goods and the emergence of new producers for traditional products. As a result, certain regions of Member States have seen their industrial base decline, with resulting job losses and depression. This has been the case in particular in the iron and steel industry and in shipbuilding, two sectors that were in the forefront of the Industrial Revolution and were major employers in Western Europe but

which have seen a great decline since the 1970s due to falling demand and cheaper competition from other regions of the world.

Although the impact has been local in many cases, with particularly hard hit regions such as Wallonia in Belgium, north-East England, Nord-Pas-de-Calais in France and the Basque country in Spain, both industrial sectors have benefited from policies and funding at EU level. In addition to the ECSC Treaty, which has provided the framework for the restructuring of the steel industry, there have also been specific programmes for the shipbuilding industry and the regions worst affected have been regular beneficiaries from the Structural Funds.

Iron and Steel Industry (See Figure 7.4)

The iron and steel industry was the first sector of the economy to come under EU control

Plate 7.1 Abandoned mines in Andalucía, southern Spain

Plate 7.2 Older industrial constructions: pot banks in Longton, Stoke-on-Trent, England. As recently as the 1950s, hundreds of these furnaces smoked prolifically

through the ECSC Treaty in 1952. Originally based on Franco-German cooperation in the coal and steel producing regions, it also included the Benelux countries and Italy in its founding membership. According to the ECSC Treaty:

> The Community shall progressively bring about the conditions which will of themselves assure the most natural distribution of production at the highest level of productivity, while safeguarding continuity of employment and taking care not to provide fundamental and persistent disturbances in the economies of the Member States.
>
> (Art. 2 ECSC).

The ECSC Treaty went on to establish a common market for steel products and 'to ensure an orderly supply to the common market, taking into account the need of third countries' (Art. 3(a) ECSC).

The crisis in the industry, which lasted from 1974 to 1987, led to a drastic reorientation of policy, aimed at saving the industry from fast

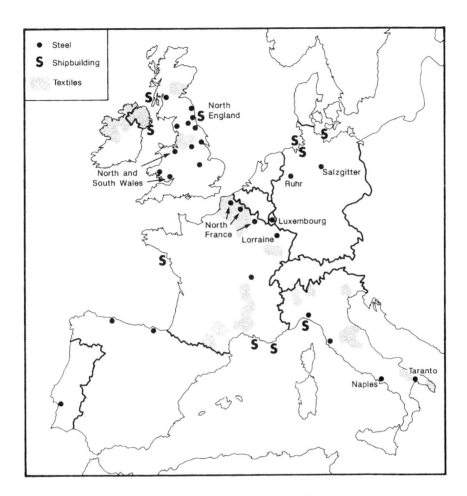

Figure 7.3 Selected areas of the European Union affected by the decline of the steel, shipbuilding and textile industries. The map does not include Austria, Sweden or Finland
Source: based on map in COM 87–230 (1987), map 2.2.3-B4, p. 84 of Annex

decline rather than promoting its expansion. The crisis was caused by both internal and external factors, with a fall in consumption and in prices for steel products, as well as competition from goods with lower production costs from newly industrialised countries. During the period in question, Community production fell from 156 million tonnes to 113 million tonnes and 500,000 jobs were lost in the industry, including in the UK and Spain, which had become Member States.

Although demand for steel began to rise

again at the end of the 1980s, the economic recession at the start of the 1990s caused production to drop further by 12 per cent between 1991 and 1993, although demand began to recover again in 1994. Production levels in the EC were also raised by the addition of the former GDR in 1991, which accounts for around 5 million tonnes of production. Sweden and Austria, which joined the EU in 1995, are also major producers of steel. It is estimated that crude steel output will stabilise at around

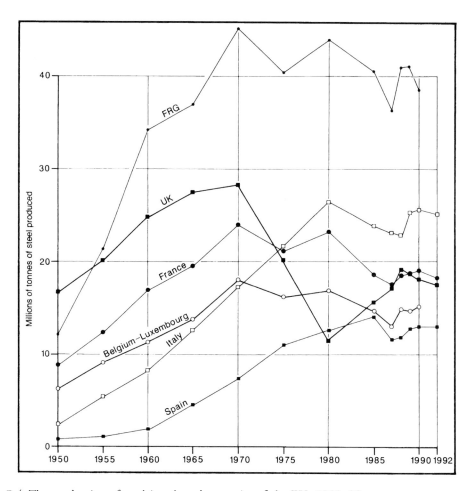

Figure 7.4 The production of steel in selected countries of the EU, 1950–92
Source: various years of *United Nations Statistical Yearbook* and numbers of *Eurostatistics*

140 million tonnes per year for the EU in 1996 and 1997 (Commission of the EC 1995a).

The industry still has overcapacity, in particular in Italy, where restructuring has lagged behind that in other main steel-producing Member States. The failure of the Commission to achieve further cuts in capacity in 1994, with the rejection of its restructuring plan by Member States, means that the competitiveness of the sector remains weak. The sector has also had to cope with increasing pressure to respect higher levels of environmental protection, in particular with regard to CO_2 emission, recycling and waste disposal.

At international level, the EU remains the leading manufacturer of iron and steel products and it is still a net exporter, although the balance is becoming less favourable due to cheap imports. In 1993, its production of 132 million tonnes was far ahead of that of the USA (100 million tonnes) and Japan (90 million tonnes). Production in the former Soviet Union (96 million tonnes) and China (89 million tonnes) is a significant source of cheaper competition for the future, and the EU must pay particular attention to the total production of 125 million tonnes in the countries of Central and Eastern Europe. If major steel producers such as Poland and the Czech Republic become Member States of the Union, not only will their production compete with that of the existing Member States due to cheaper wage costs, but the problem of overcapacity in the EU will become even more acute.

The iron and steel industry is highly concentrated, with the ten major steel-producing countries accounting for about 80 per cent of production. Germany is the largest EU producer of steel, although its economic importance is far greater in Luxembourg where it accounts for 40 per cent of total production. By contrast the total contribution of steel to total manufacturing output in Denmark, Portugal and Ireland is around 1 per cent.

According to the *Panorama of EU Industry 95/96* (Commission of the EC 1995a), Usinor Sacilor in France is the largest producer, with an output of 18 million tonnes in 1994, followed by British Steel in the UK (12 million tonnes) and Thyssen (Germany) with 9.6 million tonnes. Altogether there are around 400 steel plants, many of which are smaller 'mini-mills' located in northern Italy and Spain and which are more flexible and able to adapt production to market demands. As far as regional location is concerned, steel plants used to be situated inland, usually near coal or iron-ore fields, from where supplies were obtained, or near major steel consumers. More recently, however, they have been located on the coast, such as near Marseilles in France and at Taranto in Italy, where they have easier access to imported raw material and for exporting in international markets.

The iron and steel industry in the EU remains in difficulty, and as a result of the crisis in the industry the ECSC has reoriented its focus to three main policy goals: to regulate markets, to restructure the industry to meet changing market conditions, and to provide regional and social aid for steel-producing regions. The RESIDER programme has been aimed at assisting those regions hardest hit by the restructuring in the sector by providing opportunities for new investment, retraining and the establishment of small and medium-sized enterprises (SMEs). The European Commission is also active at the international level, negotiating agreements with other steel exporters and importers in the framework of the WTO to improve market access and avoid trade conflicts.

The Shipbuilding Industry

The shipbuilding industry has also shown the characteristics of a traditional industry in decline throughout the last two decades, with employment falling from over 300,000 jobs in the boom years of the 1970s to around 130,000 in 1993 for the countries of the Association of West European Shipbuilders (AWES), the EU plus Norway. Problems of overcapacity and competition from the Far East, in particular Japan and South Korea, have been the key factors in the decline, the worst hit Member State being the UK, where the labour force is now only 10 per cent of what it was in 1975.

According to *Panorama of EU Industry 95/96* (Commission of the EC 1995a), capacity in Western European yards decreased by 60 per cent between 1976 and 1990. Many shipyards have closed during the crisis and those remaining open have had to modernise their production and management methods and reorientate their production towards more technologically advanced vessels. Those Member States with the most capacity and employment in shipbuilding are Germany, in particular following unification, Spain, Denmark and Italy.

The European Commission has been responsible for overseeing restructuring in the shipbuilding sector in particular due to the different levels of government subsidies between Member States. In order to avoid distortion of competition, the EC adopted a series of directives on aid to shipbuilding through which the Commission has been responsible for setting a common ceiling for state aid based on the difference in production costs between the most efficient EU yards and their Far Eastern competitors. EU Structural Funds through the RENAVAL programme are also used to assist those regions of the EU that have suffered most from the closures, such as Northern Ireland and Western Scotland, Northern Germany and the South and West of France.

The EU shipbuilding industry has better prospects following an increase in orders in 1993 and 1994 as well as an improvement in the longer term outlook with environmental protection and marine safety becoming increasingly important. As the world fleet becomes more aged and more vessels need to be scrapped, new legislative requirements for the design of vessels will favour the higher levels of expertise of EU shipyards against their competitors in South Korea, Japan and, increasingly, the People's Republic of China.

BOX 7.1 DISTORTIONS IN THE SINGLE MARKET

Even though the Single Market has been in existence since 1 January 1993, with the free movement of goods, services, people and capital, the reality of free trade in goods in the European Union remains incomplete. The more blatant examples of restrictions on the circulation of goods are automobiles and telecommunications equipment, but there are also conditions that distort the internal market in a variety of areas, in particular in the case of consumer goods. Rates of VAT, excise duties and other indirect taxes have not yet been fully harmonised at EU level, and even in cases where consumers are free to purchase goods anywhere in the EU, major distortions create unfair conditions on the market, affecting consumers and industry negatively.

One example of a major consumer product for which national levies vary widely,

Table 7.3 National levies on blank recording tape in EUR 15, 1996

	E240 video tapes %	C90 audio tapes %
Belgium	8.3	7.2
Denmark	50.0	35.0
Germany	13.2	7.7
Greece	n.a.	n.a.
Spain	38.5	25.3
France	34.3	25.1
Ireland	0.0	0.0
Italy	5.0	6.0
Luxembourg	0.0	0.0
Netherlands	31.3	28.3
Austria	34.5	13.8
Portugal	n.a.	n.a.
Finland	n.a.	n.a.
Sweden	0.0	0.0
UK	0.0	0.0

Source: European Tape Industry Council 1996
Note: n.a. not available

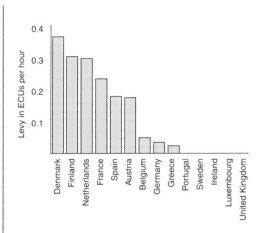

Figure 7.5 Recording copyright levies for EU Member States 1996
Source: European Tape Industry Council 1996

affecting sales prices to consumers, is the blank recording tape. Table 7.3 shows the percentage of levy on sales price in ECU for the different Member States, with enormous variations between Denmark,

at 50 per cent, and the UK at 0 per cent. There is clear incentive for consumers to purchase their tapes in a neighbouring Member State, in particular for citizens who live in border areas. The tape industry itself complains that the trade distortions are intolerable, and should not exist in a true Single Market. Recording copyright levies shown in Figure 7.5 show similar variations discouraging recording in countries such as Denmark or Finland to the benefit of countries such as the UK and Ireland.

Such distortions in trade continue to exist because Member States are able to justify special duties or other trade restrictions on grounds of environmental protection or public health under the provisions of the TEU. Until further reforms are made, the reality of a Single Market for goods will remain as incomplete as the concept of the free movement of persons within the EU.

MAJOR INDUSTRIAL SECTORS

In the first section of this chapter it was shown how the policy of the European Union on industry set as one of its key objectives the improvement of competitiveness. In *Panorama of EU Industry 95/96* (Commission of the EC 1995a), the industrial competitiveness of EU manufacturing sectors is analysed in a special article, and one of the main conclusions is that, compared with the USA and Japan, EU industry, in particular as far as exports are concerned, is less specialised than either of its two main competitors, and that where it is specialised, such as the textile industry, the sector concerned is a poor performer. The EU has only one high-technology sector, pharmaceuticals, among its specialisations. Its largest export sec-

tors, mechanical engineering and the automobile industry, have constantly lost competitiveness, reflected in their diminishing market share. EU industrial competitiveness is also hampered by lower R&TD investment than that in its main competitors.

This section analyses certain key industrial sectors of the EU, in particular textiles, engineering, automobiles and chemicals.

The Textiles Industry

The textiles industry, including clothing, leather and footwear, accounted in 1993 for 2.4 million jobs, or about 10 per cent of manufacturing employment in the EU. The sector also accounts for about 5 per cent of the manufacturing industry's value added. In 1987 the sector employed more than 3.5 million and the job losses have affected women in particular, and have concentrated in areas with heavy industry such as northern England, Nord-Pas-de-Calais in France and Nordrhein West-falen in Germany, where few job alternatives exist. As in the case of steel and shipbuilding, many of the job losses have been caused by the influx of cheaper imports, although automation in the industry has also had a considerable impact on jobs. More recently, the sector has been hit by economic recession in the 1990s.

International trade in textiles is of particular importance to the EU since it has a serious trade deficit, which was estimated at 14 billion ECU in 1994 (Commission of EU 1994b) in spite of the fact that its annual exports have grown at an average of 4.4 per cent. Imports of textiles and clothing have, however, increased by an average of over 10 per cent during the same period, mainly from China, Turkey, Hong Kong and India. The inclusion of the textiles sector in the GATT Agreement establishing the WTO in 1994 has meant that the bilateral quotas that existed under the Multifibre Agreement are being phased out and the sector is now being governed progressively by the stricter GATT rules and disciplines. Thus the European Commission has a more influential role to play in trade agreements negotiated for the sector.

The Commission also contributes to the sector through the RETEX initiative that promotes the diversification of economic activities in those regions that are dependent on the textiles industry, providing for training, cooperation between firms and the introduction of new technologies. The sector also benefits from specific EU support for SMEs; it has a high proportion of small firms, with the average size in the clothing industry being 17 people per firm, as against 27 in the textile industry. Many textiles firms are in Objective 2 regions (see Chapter 12), such as the cotton and woollen industries of Lancashire and Yorkshire in the UK, the linen industry in Northern Ireland and the woollen industry around Biella and Vicenza in Italy. Several of the textiles regions in difficulty also have an iron and steel industry, such as Nord-Pas-de-Calais and southwest Belgium.

The Engineering Industry

Mechanical engineering is one of the largest sectors in EU industry, responsible for 8 per cent of industrial output in 1993. The industry is concentrated in Germany, which has nearly 50 per cent of the sector's output, followed by the UK (15 per cent), Italy (13 per cent) and France (10 per cent). The accession of Austria, Finland and Sweden added around 10 per cent of output in the sector. While the sector has not suffered from external competition to the same extent as others, it is still the victim of cyclical fluctuations in demand and was also affected by the recession of the early 1990s.

The industry is concentrated in particular regions of the EU, which tend to coincide with historical centres of industrial activity, since producers of machinery have traditionally been located near their main customers since the start of the Industrial Revolution. The main production centres are to be found in Germany in Baden-Württemberg, Bavaria, Hessen, Lower Saxony and Nordrhein Westfalen, in the UK in the Midlands, in Lombardy and Emilia-Romagna in Italy, in the Ile de France and the Bassin Parisien in France, and in Catalunya and the Basque Country in Spain. Although transport and communication links have improved, much of the manufacturing capacity remains in these traditional centres.

The mechanical engineering sector is less export-oriented than the other key sectors, although it remains in surplus. Since much of EU production is for EU markets, it is a sector which has benefited particularly from the harmonisation of rules and technical standards established through the Single Market. The sector as a whole is expected to grow in the second half of the 1990s, benefiting also from

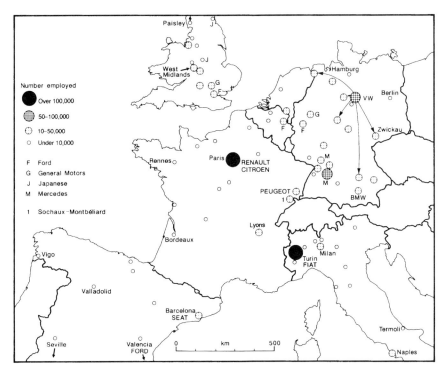

Figure 7.6 Major centres of motor vehicle manufacture in the countries of EUR 12
Main source: *Calendario Atlante de Agostini 1990* 1989

export opportunities in the expanding markets of Central and Eastern Europe, to which it already has privileged access through the Europe Agreements. Issues such as training and R&TD are crucial to the sector, over which the EU has an increasing influence through the various programmes of finance.

The Automobile Industry (See Figure 7.6)

The EU automobile industry suffered a severe recession during the early 1990s, felt throughout the Member States and leading to job losses in a major source of employment in EU industry. Jobs in the vehicle assembly industry fell from 1.23 million in 1990 to 1 million in 1994, due mainly to restructuring and rationalisation. The industry is a net exporter, although the EU's trade surplus has declined substantially between 1985 and 1993. Although the world market itself has declined there has been a particularly severe reduction in demand on the EU market itself. Car production recovered in 1994 to 12.8 million units from 11.3 million in 1993, although EU sales remain slow at around 12 million units per year.

The motor vehicle market is dominated by companies from four Member States, with the German industry accounting for the largest single share. In 1993, market share in the EU was as follows:

VW Group	16.4%
GM Group	12.7%
PSA	12.6%
Ford Europe	11.7%
Fiat Group	11.6%

Renault	10.8%
BMW	6.7%
Other	17.4%

It is significant that there is no UK company among the 'big 7' in the industry, unless Rover is counted after its takeover by BMW. The Japanese and Koreans still have a reduced market share. At national level, market share varies enormously, with the German and French markets, in particular, dominated by national producers.

The automobile industry has yet to benefit fully from the impact of the Single Market, since lack of harmonisation of tax and excise duties as well as restrictions under car registration schemes still make it difficult to purchase a car in one Member State and register it in another in order to benefit from the large differentials that apply between Member States. The industry is also coping with more stringent environmental emission requirements established under EU law and higher taxes on fuel which are reducing the use and sales of new vehicles in many Member States.

The Chemicals Industry

The chemicals industry in the EU is another principal source of employment, accounting for 1.58 million jobs in 1994. It constitutes not only basic organic and inorganic chemicals and petrochemicals, but also downstream products such as pharmaceuticals and agrochemicals. Germany is the largest single producer, with over 30 per cent of value added in the sector, followed by the UK and France with around 16 per cent each. As a percentage of GDP, however, Spain, the Benelux and Ireland have the largest chemicals industries. The largest companies in the EU in terms of turnover in 1993 were as follows:

Hoechst, Bayer, BASF, Henkel (Germany)	450,000 employees
Rhône Poulenc, L'Oreal (France)	120,000 employees
ICI, SmithKline Beecham, Glaxo (UK)	190,000 employees
Akzo Nobel (Netherlands)	60,000 employees

The export market is extremely important, since the EU is the world's leading exporter of chemicals, with over 21 per cent of production exported in 1993. It is a sector with particular problems in terms of environmental protection, both for pollution and production and the consumption and disposal of end products. Stricter EU legislation in these areas requires considerable investment on the part of the chemicals industry over the coming years.

NEW AND EMERGING INDUSTRIES
(See Figure 7.7)

The 1980s were characterised by a race between the USA, Japan and the countries of Western Europe in the development and marketing of new products, in particular in the areas of consumer electronics, information technologies, robotics, telecommunications and biotechnology. The emerging industrial sector has become even more prominent in the 1990s, given the importance it represents in strategic terms for manufacturing industry and the services sector in the future. Although the sector has grown in real terms, with a doubling of consumption of electronics components between 1984 and 1994, it is imports from the USA and Japan that have benefited most, and the Single Market has not changed the market reality of the 1980s when in 1984 there was a clear trade deficit in electronics products for the EU of 10 billion ECU. This has now trebled to a forecast level of 33 billion ECU for 1997 (Commission of EU 1995a). Employment in the sector, which has a high value added, has remained stable in the EU at around 1.4 million jobs.

When broken down into subsectors, the sector shows different levels of performance, with the telecommunications sector doing better than those of components, computers and

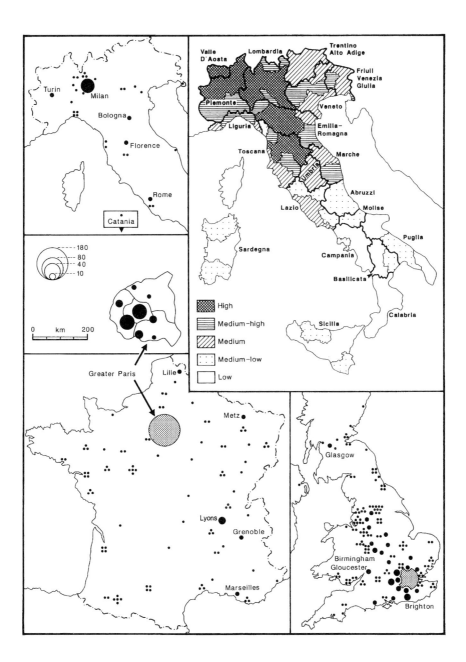

Figure 7.7 The location of a sample of new and high-technology industries in Italy, France and the UK, and the level of development of Italian provinces
Note: The key appearing in the Paris map also relates to the other maps in the figure
Source: data for Italian development: Carazzi and Segre 1989: 197

[Plate 7.3] The US company Hewlett Packard has opened a plant in Stuttgart, in Southwest Germany (picture 1990), alongside Daimler Benz, in the suburb of Sindelfingen. The NUTS level 2 region of Stuttgart is the most highly industrialised in the whole of the EU in terms of percentage of economically active population in the sector

Plates 7.3 and 7.4 New industries

[Plate 7.4] Since the Second World War Dunkerque on the North Sea coast in the extreme north of France has become a centre of heavy industry, with an integrated iron and steel works and a nuclear power plant. An aluminium plant under construction in 1990 uses imported bauxite

consumer electronics. There are at least three EU companies among the top fifteen in each subsector, as shown below:

Components:	Philips (Netherlands)
	SGS Thomson (France)
	Siemens (Germany)
Computers:	Siemens (Germany)
	Olivetti (Italy)
	Bull (France)
Telecommunications:	Alcatel (France)
	Siemens (Germany)
	Ericsson (Sweden)
	Nokia (Finland)
	Bosch (Germany)
Consumer electronics:	Philips (Netherlands)
	Thomson (France)
	Nokia (Finland)

The UK no longer has any leading manufacturer in the various subsectors concerned. ICL,

Plate 7.5 Industry on the periphery of the EU: ship repairing and maintenance at Falmouth, Cornwall, Southwest England

the last major British computer manufacturer was sold to the Japanese company Fujitsu.

The electronics sector has been given particular emphasis in EU policy, with a succession of programmes for the promotion of investment, cooperation and R&TD in the industries concerned since the 1980s. The fourth EU Framework Programme for R&TD is providing 13 billion ECU for the period 1994–8 on four action areas:

1 Specific R&TD programmes in seven sub-sectors:

Information and communications technologies (3.4 BECU);

Industrial and material technologies (1.995 BECU);

Environmental technologies (1.08 BECU);

Life sciences and technologies (1.57 BECU);

Energy technologies (2.26 BECU);

Transport technologies (0.24 BECU);

Applied socio-economic research (0.14 BECU).

2 International cooperation (540 MECU): assistance to neighbouring or developing countries.

3 Diffusion and promotion of research results (293 MECU).

4 Mobility of researchers (744 MECU).

The EU also promotes training or retraining through the Structural Funds, much of which is focused on adaptation of the workforce in less privileged regions to new technological requirements. Many of the SMEs assisted through EU programmes are also active in the new technologies, whether it be in manufacturing or in the applied services sector. As in the 1980s, however, it remains to be seen whether EU funding to promote cooperation and R&TD in the new technologies will assist in redressing the negative balance of trade that has always characterised the sectors concerned. Not only is investment in R&TD much higher in the USA and Japan, but EU industry also has to compete increasingly with emerging high-technology industries located elsewhere in South and East Asia, such as China, India, Malaysia and the Philippines.

The significance of the new industries is clear and EU industry is continually attempting to catch up with the advanced position of the USA and Japan as well as watching the emerging industries in the Far East, where wage costs are much lower and where technology transfer is providing increasingly advanced levels of know-how. Development in the sectors concerned has a far more widespread and longer-term impact on EU industry and services for the future and it is a crucial area for EU policy in terms of R&TD and training, as well as measures that provide a favourable climate for the industry to compete on a fair footing. Although the EU must continue to assist those industrial regions whose activities are in decline, it must concentrate its limited resources on the industries of the future in order to provide the competitiveness and growth that can assist in helping its regions to emerge from the economic recession and high unemployment of the early 1990s.

FURTHER READING

Commission of the EC (1995) *Panorama of EU Industry 95/96* Luxembourg: Office for Official Publications of the European Community.

Cooke, P. (ed.) (1995) *The Rise of the Rustbelt*, London: UCL Press. The regeneration of three major industrial areas, one in North America, two in the EU: the Ruhrgebiet of North Rhine-Westphalia, and South Wales.

Dyker, D. (1992) *The National Economies of Europe*, London: Longman.

Eurostat (1995) *Europe in Figures*, 4th edition, 'Industry', pp. 286–315.

Keeble, D. and Wever, E. (1986) *New Firms and Regional Development in Europe*, London: Croom Helm.

Larson, E. D., Ross, M. H. and Williams, R. H. (1986) 'Beyond the era of materials', *Scientific American*, 254 (6): 24–31.

8

SERVICES

'The services sector becomes more capital-intensive than manufacturing as advanced economies rely more on knowledge than on manual skills.'

(Naudin 1996: 22)

- Services now account for well over 60 per cent of all the economically active population in the EU.

- Services account for almost four times as much investment as manufacturing in the EU.

- Healthcare takes almost 8 per cent of the total GDP of the EU.

- Education takes 5–6 per cent of the total GDP of the EU.

- Tourism makes a strong impact on the economy of certain EU countries and on particular regions.

- The budget of the defence sector has diminished and manpower in the armed forces has been reduced in the 1990s in most EU countries, adversely affecting the economic prospects of certain locations.

The services sector does not have a specific EU policy but is covered by a number of sectoral policies as well as being governed by the provisions of the Single Market. Title III of the EC Treaty lays down the principle of free movement and its Chapter 3 establishes the progressive removal of restrictions on the freedom to provide services, in particular services of an industrial or commercial nature and the activities of craftsmen and the professions. It is also recalled that transport services are governed by a specific Title in the Treaty and that banking and insurance services are governed by the chapter on the liberalisation of the movement of capital. In other parts of the Treaty there are specific provisions for culture, public health and consumer protection, which have an impact on the services sector. In most Member States the services sector has traditionally been more protected from foreign competition than industry, and liberalisation has been more difficult to establish due to a reluctance on the part of governments to allow for markets to be opened up. Air transport, postal services and financial services have been particularly sensitive areas, and the spheres of education and healthcare remain limited to cooperation rather than a comprehensive EU policy.

EMPLOYMENT IN THE SERVICES SECTOR

While the definition and function of the agricultural and industrial sectors in the EU is reasonably straightforward, the tertiary or services sector consists of a wide variety of types of employment. In the USA, the first two sectors are sometimes distinguished as producing goods whereas the services sector is referred to as 'non-goods'. Nevertheless, many workers in such branches of services as transport and administration are directly connected with the two 'productive' sectors. Eventually the distinction between the three sectors may become so blurred that their usefulness will disappear, but at present they are widely used. In this book, transport and communications, a major element in the services sector, and one of special geographical and spatial interest, has already been covered in Chapter 4, thus recognising that the service sector does not have a clear-cut functional base.

According to COM 90–609 (1991), in 1987 the services sector accounted for 59.2 per cent of total employment in the EC, while in 1986 (Eurostat 1989) the figure of 60.1 per cent of GDP was given. *Regional Profiles* (1995) gives 62.8 for the percentage of employment in services in EUR 15 in 1993, 31.6 for industry and 5.6 for agriculture. In several highly urbanised and territorially small regions at NUTS level 2, employment in the agricultural sector is at a very low level. Again, in many EU regions, employment in industry is modest in scale. In contrast, every NUTS level 2 region has some people working in services such as local government, healthcare, education, retailing and transport. Some regions have administrative, financial and other functions that extend, nationally or even internationally, far beyond their limits. The lowest proportion of employment in services in any NUTS level 2 region in the EU is 35 per cent in one of Greece's poorest regions, Anatoliki Makedonia, Thraki. The highest is in Spain's North African territory of Ceuta y Melilla, with 89 per cent in services, followed by Brussels, with 83 per cent.

Figure 8.1 shows the uneven distribution of employment in the services sector of the EU. The data for NUTS level 1 regions are given in Table A1 (Appendix). Three reasons for having a high level of employment in services stand out at NUTS level 1: the nature of the administrative area used (e.g. Hamburg, Brussels), the presence of the national capital (e.g. Lazio (Italy), Madrid), and a strongly developed tourist industry (e.g. Canarias (Spain), Mediterranée (France)). The NUTS level 1 regions in which services are least prominent in terms of employment are either those with a high level of industrialisation (e.g. Lombardia (Italy)) or those with a large agricultural sector (e.g. Noroeste (Spain), Nisia (Greece)). Table 8.1 shows the NUTS level 2 regions with the highest levels of employment in services. Development in the conventional post-Second World War sense is associated among other features with the growth of the service sector, and it is therefore interesting to note that the average for the EU as a whole, 62.8 per cent of all employment, is well below the level of 72 per cent in services in the USA and Canada (cf. Australia 70 per cent, Israel 74 per cent) but higher than in Switzerland, with only 60 per cent, and Japan with 59 per cent. In general the service sector is less prominent in providing employment in the southern countries of the EU than in the northern ones, although even Portugal and Greece are well above the level in most developing countries (e.g. 27 per cent in India, only 6 per cent in the lowest in the world, Nepal and Burundi).

A much greater range of employment in services is found in the EU at NUTS level 2 than at national level. Twelve NUTS level 2 regions have over 75 per cent of their labour force in services (see Table A1, Appendix). Ceuta y Melilla is a small overseas region of Spain with very little agriculture or industry. Six of

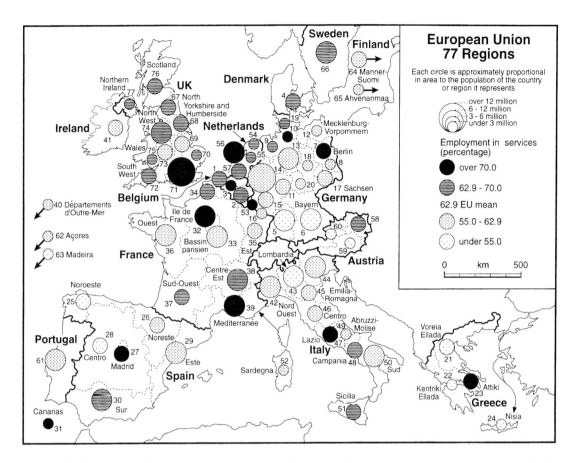

Figure 8.1 Employment in services as a percentage of total employment in the EU at NUTS level 1. See Table A1 (Appendix) for key to the numbering of unnamed regions
Source: *Regional Profiles* 1995

the remaining 11 include national capitals (2, 4, 7, 8, 11, 12), while five are in the Benelux (2–6). Corse and Canarias have limited agricultural resources, little industry, but in relation to their population sizes, large tourist industries.

At the other end of the scale there are 18 NUTS level 2 regions with less than half of their employed population in services. Greece with eight, Spain with three and Portugal with two, contain regions with a comparatively high proportion of their economically occupied population in agriculture, thus reducing the role of the service sector. Five NUTS level 2 regions of southern Germany have relatively

weak service sectors for a different reason. Here employment in industry is at its highest in the EU, again reducing the role of services, although in this case still ensuring a high level of provision.

Contrasts in employment levels in services are even greater at district level than at NUTS level 2, as shown by examples from the UK where services approach and even exceed 90 per cent of all employment in some regions, including many London boroughs (e.g. Kensington and Chelsea 97 per cent), some coastal resorts (e.g. Brighton 91 per cent, Bournemouth 90 per cent) and some smaller cities (e.g. Exeter 92

Table 8.1 Regions with the highest levels of employment in services in NUTS level 2 regions of the EU in 1993 (percentage)

1 Ceuta y Melilla (Spain)	89.2	7 Hovestadsregion (Denmark)	77.5
2 Brussels (Belgium)	83.1	8 Ile de France	76.4
3 Utrecht (Netherlands)	81.6	9 Corse (France)	75.9
4 Noord Holland (Netherlands)	78.1	10 Canarias (Spain)	75.8
5 Brabant (Belgium)	77.8	11 Lazio (Italy)	75.6
6 Zuid Holland (Netherlands)	77.6	12 South East (UK)	75.2

Source: Regional Profiles 1995

per cent, Chester 90 per cent). Such extreme cases of dependence on services are less likely to occur outside the UK and Belgium, because nowhere else does agriculture account for such a low share of employment.

Most of the assistance to the poorer regions of the EU has been targeted at regions with a large agricultural sector, the smallest employer of labour of the three sectors, and to some types of mining and industrial region. The services sector now accounts for over three-fifths of all employment in the EU and includes many very poorly paid types of work, but it hardly figures at all in regional objectives. Since employment in agriculture and industry in the EU is declining in many regions, the services sector will have to expand if unemployment is not to increase. Unfortunately, the data available for the EU on services do not give a precise indication of where new employment in the sector is needed or is likely to be created. Apart from the relocation of offices by governments or private companies, usually from the capital to the provinces, the growth of and changes in the services sector tend to be gradual. New technologies are already leading to job losses in such services as banking and retailing, as they have been doing for some decades through mechanisation in agriculture and automation in industry. On the other hand, a reduction in working hours or a change in the ratio of full-time to part-time jobs could sustain employment in the industrial and services sectors in the future.

In the rest of this chapter, various branches of

the service sector will be discussed. While all are vital to the Member States of the EU and to the EU as a whole, in general they have been less influenced in their organisation and development by EU policy than have the agricultural and industrial sectors. Nevertheless, marked regional disparities also occur in the service sector, with differences in the quality and quantity of services available.

HEALTHCARE

Virtually all the administration and financing of healthcare in the EU remains in the hands of the Member States, and EU policy is therefore largely confined to the funding of programmes and information campaigns and the intra-EU recognition of medical qualifications. A precise statement about the availability and regional distribution of healthcare provision and the occurrence of illnesses and causes of death throughout the EU cannot be made because of a lack of comparable data sets. In Eurostat publications there are notable gaps in information about various aspects of healthcare, while definitions vary from country to country, and health services are organised in different ways. There are, however, sufficient data about healthcare in the EU at both national and regional level to show marked disparities in the provision of services and in the main causes of ill health and death both between and within EU Member States.

According to Eurostat (1995a), medical and

other health personnel, including dentists and veterinary surgeons, account for approaching 5 per cent of total employment in the EUR 15 or 7.5 million people. About 7.7 of the total GDP of the EU is spent on all aspects of healthcare, compared with 13.3 per cent in the USA and 6.8 per cent in Japan. The situation at national level is shown in Table 8.2. Each variable will be discussed in turn.

Expenditure in Relation to GDP (Cols (1) and (2))

It is common to express expenditure on health as a percentage of total GDP. In 1960, when only six of the EUR 15 countries in Table 8.2 were actually members of the EC, far more was spent on health in Germany, the Netherlands and Sweden than in Spain. A comparison of the figures for 1960 and 1991 shows an increase in the share of GDP devoted to health in all EUR 15 countries, as also in Norway, Switzerland, the USA and Japan. The increase is the more remarkable because in real terms total GDP has itself grown during 1960–91. In 1991, however, a marked disparity remained in the EU, with Greece, Portugal, Spain and the UK lagging behind the other countries. The disparity may to some extent relate to needs, with the larger proportion of elderly in some countries than in others, requiring more facilities.

Population per Doctor (Col. (3))

Population per doctor illustrates disparities among EUR 15 countries in a more specific way than expenditure in relation to GDP. There is little correlation between the level of expenditure and the availability of doctors, although a more detailed breakdown according to the specialisations and qualifications of doctors might explain the large numbers in the southern countries.

Data for hospital beds (per 1,000 population) are incomplete, but data in Eurostat (1994a,

Table VII.1) show marked disparities between and within countries. Moreover, the availability of hospital beds does not correlate with the availability of doctors, as shown by the fact that the Netherlands has 11.4 per 1,000 population, Greece only 5.0 and Portugal 4.3. Within the northern Member States (Netherlands, Germany, France) regional disparities in the availability of hospital beds exist, but are not large. On the other hand, in both Italy and Greece, with poor provision compared with the EU average, regional disparities are very marked. Thus in Italy (national average 6.8), Lazio (8.0) and Nord Est (7.9) have much higher scores than Sicilia (5.5) and Campania (5.2). In Greece (average 5.0) the contrast is even sharper. Attiki (7.0), which includes Athens, contrasts with Kentriki Ellada (2.9), within which, at NUTS level 2, Sterea Ellada has only 2.0, worlds apart from Utrecht with 13.3 (West-Nederland). In so far as hospital beds represent all support facilities in the health service, then the northern countries tend to have a superior infrastructure to that of the southern countries, while having fewer actual doctors.

Public Funding (Col. (4))

Whether or not there is a large element of private funding in the health service (contrast Norway's 2 per cent with 61 per cent in the USA), concern over the cost of maintaining the health service is now expressed throughout the industrial countries of the world. The trend in most EU Member States is towards a reduction in the share of healthcare provided from public funding. The result could be that the healthy no longer contribute as much as the unhealthy, and the better-off afford care superior to that obtained by the less well-off. The growth in the proportion of the elderly (see Chapter 9), has implications for the above problem. Mostly the over 64s are or become less healthy than average and poorer economically.

Table 8.2 Healthcare data

	(1) Total expenditure on health as % GDP 1960	(2) Total expenditure on health as % GDP 1991	(3) Population per doctor 1988–91	(4) % public 1990	(5) Life expectancy in years	(6) Infant mortality per thousand	(7) Heart disease	(8) Cancer	(9) AIDS
Belgium	3.4	8.1	298	n.a.	75.7	8.5	26	23	2.2
Denmark	3.6	7.1	360	n.a.	75.4	6.5	32	22	4.6
Germany	4.9	9.1	n.a.	77	75.7	6.9	35	21	3.1
Greece	2.6	4.8	313	n.a.	77.4	9.0	32	16	1.7
Spain	1.6	6.5	262	69	77.0	7.6	23	19	14.0
France	4.3	9.1	333	74	77.7	7.3	23	23	9.9
Ireland	3.8	8.0	633	n.a.	75.0	7.6	34	20	1.9
Italy	3.6	8.3	211	79	77.2	8.0	26	22	8.0
Luxembourg	n.a.	6.6	n.a.	n.a.	76.1	7.2	n.a	n.a.	5.4
Netherlands	4.8	8.7	398	71	77.3	6.5	28	24	2.9
Austria	4.4	8.5	n.a.	65	77.0	6.2	38	21	2.9
Portugal	2.3	6.2	352	60	74.4	9.3	18	18	4.5
Finland	3.8	8.9	405	n.a.	76.0	4.4	37	18	0.7
Sweden	4.7	8.8	395	91	78.0	4.8	38	19	2.1
UK	3.9	6.6	n.a.	85	76.4	6.6	31	23	2.8
Norway	3.2	8.4	309	98	77.0	5.8	33	19	1.5
Switzerland	3.3	8.0	585	67	78.0	5.6	33	22	10.4
USA	5.3	13.3	n.a.	39	76.0	8.0	38	21	25.4
Japan	3.0	6.8	n.a.	70	79.0	4.3	23	20	0.1

Sources: (1)–(3), (7)–(9) UNDP 1995, Table 24; (4) Laurance 1995: 8; (5), (6) WPDS 1995 (PRB 1995); (7) and (8) likelihood of dying after age 65 per 1,000 people, average for males and females, 1990–2; (9) AIDS cases per 100,000 people, 1993
Notes: n.a. not available

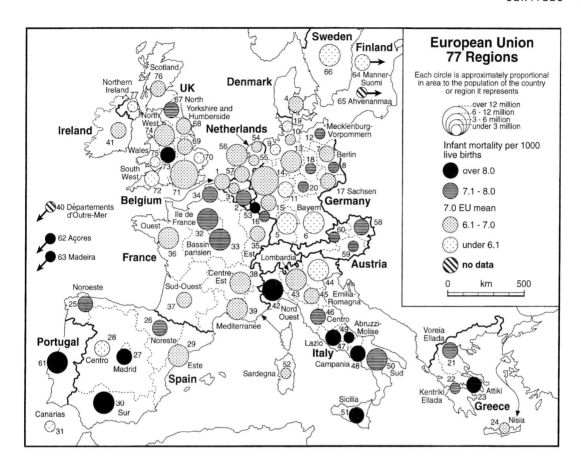

Figure 8.2 Deaths of infants under the age of one year per 1,000 live births, 1991. See Table A1 (Appendix) for key to the numbering of unnamed regions
Source: Eurostat 1994a: 2–9

Life Expectancy and Infant Mortality
(Cols (5) and (6))

Average life expectancy (at birth) and infant mortality (deaths of infants under the age of 1 per 1,000 live births) are general indicators of the availability, quality and sophistication of healthcare services in a country or region. In a global context, average life expectancy is very similar throughout the EUR 15 countries (extremes Sweden 78 years, Portugal 74.4 years) when compared with that in many African countries, where levels are between 40 and 50 years (the lowest is Sierra Leone, 39 years).

Infant mortality rates in the EU are very low by world standards, and disparities at national level (see Figure 8.2) are not great (Finland 4.4 per thousand, Portugal 9.3). Indeed, 4–5 per thousand is probably as low as the level can realistically be expected to fall. In contrast, more than ten countries in Africa south of the Sahara have levels of over 200, still common in much of Europe well into the nineteenth century. With infant mortality rates of 55 and 57 respectively, Algeria and Morocco have levels similar to that of Sweden around 1920.

Since the number of infant deaths per 1,000 live births is so small, the absolute number in

Table 8.3 Contrasts in level of infant mortality between and within selected EU countries

	National mean	Lowest region		Highest region	
Germany	8	Mittelfranken	5	Dessau	12
France	8	Centre-Est	6	Corse	23
Spain	7	Asturias	5	Navarra	11
Italy	8	Nord-Est	5	Campania	11
UK	7	Essex	5	West Yorkshire	10

Source: Eurostat 1994a: 2–11

Table 8.4 Infant mortality rates in selected regions of Italy, 1961–91 (per 1,000 live births)

Region	1961	1971	1991
Lombardia (North)	35	23	7
Emilia-Romagna (North)	31	23	6
Lazio (Centre)	35	24	8
Campania (South)	55	40	11
Sicilia (Islands)	47	34	11

Source: various years of *Annuario statistico italiano* (ISTAT Rome)

any region is itself too small for the figures each year to be consistent. Nevertheless, given the limitations, data for 1991 in Eurostat (1994a: 2–11) show great internal disparities, reflecting both the level of appropriate healthcare facilities and general environmental conditions (Table 8.3).

Progress in reducing infant mortality levels in the EU may be illustrated by reference to selected Italian *regioni* (see Table 8.4). The picture is one of universal rapid decrease and also of regional convergence. The decrease has come more quickly in the poorer regions. Even allowing for fluctuations due to small numbers of births at district level, marked disparities occur at the smallest scale of region for which data are aggregated within countries. A UK study (National Audit Office 1990) of the occurrence of perinatal deaths (still-births plus deaths in the first week of life), per 1,000 births, shows a range between 13.5 in

Bradford (Yorkshire) and 6.5 in west Surrey and northeast Hampshire.

Diseases (Cols (7) and (8) Table 8.2)

In the developed/industrial countries of the world, deaths from infectious and parasitic diseases are now rare. Deaths from accidents, suicide and other 'man-made' causes such as war have also been limited in scale since the Second World War. In the early 1990s they accounted for about 6 per cent of deaths of men, 3.6 per cent of those of women. In most developed countries, death from cardiovascular (heart) diseases and malignant neoplasms (cancer), account for between two- and three-fifths of all causes of death. Infectious and parasitic diseases can largely be controlled and their treatment, when they do occur, tends to be low in cost. In contrast, degenerative diseases often need costly medical treatment and the extended use of hospital facilities. Hence the growing concern over the growth of the elderly population (see Chapter 9).

AIDS (Col. (9))

Since the advent of AIDS among humans, an affliction only recognised in the early 1980s, there has been less complacency than previously among the public of the developed countries about the control of infectious diseases. Although compared with some other contagious diseases the occurrence of AIDS is very rare in the developed countries, again differences among the EUR 15 countries are great. Spain, France and Italy have the highest levels of AIDS, Ireland, Greece and Finland (and Norway) the lowest levels; apparently for once location on the periphery could be an advantage. On a global scale the remarkable isolation of Japan (0.1 cases per 100,000 people) contrasts even more strikingly with the open and cosmopolitan nature of the USA (25.4). The EU countries still have a long way to go to reach Zambia's score of 239 (see also Chapter 9).

Even without complete, standardised data for the measurement of the availability of healthcare facilities and the occurrence of various causes of death, some generalisation can be made about healthcare in the EU. Only in North America, Australia and New Zealand, and Japan is it matched over comparatively large populations. Within the EU itself, however, there are differences. The widest range of facilities can be reached quickly by people living in and around most of the large urban agglomerations, especially the national capitals, as well as in retirement areas of the more affluent (e.g. Provence in France, southwest England). Medical specialists, medical schools and research centres tend to be concentrated in larger cities. At the other extreme, healthcare conditions are well below the EU average in much of Portugal, Greece and southern Italy. People living in thinly populated areas, remote from large urban centres, also tend to be comparatively poorly provided with easily accessible healthcare services. Those areas designated for EU Structural Funds as Objective 5b (see Chapter 12), are mainly rural and thinly populated.

The impressive advances in medical science and their success in reducing degenerative diseases among the 'middle-aged' population mean that more and more people move into age groups in which the cost of medical treatment, nursing care and residential care is highest (see Chapter 9). Greatly increased expenditure on healthcare since the Second World War throughout the EUR 15 countries has not closed the 'statistical' regional gaps between the quality and the quantity of medical facilities. The needs differ from one region to another as a greater understanding of diets and life-styles is showing. For example, the reduction of smoking has been more successful in some countries than in others. The effects on health and on the causes of death of high or low levels of consumption of alcohol, certain fats and various other beverages and food are still being assessed. Different types of employment, as well as environmental pollution, are also complex factors with different regional impacts. Cohesion is a long way off in the world of EU healthcare, and convergence, difficult enough to measure, does not appear to be occurring.

EDUCATION (See Figure 8.3)

As with healthcare, policy on education at European Union level is limited in scope. The main influence on education at the EU level so far has been to ensure mutual recognition of examinations and qualifications between countries rather than the imposition of harmonisation and uniformity. The EU also funds training programmes through the Social Fund to assist unemployed people, vocational training, and exchange and education programmes between establishments in different Member States. As

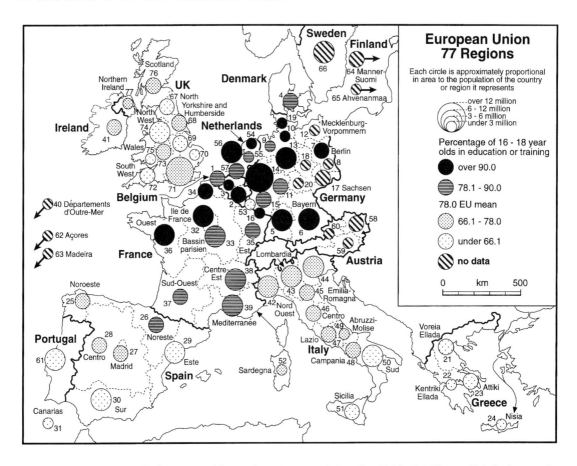

Figure 8.3 Percentage of 16–18 year olds in education or training. See Table A1 (Appendix) for key to the numbering of unnamed regions
Source: Eurostat 1994a: 2–9

with healthcare, the availability, accuracy and recency of EU data on education are far from satisfactory. There are considerable differences between Member States in the ages at which compulsory education starts and finishes, while participation in non-compulsory schooling (mostly 16–18 years) and higher education varies greatly (see Eurostat 1995a: 162–9).

In the early 1990s, the whole educational sector of the EU accounted for about 5.5 per cent of total GDP and employed about 8.5 million people, of which only about half were actually engaged in teaching, the remainder being in administrative, maintenance and other non-teaching duties. In 1991/2 there were about 71 million pupils and students, 19 per cent of the total population. About 60 per cent were in compulsory education, 25 per cent in post-compulsory secondary education and the rest in higher education.

One of the most remarkable demographic trends in Western Europe in the last three decades of the twentieth century has been the sharp fall in fertility rates. This is reflected in a sharp drop in the proportion of under 15 year olds in the total population. According to *WPDS 1975* and *WPDS 1995* (PRB 1975 and 1995), the percentage of under 15s in

Table 8.5 Education data for the EU at national level

	(1) Education as a percentage of GDP	(2)	(3) % of population aged 5–24	(4) % spent on tertiary	(5) Science as % of tertiary exp.	(6) % of women in higher education
	1960	1991	1991/2	1991	1990–1	1991
Belgium	4.8	5.4	80	19	39	52
Denmark	4.0	6.1	72	21	34	53
Germany	2.4	5.4	71	22	42	43
Greece	2.0	3.0	65	20	40	53
Spain	1.1	5.6	74	19	29	57
France	3.6	6.0	72	18	31	48
Ireland	3.0	5.9	68	24	34	47
Italy	4.2	4.1	64	10	31	51
Luxembourg	n.a.	5.8	55	n.a.	n.a.	n.a.
Netherlands	4.9	5.8	87	30	18	43
Austria	2.9	5.4	66	23	31	48
Portugal	1.8	5.5	63	17	28	61
Finland	4.9	6.6	79	24	52	57
Sweden	5.9	6.5	65	18	43	58
UK	3.4	5.3	76	21	39	49
Norway	4.6	7.6	75	20	27	54
Switzerland	3.3	5.4	69	23	34	n.a.
USA	5.3	7.0	86[1]	34	14	55
Japan	4.9	5.0	73[1]	21	26	42

Sources: UNDP 1995, Table 25 (1), (2), (4), (5); Eurostat 1994b: 132–5 (3); Eurostat 1995a: 164–5 (6)
Notes: 1 per cent of age 6–23
    (1), (2) Total expenditure on education as a percentage of GDP
    (3)   Percentage of total population aged 5–24 years in all levels of education
    (4)   Expenditure on tertiary education as a percentage of expenditure on all levels
    (5)   Tertiary natural and applied science enrolment as a percentage of total tertiary expenditure
    (6)   Women as a percentage of all graduates

Western Europe dropped from 24 per cent in the mid-1970s to 18.5 per cent in the mid-1990s. Since total population increased only slightly during those two decades the absolute number of children in compulsory education actually decreased in EUR 15 from about 50 million to about 40 million. During the same period, however, the numbers in upper secondary and higher education rose from about 20 million to 30 million.

The data in Table 8.5 highlight some of the contrasts between the educational systems of the EUR 15 countries. Each set of data will be considered in turn.

Education as a Percentage of GDP
(Cols (1) and (2))

On average, somewhat more of total GDP is spent on education in the EU than in Japan, but considerably less than in the USA. In column (1) it can be seen that in 1960, when neither country was in the EU, Sweden spent far more of its GDP on education than Spain, while among the original EUR 6 countries the difference was considerable (Netherlands 4.9 per cent, Germany 2.4). Between 1960 and 1991 the proportion of GDP spent on education rose in every EUR 15 country, with the level picking up sharply in Spain and Portugal, leaving Greece with only 3.0 per cent at the lower end and the Nordic countries, typically, at the upper end, with over 6 per cent.

Percentage of Population aged 5–24 in Education (Col. (3))

Since compulsory schooling starts at different ages in different EU countries (e.g. at 7 years in the Nordic countries, at 4 years in Luxembourg and Northern Ireland) a comparison of figures for the number of pupils in compulsory education in relation to population aged 5–24 is not straightforward. The proportion continuing beyond the compulsory age limit (mostly 16 years) also varies. For what they are worth, the data in column (3) show marked differences overall, with Portugal (63 per cent) and Italy (64), having much lower levels than the Netherlands (79 per cent).

Expenditure on Tertiary Education (Col. (4))

Column (4) shows that there are also variations within the EU with regard to the percentage of education funding devoted to the tertiary level (19–24 year olds). In spite of its distinction of having some of the oldest universities in Europe, modern higher education in Italy appears to be badly underfunded, with only 10 per cent going to the sector. There is broad conformity if not homogeneity among most EU Member States, but none approaches the US level of 34 per cent.

Science as Percentage of Tertiary Expenditure (Col. (5))

Increasing emphasis has been placed on science and technology in education in the developed countries since the Second World War. As measured by the percentage of graduates enrolled in tertiary natural and applied science,

some EUR 15 members have the highest levels among Western countries (Finland 52 per cent, Sweden 43, Germany 42) but the Netherlands, with only 18 per cent, falls well below the EU average yet is still higher than the USA, with a surprisingly low 14 per cent.

Women in Higher Education (Col. (6))

The place of women in higher education was an issue earlier in the twentieth century but the data in column (6) show that in most EUR 15 countries women are now well represented and in some countries 'over-represented'. There remains a marked gap between Germany and the Netherlands (and Japan), where women appear to miss out in the tertiary sector, and the Nordic and Iberian countries (and the USA), where they predominate. In the EUR 15 as a whole the percentage of women is 49, but they greatly outnumber men in medical sciences, healthcare and related areas (71 per cent) and humanities (64 per cent), while coming well behind men in natural sciences and mathematics (35 per cent) and far behind in engineering, architecture and transport (16 per cent). The position of women in various aspects of life in the EU is further covered in Chapter 9.

In conclusion the following points may be noted:

- By world standards, educational levels in the EUR 15 countries are very high in all respects and are broadly on a level with those in the USA and Japan.
- Nevertheless, marked differences exist between EU Member States, mainly at post-compulsory school and higher levels, and even if the EU takes a greater role in educational policy than it has up to now, it is unlikely that homogeneity will be reached in the foreseeable future.
- Greater cohesion and integration of the EU is

one aim of policy, and one way of achieving greater understanding among EU Member States is to encourage the teaching of foreign languages. According to Eurostat (1995a), in 1991/2, 83 per cent of all pupils in general secondary education (outside the UK and Ireland) learned English as a foreign language while 31 per cent learned French and 17 per cent German. Before the Second World War French and German were widely used in southern and eastern Europe. Denmark and the Netherlands are now the only countries in the EU in which over half of the pupils learn German. With time, the use of English in international communication in the EU seems set to grow. It is ironical that while in Greece every important road sign in the native language is duplicated in English, in Wales, Welsh is now shown alongside English although virtually no visitors to Wales can understand Welsh and the number of monoglot Welsh is virtually zero.

- In the 'near abroad' of the EU, participation rates in the various stages of education in Central Europe and the former USSR are roughly comparable with those in the EU, although until the 1990s the content of teaching and the function of education itself were very different. On the other hand, in the Maghreb countries and Turkey participation

levels and the quality of the backup to education are far below those in the EU.

FINANCIAL SERVICES

Such has been the expansion of services connected with various aspects of finance that more people are now employed in the EU handling finances than growing food. In the early 1990s about 10 million were employed in EUR 12 in a broad sector covering banking and finance, insurance, business services and renting. Table 8.6 shows the percentage of the economically active population engaged in financial services according to two somewhat different definitions in 1991 (column (1)) and in 1992 (column (3)). The absolute number employed is shown in column (2). Although pay is not particularly high in the general run of banks and offices, there is a considerable number of highly qualified and well paid jobs in the main centres of finance, principally the larger cities of the EU, and the whole sector probably accounts for some 10 per cent of the total GDP of the EU.

As with most activities and types of production of goods and services in the EU, the proportion of total employment accounted for by financial services varies sharply among Member States. The freak case of Luxembourg apart, the

Table 8.6 Employment in banking and finance, insurance, business services and renting in EUR 12, 1991 and 1992

	Per cent 1991	Thousands 1992	Per cent 1992		Per cent 1991	Thousands 1992	Per cent 1992
Belgium	8.2	256	8.3	Ireland	8.5	81	9.3
Denmark	9.1	216	9.2	Italy	4.7	1,057	7.1
Germany	8.4	2,429	7.4	Luxembourg	13.0	19	12.8
Greece	5.3	119	6.1	Netherlands	11.1	591	10.1
Spain	5.8	601	6.6	Portugal	4.5	226	6.7
France	9.7	1,851	9.9	UK	11.4	2,403	10.9
				EU 12		9,848	8.5

Sources: 1991: Commission of the EC 1994b: section 23, p. 2; 1992: Eurostat 1994b: 140–1
Note: Employment in the financial sector as a percentage of total employment

range is approaching 2:1, with over 10 per cent in the UK compared with about 6 per cent in the four southern countries, Italy, Greece, Spain and Portugal. In the latter countries the nation-wide financial infrastructure is not so developed as in the UK, the Netherlands and France, while the intra-EU and global functions of financial institutions are highly concentrated in a few cities such as London, Paris, Frankfurt and Amsterdam.

Since the creation of the EU, and especially since the 1970s, great changes have taken place in sectors providing financial services, as also in agriculture and manufacturing. According to *Panorama of EU Industry 94* (Commission of the EC 1994b):

> The traditional boundary lines between sectors providing financial services are becoming blurred in Europe. Whereas in the USA and Japan, these demarcation lines still exist, in Europe deregulation and liberalisation stimulate the emergence of integrated markets. Disintermediation, securitisation, the growth of the market for derivatives, 'Allfinanz' mergers and alliances, electronic banking and electronic markets are examples of important trends in the financial services sector.

Just as the demarcation lines between the providers of financial services have become fuzzy, so some services are increasingly pro-vided on an EU rather than a national level, a trend that could accelerate with the elimination of the remaining obstacles to cross-border trans-actions, required by the Single Market, for banking, insurance and other financial services. The introduction of the single currency will further facilitate financial transactions of all kinds among EU Member States, but those countries remaining outside the single currency would be at a marked disadvantage, including possibly the UK and certainly for a time non-members, notably Switzerland and Norway.

One of the central features of the EU financial services is the presence of 32 stock exchanges in the EU and EFTA. According to Butler (1995a) by the turn of the century, the number could drop to eight, the number currently in the USA. Seventeen exchanges or groups of exchanges are listed in Table 8.7, including eight in Germany and three in Switzerland. The eight surviving exchanges could be London, Paris, Frankfurt, Amsterdam, Zurich (with Milan), Madrid, Stockholm (for the Nordic countries) and Vienna, the last an attractive location for firms from Central and Eastern Europe. Although the above prospect is spec-ulation it does underline the need for various EU Member States to pool their resources, and

Table 8.7 Stock Exchange capitalisation

	European market capitalisation – domestic equity			European market capitalisation – domestic equity	
	Billions of ECU	Thousands of ECU per head		Billions of ECU	Thousands of ECU per head
London	982.1	16.8	Copenhagen	42.4	8.2
German Exchanges	435.7	5.3	Helsinki	40.0	7.8
Paris	388.9	6.7	Oslo	32.9	7.7
Swiss Exchanges	262.3	37.5	Vienna	24.0	3.0
Amsterdam	206.5	13.3	Luxembourg	20.8	52.0
Italian Exchanges	142.2	2.5	Dublin	17.8	4.9
Madrid	138.7	3.5	Lisbon	13.6	1.4
Stockholm	121.4	13.6	Athens	12.7	1.2
Brussels	73.7	7.2			

Source: Butler 1995b: 19

is similar to the prospect, discussed in Chapter 13, that especially if a dozen new countries join the EU, there might need to be some pairing of Member States to prevent the proliferation of commissioners and other representatives in the EU administration.

Those financial centres of the EU already at the top are increasingly competing for enhanced status and even more business. For example, although Frankfurt is the headquarters of the European Union's monetary institutions, Naudin (1994c) questions whether that will turn it into the financial capital of Europe: 'Bankers and officials assembled in the city recently and agreed that while it concentrates most of the *deutschmark* business, it will take more effort to enhance the city's role as *Finanzplatz Deutschland*, let alone *Finanzplatz Europa*.' Meanwhile there is hope in Paris that if the UK stays out of European Economic and Monetary Union it will be able to freeze the City of London out of doing business in the new single currency.

At present London remains pre-eminent as a financial market in the EU and indeed in the world, the volume of its transactions and assets matched in Europe only by those of Switzerland (Zurich, Basle and Geneva). According to Evans (1994) more than 500 banks from abroad operated in London in 1994, compared with fewer than 250 in Paris and 180 in Frankfurt. The 514 in London included 54 Japanese and 49 US banks. Evans notes: 'the biggest single factor behind London's success is its "critical mass", with the huge number of banks and financial institutions and wide range of financial markets available'. One should not overlook, also, the positive but unquantifiable convenience of working in English, the favourite second language in nearly every country in the world. It would be ironical if, after surviving physical destruction by the Germans in the Second World War, the more recent activities of IRA bombers, and scandals in the form of BCCI and Barings, the City of London should slip on the single currency banana skin.

In conclusion, some trends may be noted:

- Job losses are expected in financial services over the next few years, including in Switzerland.
- Many of Europe's regional business centres, including several EU capitals, notably Athens, Copenhagen, Dublin and Lisbon, each with the headquarters of more than ten banks, together with regional centres of distinction such as Barcelona, Stuttgart and Manchester, are likely to lose out against the key stock exchanges.
- Emerging markets in the EU's 'near abroad' can be expected to bring further businesses to EU financial institutions, especially from Poland, the Czech Republic, Slovenia and Israel.
- As the EU becomes more integrated and consolidated, Verchère (1994a) sees a threat even to the largest and most prestigious financial centres in the EU:

> there are signs that companies with their sights set on the Union's Single Market – notably US and Japanese companies – may eventually leapfrog national centres and relocate their head offices in or near Brussels. An example is Pilkington, Britain's largest glass manufacturer, which has moved its corporate headquarters from St Helens in Lancashire [to Brussels] and skipped the traditional migration to London.

TOURISM

The importance of tourism in the EU is underlined by several different measures of its impact. According to Eurostat (1995a: 330), in the early 1990s it accounted for about 5.5 per cent of the GNP of EUR 12. At that time it gave employment to some 9 million people, providing between 6 and 7 per cent of all jobs in the EUR 12, and it also makes a major

[Plate 8.1] Old monasteries are perched on the top of natural features with precipitous slopes. Meteora in central Greece offers the tourist spectacular scenery as well as insights into the world of the Greek Orthodox Church for a fairly modest fee

Plates 8.1 and 8.2 Tourist attractions

[Plate 8.2] On a flat area near Montpellier, southern France, man-made constructions produce a rugged landscape in which accommodation is created to capture the sun, and access to the beaches is a matter of a few minutes' walk

contribution to the balance of payments of several EU countries. In 1994, EUR 12 countries accounted for 38 per cent of all tourist arrivals in the world, and as a contribution to the balance of payments, tourism represents 28 per cent of the services exported by the EU to non-EU countries. There are, however, various reasons why the impact of tourism on the economy of the EU is difficult to quantify.

Tourism is a subset of all leisure activities and is closely related to travel and to the availability of the transport network and the services provided on it. Conventionally, it is distinguished from home-based recreation and day trips, and involves temporary movement to places beyond the normal places of residence and work. It is difficult, however, to distinguish travel in connection with work from that for leisure or pleasure. Many people working in such activities as hotels, catering, travel agencies and other services associated with tourism are also providing services for people who are

BOX 8.1 PRIME LOCATIONS IN THE EU

At local and regional level in each EU country there is a continual opening and closing of service facilities such as hospitals, schools, retail outlets and banks, corresponding to changes in the distribution of population, consumer needs and economies (or diseconomies) of scale. Such changes in the distribution of services in the EU generally take place regardless of the existence and influence of supranational guidance and funding. At the other extreme, it is vital for the success of service undertakings with the whole of the EU market in mind to be set up in a good location, if not the best. The attraction of just a few centres to the headquarters of banks and other institutions is described in the financial services section of this chapter. Here three other examples are given. Figure 3.2 is a good starter for determining roughly how many people live within given distances of any place in the EU.

Euro Disneyland

Euro Disneyland is located outside Paris. Once it was decided that there would be only one Euro Disneyland centre, the choice of location depended on finding a place that satisfied above all one criterion: how to include the largest number of people in a given radius. A Euro Disneyland advertisement in *The Times* (12 Oct. 1989, p. 31) showed there to be about 16 million people within 160 km of Paris, 41 million within 320 km and 109 million within 480 km, the last containing most of France, the southeastern UK, the Benelux and the western side of Germany. Other places in France to the east of Paris would have produced similar results, but

the special advantages of choosing Paris were the presence 'on the doorstep' of Euro Disneyland of about 10 million people, the proximity of one of Europe's busiest airports, with national and intra-EU flights to most places of consequence, and the image of Paris, worth visiting for its own sake by visitors to Euro Disneyland. In the event, Euro Disneyland almost ended in financial disaster, one factor blamed being the general indifference if not antipathy of many French to American culture. It would be unrealistic now to re-locate Euro Disneyland, but this experience underlines the need to examine very critically the feasibility and financial viability of large enterprises serving all or a large part of the EU. One is left to speculate whether or not London (culturally more sympathetic to the USA), Amsterdam (entertainment 'capital'), or Cologne (with a massive population within a small radius) might have been more successful. In a similar fashion, it can be seen with hindsight that the Channel Tunnel would never capture enough traffic to make it a financial success.

Sophia Antipolis

Europe's first scientific park, Sophia Antipolis, was developed in the 1960s and opened in 1969. It is situated in pleasant wooded hills behind the French Mediterranean coast between Cannes and Nice, occupying some 2,300 hectares. There are almost 1,000 enterprises, employing more than 15,000 people from 50 countries, and some 18,000 additional jobs indirectly created by on site activities.

As Max (1994) states: 'the park is

Table 8.8 The world's top conference cities, international conferences per year (to nearest 5)

EU	Paris	230	EU	Berlin	85
EU	Vienna	180	E	Budapest	70
EU	Brussels	175		Singapore	70
EU	London	155	EU	Rome	65
E	Geneva	125	EU	Lisbon	65
	New York	100	EU	Madrid	65
EU	Copenhagen	100	EU	Helsinki	60
EU	Amsterdam	95	E	Prague	60
EU	Strasbourg	95	EU	Barcelona	55
	Washington	90		Bangkok	50

Source: Tillier 1995
Note: EU – European Union   E – Other Europe

geared towards contented employees; two thirds of its land is a nature reserve, and 150 hectares are dedicated to sports and leisure'. Many of the early non-French arrivals were US companies, whose employees would find the climate and other environmental conditions attractive. Sophia Antipolis does not have such a large population in a given radius as Paris, but the cost of delivering its products is negligible so, unlike many manufacturing firms or places of entertainment such as Euro Disneyland, its somewhat remote location is a minor drawback. Nice has France's second busiest airport, thanks to the tourist activities of the Côte d'Azur, and a new motorway link will put this part of France in close contact with north-west Italy. Moreover, the University of Nice, opened only in 1965, now has some 25,000 students, and it is planned to double that number. Other places on the Mediterranean coast appear to be equally favourable locations for similar scientific parks, for example Genoa (although there is little spare land there), Barcelona and Valencia, and it could be that it was the early initiative of the Nice-Côte d'Azur Chamber of Commerce, against indifference from Paris, but with

some of Europe's favourite coastal resorts to hand, that accounts for the success of this particular location.

Conference Centres

Almost anyone who has attended a conference on whatever subject will agree that the choice of venue is a consideration in determining whether they will attend, if their presence is not essential. The attractiveness of the venue to an individual may depend on the actual cost of the conference (travel plus accommodation) and on features of the locality itself (historic associations, scenery, climate). The Brussels-based Union des Associations Internationales ranks the cities of the world in various ways according to their importance as conference centres. Table 8.8 shows the 20 top conference cities of the world according to the number of international conventions and congresses attracting a minimum of 300 people, with 40-plus per cent foreign participation, from at least five countries. Of the top 20, 13 are in the EU, three more elsewhere in Europe. Of the 13 in the EU, 11 are capital cities, only two regional centres, Strasbourg and Barcelona.

The impression is that in addition to their generally good airline connections, capital cities rather than other cities have adequate facilities for comparatively large conferences. Among the present EUR 15 capitals, Paris and London are much more attractive than Berlin or Rome, but Vienna, with a special image and, for Europe as a whole, a good location, has a prominent place, as, for obvious reasons, does Brussels.

not tourists. The problem is summarised in Eurostat (1995a: 330):

> The tourism sector comprises a wide variety of services and industries: not just accommodation, catering and travel services but also craft industries, the entertainments industry, banking, business services and local authority activities connected with national parks and historic monuments. It is therefore difficult to say exactly where the tourism sector begins and ends.

Much work is done in the tourist industry by unregistered workers, particularly during seasonal peaks, while family members may also assist.

In 1993, tourism receipts and expenditure for the EUR 12 countries were almost identical, each almost 93 billion ECU. This fact underlines the nature of tourism as a kind of trade, whereby a number of countries, and in particular certain regions within them, 'export' their attributes to people from other countries and regions, whether natural features such as sun, snow, beaches and mountains, or man-made features such as historic buildings and cultural attractions. Arguably the most positive aspect of tourism in the EU from the point of view of cohesion and convergence is the fact that there is a net flow of tourists and spending from richer countries and regions to poorer ones. In reality, however, certain privileged regions of the EU such as Mallorca, the Canarias, the Algarve, the Austrian Alps, attract a disproportionate share of all tourists, while many areas, some nearby, are hardly visited at all. Moreover, some of the money spent in tourist areas themselves ends up with regional or national governments, airlines, and shareholders in companies owning hotels.

Table 8.9 shows the receipts from tourism in 1984 and 1994 in the countries of EUR 15 and the number of tourists arriving. The data in columns (3) and (5) showing receipts and visitors in relation to the population size of the countries give a better idea of the importance of tourism to each country than the absolute amounts in the other columns. The numbers include receipts and tourists from non-EU as well as EU countries. Again, average length of stay is not clear from the data, but varies considerably, with short-stay movement between neighbouring countries of the EU for shopping and transit purposes, much of it unlikely to be officially recorded. From the data in the table, very marked differences can be seen in both receipts and visits per population. Germany languishes unloved at the lower end of the scale whereas Austria has by far the highest receipts, an asset to be added to its great locational advantage as the gateway to Central Europe and the Balkans.

Table 8.10 shows the contribution of tourism to employment and GDP in EU countries for which data are available. Austria, Greece and Spain lead the field while Germany apparently does better than on the criteria used in Table 8.9. Table 8.11 shows that Hungary, Poland and the Czech Republic now figure among the principal destinations of tourists, a warning that cheaper European holidays can now be more readily obtained outside the EU, at least while the novelty lasts and until prices approach more closely those in the EU

Table 8.9 Receipts from tourism and number of tourists visiting EU countries

	(1) (2) Receipts in billions of ECU		(3) Receipts per capita 1994	(4) Tourists arriving in millions 1994	(5) Arrivals per thousand total population 1994
	1984	1994			
Belgium	2.1	4.3	422	3.3	324
Denmark	1.6	3.0	577	1.6	308
Germany	5.4	9.0	110	14.5	177
Greece	1.7	2.8	267	10.1	962
Spain	9.8	18.0	460	43.2	1,105
France	9.6	21.1	363	60.6	1,043
Ireland	0.6	1.5	417	4.2	1,167
Italy	10.9	20.1	348	27.3	473
Luxembourg	0.1	0.2	500	0.9	2,250
Netherlands	2.1	4.7	303	6.2	400
Portugal	1.2	3.8	384	9.1	919
UK	7.8	11.8	201	19.7	336
EUR 12	53.0	100.4	286	200.7	573
Austria	6.4	11.1	1,370	17.9	2,210
Finland	0.6	1.1	216	0.8	157
Sweden	1.4	2.4	270	0.6	67

Source: Commission of the EC 1995a: 21–2

Table 8.10 Importance of tourism in relation to national economies of selected EU countries

	% share of total employment	Tourism as % of total GDP		% share of total employment	Tourism as % of total GDP
Belgium	2.0	4.0	Luxembourg	6.3	5.5
Denmark	2.7	2.5	Netherlands	2.9	2.0
Germany	6.8	5.5	Austria	14.0	14.0
Greece	10.0	8.0	Portugal	5.5	8.0
Spain	9.0	8.0	Sweden	3.3	6.0
France	5.0	2.6			

Sources: The European (1994) *Focus*, 27 May–2 June, p. 25

countries. The main countries of origin of tourists to the EUR 15 countries (except Luxembourg) are shown in Table 8.12. Germany is by far the largest source of tourists to other EU destinations, the UK the second. France is roughly in balance, while Spain and Italy receive far more tourists than they provide. In the end, about 100 billion ECU is both spent and received by EUR 15 countries, and the whole industry represents a redistribution of income and spending, generally to the benefit of the poorer EU Member States but in the last resort a case of 'taking in each other's washing', the result of the increasing tendency for more and more EU citizens to take their holidays abroad, which has also resulted in an increase

Table 8.11 Top tourist destinations in Europe, 1994

	Receipts			International arrivals in millions
	$ mln	$ per capita		
1 France	25.0	430	France	60.6
2 Italy	23.9	415	Spain	43.2
3 Spain	21.4	548	Italy	27.3
4 UK	14.0	239	Hungary	21.4
5 Austria	13.2	1,625	UK	19.7
6 Germany	10.7	130	Austria	17.9
7 Switzerland	7.8	1,114	Poland	17.6
8 Poland	6.2	159	Czech Republic	17.0
9 Netherlands	5.6	362	Germany	14.5
10 Belgium	5.2	505	Switzerland	12.6

Source: 'Why tourism is in need of some bright ideas', *The European* 24–30 Aug. 1995, p. 20

Table 8.12 Main origins of tourists to EUR 15 countries, 1992

Destination of tourists	Origin of tourists (percentage)					
	First		Second		Third	
Belgium	Netherlands	33	Germany	18	UK	12
Denmark	Germany	39	Sweden	22	Norway	10
Germany	Netherlands	21	C and E Europe	10	USA	10
Greece	Germany	31	UK	18	Italy	7
Spain	Germany	34	UK	25	France	10
France	UK	19	Italy	14	Germany	14
Ireland	UK	36	USA	20	France	10
Italy	Germany	40	France	7	UK	6
Netherlands	Germany	50	UK	11	Belgium	6
Austria	Germany	60	Netherlands	9	UK	5
Portugal	UK	29	Germany	19	Spain	10
Finland	Sweden	24	Germany	18	USA	6
Sweden	Germany	26	Norway	20	Denmark	9
UK	USA	15	Germany	10	France	9

Source: Eurostat 1995a: 333

in tourism outside Europe altogether at the expense of fellow EU countries, thanks to relatively cheap charter flights and package tours to almost every corner of the world.

Given its economic and social importance, there is increasing pressure for the development of a clearer EU policy on tourism. In Cornelissen (1993) it is stated that 'tourism should be identified as a strategic economic development priority and . . . it should be given a specific status as an industry and funded in the Community budget, corresponding to the importance of tourism for economic growth, employment and social and economic cohesion'.

THE DEFENCE SECTOR

Following the failure in the 1950s to establish a European Defence Community, defence and military affairs have long remained bastions of national sovereignty in spite of EU integration in many industrial and services sectors.

Cooperation between Member States in the defence area has been restricted to action within broader international entities such as NATO, and there are several Member States with long traditions of neutrality. Nevertheless, the fall of the Berlin Wall, the removal of the Iron Curtain and the collapse of the Soviet Union led to increased calls within the EU for greater EU control over the defence of Europe, with a corresponding cut in the role of bodies such as NATO and of the USA in particular. Defence budgets were also cut back in most Member States as a result, contributing to the concept of more efficiency through the pooling of military resources, both in terms of production and procurement.

It was against this background that the negotiators of the Treaty of Maastricht began to revitalise the concept of a shared defence entity, and the Treaty on European Union, in its section on a Common Foreign and Security Policy, calls for 'the eventual framing of a common defence policy, which might in time lead to a common defence' (Art. J 9 TEU). Member States such as the UK and France are, however, extremely protective of their own defence identity and sovereignty over decision-making in this area, and are reluctant to hand anything over to an EU level. The existence of NATO and the Western European Union (WEU) as entities that have functioned satisfactorily in the defence of Europe renders it more difficult for any new or parallel body to be created. The more ambitious Member States talk of the WEU becoming the defence arm of the EU, but several Member States are not members, and a number of NATO Member States are not in the EU.

Cuts in military spending due to the global trend towards cutting public expenditure deficits and the end of communism have led to a serious recession in the defence industry. The UK has cut its annual defence spending from 14 billion ECU to 9.5 billion ECU over the past five years, with Germany and France following suit. There are calls for the pooling of military expenditure in the form of procurement with the establishment of a European Armaments Agency (EAA) which would actively procure projects for a European defence policy. In this area France and Germany are moving ahead faster than other Member States, in particular thanks to Eurocorps, the defence corps the two countries have formed together and which became operational at the end of 1995, together with forces from Belgium and Luxembourg. Its main tasks are participation in collective defence, peace-keeping and peace-making operations, and humanitarian actions. Further progress in this area is bound to be slow because of the political importance attached to defence for historical reasons in many European countries. The role of the EU in out-of-area conflicts such as the former Yugoslavia is, therefore, still unclear, and it remains the role of the USA to provide most of the peace-enforcement effort even within the confines of Europe.

The defence industry suffered considerably in the 1990s due to the defence budget cuts referred to above, and traditional centres of employment, such as Roanne in France, where military tanks have been built, and the Northwest of England, where British Aerospace has several factories, have been hard hit, especially since little alternative employment exists in the zones concerned. One of the programmes under Objective 2 of the Structural Funds, CONVER, is aimed at assisting the zones in the EU where the defence industry is in greatest decline.

FURTHER READING

Daniels, P. (1995) 'Services in a shrinking world', *Geography*, 80 (347), Part 2, April: 97–110. The explanations for the growth of the service sector are examined. Examples from situations in Western Europe are included.

Eurostat (1991) *Le tourisme en Europe, tendances 1989*, Brussels/Luxembourg, ISBN 92–826–1852–0.

Eurostat (1995) *Europe in Figures*, 4th edition, 'Health', pp. 228–35. Background to main causes of death in EUR 12.

9

THE SOCIAL ENVIRONMENT

'Since the late 1970s a significant and disturbing shift has been taking place in the distribution of income and wealth in the U.S. The shares of total income going to different segments of the population have changed in such a way that the rich are getting richer, the poor are increasing in number and the middle class has trouble holding its own. The trend can be described as a surge towards inequality.'
Is this true also of the EU?

(Thurow 1987: 26)

- In general, given the application of the principle of subsidiarity, social issues and problems in the EU are largely left to national and local governments and to institutions such as the Church and private charities to handle and resolve.

- Nevertheless, working conditions, unemployment and poverty have been referred to in EU policy documents, albeit often more in vague and pious terms than in practical ways.

- Life-styles vary greatly from one part of the EU to another, partly the result of environmental and cultural differences, partly of differences in income.

- Negative aspects of life in different parts of the EU, of marginal concern at EU level, include the excessive consumption of alcohol, smoking, drug use, crime, and road deaths and accidents.

- The status and life-style of women in EU countries is an issue of growing concern.

- The relative and absolute increase in the number of elderly citizens throughout the EU will require changes in social security and pensions funding.

The quality of life is an elusive concept measured tentatively by such variables as levels of crime, environmental pollution and care of the elderly as well as by more conventional economic indicators such as GDP per inhabitant. Much of what is said and done about the quality of life relates to personal preferences and opinions, as is dramatically illustrated by the almost equal number of votes in the referendum for and against allowing divorce in the Republic of Ireland in November 1995. In this chapter the social environment of the EU is investigated. How have economic achievements actually affected the quality of life of the citizens of the EU? The choice of subjects covered has depended on both the availability of information and the constraints of space. For some aspects of the quality of life the spatial breakdown at national level is all that needs to be or can be discussed. Thus, for example, minimum school leaving age is a standard set at national level. On the other hand, infant mortality rates in the larger EU countries vary considerably between regions around the national average. A number of themes are developed in this chapter: material achievements measured by availability of information and car ownership, human distress, the status of women, prospects for the elderly, and the situation of non-EU immigrants.

INTRODUCTION TO SOCIAL POLICY

The establishment of a social policy at EU level has been an objective dating back to the founding treaties. The original provisions of the ECSC, Euratom and EEC Treaties contained social provisions that concentrated on the improvement of working conditions, in particular safety and hygiene, and on cooperation in the field of vocational training, for which the European Social Fund was set up. They also laid down the principles of the right of association and collective bargaining and the principle of men and women receiving equal pay for equal work. The Single European Act of 1987 added the concept of 'economic and social cohesion' as a fundamental objective, but added few specific provisions of a social policy nature.

It was only with the negotiations for the TEU that efforts were made to develop a more comprehensive social policy for the EU, in particular one based on the Social Charter adopted in 1989. The opposition of the United Kingdom to the inclusion of a Treaty chapter establishing a detailed social policy led to its adoption in the form of a protocol appended to the TEU signed by the other 11 Member States. Protocol (No. 14) on Social Policy contains a number of provisions providing for harmonised policies in areas such as the information and consultation of workers, social security and social protection of workers as well as working conditions, including working hours. The United Kingdom's opposition to what it views as continental traditions, maintaining that industry would be made less competitive as a result of such provisions, has not prevented certain directives from being adopted to implement the policy. The accession of Sweden and Austria, which have strong social policy traditions, has contributed to the further isolation of the UK on this subject.

The economic recession of the early 1990s and the continued restructuring of industry, with the resulting high levels of unemployment in Western Europe and public sector spending cuts, have reinforced calls for greater social protection and policies at EU level to promote employment and to protect what has already been achieved in social policy terms. The Intergovernmental Conference for the reform of the Treaty will be faced with further pressure from certain Member States and the trade union movement to strengthen EU policies in these areas, in particular at a time when public opinion views with scepticism the ability of the EU to address issues such as unemployment and social deprivation.

The European Commission has published numerous reports and sets of data on living conditions and social aspects of the EU. Unfortunately at the time of writing information was not generally included for the three newest members, Austria, Finland and Sweden. Nor is it customary for EU reports to include data for non-EU countries. For that reason the authors have obtained much of the material used in this chapter from Human Development Reports (HDR) of the United Nations Development Programme (UNDP), which include

Plate 9.1 Sleeping rough, location not revealed in this EU photograph. Data for the homeless are not readily available and do not figure in EU publications
Source: European Commission, Direction Générale X, Audiovisuel

material on all the countries of the world. The reader may therefore refer to the above reports to compare the EUR 15 countries with developing countries, an exercise included in Chapter 14 of the present book. In this chapter comparisons are made only with four of the most highly developed countries of the world outside the EU: Norway, Switzerland, the USA and Japan.

Before the various themes in this chapter are developed, it is appropriate to remind readers that many of the issues and problems that were prominent in Western Europe in the eighteenth and nineteenth centuries no longer exist there, but are still widespread in many parts of the developing world. Few deaths are now caused by infectious or parasitic diseases in Western

Europe but such deaths are common in the developing world. Literacy is widespread if not universal in Western Europe, whereas illiteracy is prevalent in many parts of the developing world. In *HDR 1995* (UNDP 1995) numerous problems are considered for the developing countries but not for the industrial countries, for example access to safe water and sanitation, daily calorie supply, adult literacy rate, underweight children, malaria cases. For the industrial countries the same report has information about injuries from road accidents, homicides in selected cities, adults who smoke, and the frequency of deaths from heart disease and cancer. Countries such as Mexico, Chile, the Republic of Korea and China, which are still classified as developing, are moving from the causes of human distress widely found in developing countries to those widely found or at least discussed in the EU and other developed regions. But when in Europe there is a water shortage, the outbreak of an infectious disease or a rise in mortality rates, people soon voice concern at what is unusual for them but common among poorer countries. When things go badly wrong it is not unknown for people in Western Europe to describe their own countries as Third World, hardly a fair comparison.

The matters to be discussed in the rest of this chapter cannot be seriously considered without reference to some numerical data. Thus each section has a key table, which should be referred to as each feature is discussed. It should be appreciated that while numerical data are broadly accurate, most are imprecise, since the art or science of data collecting is far from perfect, and intelligent estimates have to be made. The use or occurrence of such negative social features as drugs, murders, road accidents, smoking and the excessive consumption of alcoholic drinks are well documented in developed countries. These features are also widespread in developing countries, but data are not widely available. Other more basic issues and problems of living conditions relegate these topics to a secondary position.

ACCESS TO INFORMATION, AND CAR OWNERSHIP

The study of the world development gap and the implementation of measures to reduce it, or at least to help to keep it from widening, have focused on Gross National Product (or Gross Domestic Product, which differs only slightly – see Glossary). The GNP and GDP indices put a value to the production of goods and of services. All other things being equal, the higher the level of production of goods and services, the higher the material standard of living to be expected. The national average amount of GNP

per inhabitant does not, however, reveal differences within each country at regional, local or family level. In order to illustrate differences in the availability of material possessions and consumption in the EU at national level the subjects of availability of information and car ownership have been chosen. The items considered are produced and used throughout the EU, but there is no simple reason why, for example, Swedes and Finns should be such avid readers of newspapers whereas few Portuguese buy them.

In Table 9.1, GDP per inhabitant (in column (1) in thousands of ECU) shows great differences between EU countries according to one measure used in an official EU publication, *Regional Profiles* (1995). As would be expected, there are broad positive correlations between this variable and the remaining four, which are now discussed in turn.

Daily Newspapers

The Finns, Swedes and Norwegians stand out among the purchasers (if not readers) of newspapers. Is it a coincidence that these three countries have by far the largest stands of softwood trees per inhabitant in Western Europe, forming one of the 'newsprint capitals' of the world? Or are they particularly aware of the value of news? In contrast, the citizens of the four 'southern' countries of the EU are far less inclined to purchase newspapers.

Ownership of TV Sets

Most families in the EU now have a TV set, but Greece and Portugal again lag far behind the mainstream TV owning countries of Western Europe, while in the USA the time seems near when there will be an average of one TV set per person. TV ownership in Central Europe, the Baltic Republics and Russia is broadly on the same level as it is in the EU. On the other hand, TV ownership levels vary enormously among developing countries. Thus some small affluent

Table 9.1 Income and selected aspects of access to information and car ownership

	(1) GDP per capita	(2) Daily newspapers per 100 persons	(3) TV sets per 100	(4) Library books per person	(5) Cars per 100 persons
Belgium	16.0	31	45	5	37
Denmark	20.5	33	54	9	31
Germany	18.1	33	56	4	39
Greece	6.8	14	20	1	14
Spain	10.7	11	40	1	29
France	17.2	21	41	2	42
Ireland	10.4	19	30	4	22
Italy	15.8	11	42	1	42
Luxembourg	23.2	38	27	2	46
Netherlands	15.6	30	49	5	36
Austria	18.2	40	48	8	39
Portugal	6.6	5	19	2	19
Finland	16.3	52	51	11	39
Sweden	22.1	51	47	13	42
UK	13.8	38	44	3	33

Source: Regional Profiles 1995 for (1), UNDP 1995 for (2)–(5)
Notes: (1) Average GDP per inhabitant for 1990–2 in thousands of ECUs
      (2) Copies of daily newspapers sold per 100 people, 1992
      (3) Televisions per 100 people, 1992
      (4) Library books per person, 1988–90
      (5) Passenger cars per 100 people, 1985–89

countries such as Singapore and Oman exceed the EU average, whereas in Latin American countries the level is mostly between 10 and 20 per hundred inhabitants, and in most African countries at 2, 1 or none.

Library Books

Since reading and culture are generally (but not universally) regarded as desirable features, it is surprising to find a large disparity among the countries of the EU in the availability of books in public libraries. The Nordic countries (including Norway) make the most generous provision, while the four 'southern' countries, together with France and the UK, fall well below that level. Can one attribute some of the enthusiasm for borrowing books in the 'northern' countries to the long winter nights or is the presence of many well-stocked libraries

dispersed among the small rural settlements an influence? As with TV sets in use, the availability of books in public libraries in Central Europe and the former Soviet Union compares with the level in the better provided countries of the EU; presumably some of the stock is being replaced, some even being returned there after a rest of a few years.

Car Ownership

It is not universally agreed that the widespread use of private cars is a 'good' thing. Nevertheless, as in North America, Australia and to a lesser extent Japan, the life-style in most regions of the EU is such that the ownership of a car is now a necessity rather than a luxury. If anything, the car is more necessary for people dwelling in rural areas, and in the outer suburbs of the larger cities, than in the central parts of cities. Ever since the 1920s the USA has led the world in car ownership levels and many families own two or even three cars, although some, mainly in the central parts of large cities, do not run a car at all.

The level of car ownership in the present EU countries has risen sharply since the 1950s but Greece, Portugal and Ireland remain well behind the rest of the EU. The use of one or more cars takes up a large part of the travel expenditure of EU household budgets and the manufacture of motor vehicles is now a key branch of engineering and of industry as a whole in the larger EU countries. Yet road transport is a major cause of pollution and one of the main causes of accidents, if not deaths, among all but the elderly. Cole and Cole (1993: 181–3) calculated that for car ownership in EUR 12 to reach an average of 50 per 100 inhabitants by the year 2005, the number registered would have to rise from about 140 million in 1990 to 190 million in 2005. One can hardly expect a corresponding increase in the length and quality of roads during that time.

THE NEGATIVE SIDE OF THE SOCIAL FABRIC

Table 9.2 contains ten measures of negative aspects of EU society. Some, such as unemployment and alcohol abuse, directly affect sizeable proportions of the community, while others, such as AIDS and homicides, directly affect only very small minorities. Almost all are aspects of society in both developed and developing countries, but awareness and information about them are more extensive in developed countries than in developing ones. Each of the ten variables in Table 9.2 is discussed in turn and reference is made to non-EU Europe and to the rest of the world where it is considered appropriate.

Unemployment

Unemployment has been described, unkindly for those without jobs, as a state of mind. It might more concretely be described as the result of a particular definition and statistical exercise in each developed country. Unemployment rates are calculated for EU countries on a month to month basis, with attempts being made in EU official sources to achieve a standard definition. Like GDP per capita, the unemployment level of different regions in the EU is one of the main indicators used to determine eligibility for regional development assistance from the EU budget, whether to provide facilities for retraining or to develop infrastructure to encourage the creation of new jobs. Unemployment at NUTS level 1 and NUTS level 2 will be considered in greater detail in Chapter 11.

Unemployment levels vary greatly among the EU Member States. Indeed, in spite of standardisation in the EU, different criteria appear to be used to measure unemployment in different countries, while changes may also be made from time to time within countries. Otherwise, for example, the enormous difference between levels in Spain and Portugal is inexplicable. It would be implying a conspiracy to suggest that in some countries (e.g. Spain) the definition and counting of unemployment is actually slanted to give a high percentage whereas the less sophisticated Portuguese under-state the level. The high level in Finland in the early 1990s relates partly to the great reduction in jobs geared primarily to producing for the Soviet market.

In principle, virtually all men between official school-leaving age and retirement age, other than those in further full-time education, or carrying out national service, are eligible for employment. On the other hand, many women in the employable age band are not only not earning wages but, as housewives, not registered as unemployed unless

Table 9.2 Measures of social stress, distress, violence and crime

	(1a) Unemployment rate as % of all employment 1993	(1b) 1995	(2) Alcohol cons. per cap. in litres 1991	(3) Male adults who smoke % 1986–94	(4) Drug crimes per 100,000 1980–6	(5) AIDS cases per 100,000 people 1993	(6) Homicides per 100,000 people 1987–9	(7) Male suicides per 100,000 people 1989–93	(8) Prisoners per 100,000 people 1990	(9) Injuries from road accidents per 1,000 people 1990–1	(10) Deaths from road accidents per 100,000 people early 1990s
Belgium	9.3	14.7	9.4	35	40	2.2	n.a.	28	64	9	19
Denmark	12.4	9.8	9.9	49	176	4.6	5.7	29	47[1]	2	12
Germany	8.2	9.2	10.9	n.a.	n.a.	3.1	3.8	23	n.a.	7	15[2]
Greece	8.7	4.7	8.6	54	n.a.	1.7	1.5	6	24[1]	3	21
Spain	22.7	15.3	10.4	58	15	14.0	2.0	11	74	4	23
France	11.6	11.5	11.9	49	n.a.	9.9	4.6	30	78	4	19
Ireland	15.6	14.5	7.4	39	n.a.	1.9	n.a.	17	n.a.	3	13
Italy	11.5	11.7	8.4	46	6	8.0	4.3	11	45	4	14
Luxembourg	1.6	2.6	12.3	n.a.	n.a.	5.4	n.a.	25	n.a.	5	21
Netherlands	7.5	7.0	8.2	41	38	2.9	n.a.	14	44	3	9
Austria	3.7	4.6	10.3	33	77	2.9	2.3	32	261	8	20
Portugal	5.5	7.2	11.6	37	13	4.5	4.2	12	92	6	34
Finland	17.7	16.2	7.4	35	n.a.	0.7	2.4	45	68	3	13
Sweden	8.2	7.6	5.5	26	n.a.	2.1	1.5	22	54	4	9
UK	10.2	8.2	7.4	36	n.a.	2.8	1.6	13	77[1]	6	8[3]
Norway	6.0	5.2	4.1	42	116	1.5	0.9	21	56	3	8
Switzerland	3.7	4.0	10.7	46	129	10.4	2.5	30	54[1]	4	12
USA	6.7	5.6	7.0	30	234	25.5	8.0	20	426[1]	14	16
Japan	2.5	3.2	6.3	66	31	0.1	1.5	22	38	6	12

Sources: UNDP 1993 (6) Table 30; Eason 1994 (10) road deaths per 100,000 population; UNDP 1995 (1a) Table 27; (2), (3) Table 24; (4) Table 23; (5) Table 24; (7) Table 23; (8) Table 23; (9) Table 22; The European 16–22 Nov. 1995, p. 20 for (1b)

Notes: n.a. data not available
1 data for earlier year
2 average for West and East Germany, 12 and 23 respectively
3 England and Wales (Scotland 9.5, N. Ireland 12)

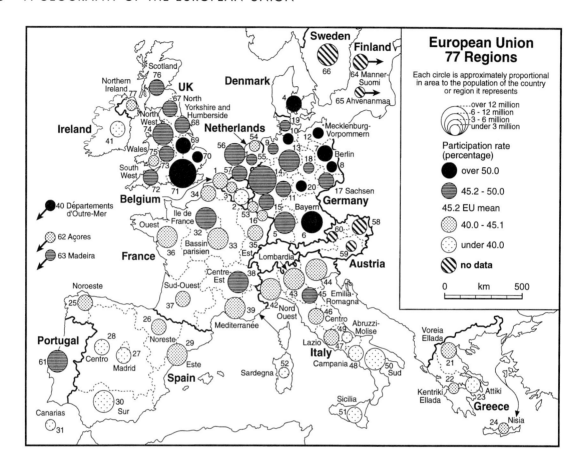

Figure 9.1 Participation rate: economically active population as a percentage of total population. See Table A1 (Appendix) for key to the numbering of unnamed regions
Source: Regional Profiles 1995

they are seeking work. The proportion of women in the appropriate age band actually working is much higher in some countries than in others, being much higher in the northern countries of the EU than in Greece and Portugal. Figure 9.1 shows the great variations among the countries of EUR 15 in the participation rate of the population. This index reflects broadly the proportion of women in employment.

The average EU unemployment level of 8.2 per cent in the mid-1990s is considerably higher than the US level of 5.6 per cent and well above the Japanese level of 3.2 per cent. It

is, however, known and widely accepted in the EU that people work in the illegal or black economy, making unemployment statistics (see also Chapter 11) far from accurate. In Central Europe, officially credited before the 1990s with near zero unemployment, the 1993 levels varied from 3.5 per cent in the Czech Republic to 15 per cent in Poland, as economies adjusted to the market economy. To compare unemployment rates in EU countries with those in most developing countries would be spurious. Information about unemployment and underemployment is incomplete, sporadic or non-existent in the rest of the world.

Alcohol Consumption

The excessive consumption of alcohol is recognised to be both a health hazard and the cause of anti-social behaviour. The (unweighted) average consumption of alcohol in the EU in 1991 was 9.3 litres per inhabitant. Apart from Luxembourg, France has the highest level, at more than twice the Swedish level. Attempts to reduce the consumption of alcohol have varied in seriousness and effectiveness among the countries of the EU. There are high taxes on alcoholic beverages in some countries, leading to extensive trans-border movements by private individuals. In Sweden (and Norway) availability is limited, while there have been anti-alcohol campaigns such as that in France in the 1950s, when French men were (in disbelief) exhorted to drink more milk by the prime minister of the time, Mendes France.

Data for Hungary, Poland and Bulgaria point to levels of alcohol consumption in Central Europe comparable to those in the EU. They are notoriously high in Russia despite attempts to cut production of alcoholic beverages during the latter years of the Soviet period. In contrast the level in Israel is on average only 0.9 litres per person per year. Perhaps the subject of attitudes and policies towards alcohol consumption should not be taken too seriously. Musto's (1996) paper 'Alcohol in American History' is summarised thus 'In the US, attitudes towards alcohol and drinking seem to oscillate between approval and condemnation over intervals of about 60 years. The medical research cited to defend each point of view tends to reflect the prevailing social opinion of the time.'

Smoking

Unlike alcohol consumption, which can be both a health and a social problem, smoking is now seen largely as a health problem, although several decades ago it was not regarded as a health problem at all and was indeed considered in some circumstances to be a social asset. The average proportion of male adults who smoked during 1986–94 in the developed countries for which data are available was 43 per cent, compared with only 23 per cent of women. With 26 per cent of adult males smoking, Sweden had the lowest rate in the EU, less than half the level of 58 per cent in Spain and 54 per cent in Greece.

To what extent the increase in smoking in the female population of many EU countries can be regarded as a measure of their emancipation is a matter of opinion. Again, however, a marked divergence occurs between Sweden and the UK on the one hand, where male and female smokers are roughly equal in proportion, and Greece, Portugal and Italy on the other, where more than twice as many men smoke as women. In Japan, the respective percentages are 66 for men against a mere 14 for women, a commentary on the gulf separating men and women or on the less stressful life-style of Japanese women.

Like the consumption of alcohol, smoking is here to stay in the EU, reasons being not only that people enjoy drinking and smoking, but also that governments derive large tax revenues from the sale of tobacco products and alcoholic drinks, while several large transnational companies and numerous small farmers have great advertising power and political influence respectively. Thus the EU budget explicitly subsidises tobacco growers in Greece and elsewhere while also supporting an anti-cancer campaign.

Drugs

Although the excessive consumption of alcohol has been described in the UK as a problem ten times as large as the use of drugs, there is great concern in the EU about the latter. Not only is drug addiction a serious healthcare problem, but drug-related crime is one of the most serious issues in many EU urban areas. For obvious reasons, precise data about the traffic in and use of drugs do not exist. The occurrence of drug crimes serves, however, as a rough guide to which countries are most afflicted by the drug problem, or at least to which countries achieve the highest rates of detection and prosecution. Among the developed countries, Australia (403 drug crimes per 100,000 population), the USA (234) and Canada (225) are the drug consuming 'capitals' of the world. Among West European countries reporting drug crime rates, Denmark and Switzerland, two of the most affluent, have the highest rates, contrasting with Italy, Portugal and Spain, which report very low levels. EU involvement in forming policy on drugs limitation and ensuring a reduction in their use has been very modest, and would itself make little impact unless carried out in consultation and collaboration with a global anti-drug effort. Ironically, according to Griffin (1995) hashish, one of the main sources of cannabis, is grown in the Rif Mountains of Morocco, not far across the Mediterranean from Spain. The European Union has traditionally turned a blind eye to the Moroccan narcotics industry, on account of the pro-Western stance of King Hassan II during the Cold War and now because he is seen as a bulwark against the growth of Islamic fundamentalism in parts of North Africa. A coherent policy in the EU on drugs is all the more difficult to establish given the extreme differences in attitude between Member States ranging from the ultra-liberal Netherlands to the highly repressive Swedes.

AIDS

Since appreciation of the nature of Acquired Immune Deficiency Syndrome (AIDS) became widespread in the early 1980s the significance and potential negative impact of this affliction has been given much publicity. Spain, France and Italy reported the highest levels of AIDS cases in the EU in 1993. The level was much lower in the northern EU countries and in Greece. The average EU level of 4.4 cases per 100,000 population in 1993 was, however, far below the US level of 25.5, but well above the Japanese level of 0.1. The latter statistic implies the virtual absence of AIDS in Japan, thanks presumably to the policy of the government to control immigration, and also to the discerning way Japanese men find sex outlets abroad. It is difficult to imagine the implementation of a strong EU policy that would reduce, if not prevent the further incidence of AIDS, until much higher levels are reached, although not, one hopes, as drastic as the highest in Africa: Zambia 239 per 100,000, Zimbabwe 86, Namibia 72, Botswana 69.

Homicides

Intentional homicide, as murder is officially defined, is a distinct if chilling description of a rare event, for which statistical information is available for most EU countries. The unweighted mean number of homicides per 100,000 for the 11 EU countries reporting was 3.1, compared with 8.0 in the USA. Since total numbers involved are small, no precise inference can be made at national level about differences in the occurrence in EU countries. Sharp contrasts do, however, occur between levels in selected EU cities (*HDR 1995*, UNDP 1995, Table 23). With 38 per 100,000 population, Amsterdam is poles apart from Madrid with 2.7 and London with 2.5. Homicide rates in the four Nordic capitals and also in Riga (Latvia) are relatively high, while by contrast Tokyo records only 1.6 per 100,000.

Male Suicides

Like homicide, successful suicide is a very rare event in relation to total population. For the record, male suicide rates for all developed countries are about three times as high as female suicide rates, a ratio roughly similar in most countries reporting suicide. In column (7), only male suicides (per 100,000 people) are included. Among developed countries, Hungary (55), Russia (53) and the Baltic Republics have the highest 'scores'. The unweighted average for the 15 countries of the EU is 21. Finland has the highest level, while the southern countries of the EU and the UK have the lowest levels. It would be inconsistent to invoke the influence of the Roman Catholic Church as a deterrent to suicide in Spain, Portugal and Italy and to ignore it in relation to the fact that these countries also have among the lowest fertility rates in the world in spite of the Church's ban on artificial contraception.

Prisoners

At first sight, the number of prisoners in relation to total population should serve as a broad guide to the intensity of criminal activity in each EU country. There is no standard procedure throughout the EU, however, for determining the application of a prison sentence or its length. Thus the enormous range between the level of 261 per 100,000 people in Austria and 24 in Greece, while a matter of some surprise, probably overstates the difference in criminal activity in the two countries. The USA is the 'prison capital' of the world, with a total of around 1.5 million people behind bars, more than 1 out of every 200 citizens.

Road Deaths/Injuries

Injuries from road accidents in the EU affect an appreciable part of the population each year and, cumulatively, the probability is that on average about one person in three will suffer some kind of injury from a road accident during her or his lifetime, whether in a vehicle or as a pedestrian. The number of deaths is of course much smaller. Car ownership varies considerably among EU countries (see Table 9.1) and is a variable that must be considered when com-

parisons are made. The UK and Norway have the lowest levels of road deaths in relation to total population, Portugal the highest. There has been an almost universal reduction throughout Western Europe over recent decades in the number of road deaths in relation to the number of vehicles in use. Indeed, in the UK the absolute number is less in the 1990s than it was in the 1940s in spite of the presence on the roads of about ten times as many vehicles. The EU Commission can fund programmes to encourage positive practices such as more rigorous regulations and the enforcement of speed limits, alcohol consumption limits, and safety features for vehicles and roads, but it cannot impose rules that would violate national sovereignty.

From the selection of variables in Table 9.2 measuring negative aspects of life in EUR 15 it is evident that there are marked differences among the countries. In the view of the authors, the inclusion in this study of many other possible variables would broadly confirm the divergence shown by the data used in this section. Some countries would score much more highly than others in, for example, illegal arms shipments, terrorist attacks, kidnapping, trafficking in stolen cars, nuclear material, babies, body parts, objets d'art, rare animals. What does emerge from the data, however, is that no particular country has high or low scores on all the variables. Thus, for example, Spain comes out badly on unemployment and frequency of AIDS but well on drug-related crimes, homicides and suicides. Austria scores favourably on unemployment but badly on drug crimes and size of prison population. The level of affluence, as measured by real GDP per capita, appears to have little overall effect on the negative side of social life in the EU. The best prospect for the future is that for each aspect the example of countries with the best standards should be emulated by the countries with the worst standards, rather than vice versa. The entry of countries from Central Europe would, however, change considerably the present social structure of the EU and complicate if not terminally damage the 'cool sequestered vale of life' along which most West Europeans are fortunate enough to pass their days.

THE STATUS AND LIFE-STYLE OF WOMEN

The disparity between the status, opportunities, obligations and rights of men and women has long been an issue in Western Europe. Earlier this century it focused, for example, on the right of women to vote. More recently there have been debates on differences between the average wages of men and women performing the same jobs, and on the right of women to be considered impartially in applications for jobs in sectors previously the preserves of men. Before some of the aspects of the gap between men and women are considered, it is appropriate to distinguish inescapable universal biological differences due to sex, sometimes referred to as gender differences, from social differences, which vary from one part of the world to another. The physical tribulations of child-bearing cannot be compared between the sexes. Again, in most countries, on average men are considerably stronger physically than women.

While there has been much speculation about mental differences between the sexes, these are difficult to establish and apparently less influential than child-bearing and physical strength. Other differences between women and men are often embedded in religious and secular ideologies and practices. Deformation of women's feet through binding was common in China up to the early part of this century, clitoral circumcision is widely practised in Africa today, and the wearing of veils by women in some Muslim countries could be regarded as discrimination. None of these practices is traditionally found in Western Europe

and all would be condemned as bad or unfair. On the other hand, it is purely a matter of opinion whether or not national parliaments should be populated equally by men and women.

The choice of variables for inclusion in Table 9.3 has been determined partly by the availability of information, partly by the desirability of obtaining a broad representation of gender issues. Before the variables in the table are examined, it is fair to point out that while the general consensus is that men get a better deal in the world than women, in reality they fare worse than women in some respects. For example, the average life expectancy at birth of women is nine years longer than that of men in Europe as a whole, and six to seven years longer in Western Europe. In the EU, about nine-tenths of the prison population is male. The suicide rate is about three times as high for men as for women. In most countries, a larger proportion of men than women smoke. In wars, almost all the deaths of military personnel are men, although among civilians there is more or less equality. It could be argued, of course, that since much more of the political and economic power has been wielded by men than by women, on the stage, at least, if not behind the scenes, men are largely to blame for their unfavourable situation.

It has been argued, also, that healthcare provision tends to be superior for the treatment of men than of women, one reason being the greater outlay on research and facilities directed to degenerative afflictions found or considered to be more prevalent among males than among females. The situation is examined by Holloway (1994) in the USA and India in particular. Such a view is difficult to substantiate conclusively without carrying out an esoteric comparison of, for example, the relative amount of research funding and effort on malignant neoplasms (cancer) unique to parts of the anatomy of females and of males.

Given the existence of basic sex differences, it

Table 9.3 The status of men and women compared

	(1)	(2)	(3)	(4)	(5)	(6)	(7)	(7a)	(8)	(9)	(10)	(11)
	Life expectancy at birth early 1990s Females	Males	Maternal mortality rate per 100,000 live births 1980–92	Average age at first marriage 1980–5	Divorce % 1987–90	Births outside marriage % 1985–9	Female school years 1990	Female–male schooling 1990	Employment of women as % of total emp. 1990	Wages of women as % of male wages 1990–1	Women in science as % of total 1988	Women in parliament as % of total seats 1991
Belgium	80	73	3	22.4	31	8	10.7	100	37	64	24	9
Denmark	78	73	3	26.1	44	45	10.3	98	40	82	23	33
Germany	79	73	5	23.6	30	10	10.6	90	38	74	28	20
Greece	80	75	5	22.5	13	2	6.5	89	31	68	28	5
Spain	81	73	5	23.1	n.a.	8	6.5	92	27	n.a.	27	15
France	82	74	9	24.3	31	26	11.7	102	37	88	n.a.	6
Ireland	78	73	2	23.4	none	12	8.8	102	27	62	30	8
Italy	80	74	4	23.2	8	6	7.3	99	31	80	32	13
Luxembourg	79	73	n.a.	23.1	37	12	9.8	95	34	65	n.a.	13
Netherlands	80	74	10	23.2	28	10	10.8	104	35	78	16	21
Austria	80	73	8	23.5	33	22	10.5	90	39	78	25	22
Portugal	78	71	10	22.1	13	14	5.2	76	41	76	35	8
Finland	79	72	11	24.6	38	n.a.	10.5	98	44	77	24	39
Sweden	81	76	5	27.6	44	52	11.1	100	47	89	25	38
UK	79	74	8	23.1	41	25	11.6	102	40	67	21	6
Norway	80	74	3	24.0	40	26	11.5	98	43	85	30	36
Switzerland	81	75	5	25.0	33	6	10.7	93	37	68	13	14
USA	79	72	8	23.3	48	27	12.4	102	42	59	n.a.	2
Japan	83	76	11	25.1	22	1	10.6	98	40	51	7	6

Sources: (1), (2) WPDS 1995 (PRB 1995); (3) UNDP 1995, Table 21; (4)–(11) UNDP 1993, Tables 28–30, 32, 33
Notes: (7) Mean years of schooling per female over 25 years of age, 1990
(7a) Mean years of schooling of females as a percentage of that of males, 1990
(10) Percentage of women in science and engineering fields (third level), 1988

is impossible in some situations to say whether women are better provided for than men. Maternal mortality rate (column (3) in Table 9.3) can be compared only between different countries, not between sexes. Again, the occurrence of divorce, whether 'good' or 'bad', cannot itself meaningfully be compared between men and women because, arithmetically, exactly the same number of men and women get divorced. On the other hand, the evidence is that, at least financially, divorced women fare worse than divorced men. The variables in Table 9.3 will now be considered.

Life Expectancy

Female and male life expectancy rates can be compared. The world average is 68 years for women and 64 years for men but the average difference is greater in developed countries, 78 and 70 years, than in developing ones, 65 and 62. In India and Pakistan there is virtually no difference in life expectancy between females and males, a fact used as evidence to show how badly women fare because they 'should' live several years longer than men. In Russia the gap has reached 13 years, a fact that might be used as evidence of the ruthless treatment of the male population during the Soviet period. The EU countries fall between these extremes, although there is some difference between Member States even here. Thus French men are relatively disadvantaged, living eight years less than women, compared with men in several other countries (e.g. Sweden and Greece each with a five year gap).

Maternal Mortality

By world standards the average maternal mortality rate in all EU countries is extremely low. Even so, concern is expressed within countries about regional and local differences. This variable reveals something about the availability and quality of health services but nothing

directly about the relative position of men and women in society. Data are not available for Central Europe or the former USSR. For what they are worth, levels in developing countries are far higher than in Europe, as for example in India and Indonesia (about 950), Ghana (1,000) and Mali (2,000).

Age at Marriage

Average age at first marriage relates partly to the general level of educational attainment and it tends to influence family size. Sweden and Denmark have the highest scores for age at first marriage in the EU, Portugal the lowest. In many parts of the developing world, age at first marriage is much lower than in Europe, reflecting the small amount of full-time education received by most women. The average age at marriage of men is generally somewhat higher than of women.

Divorce

Whether the marked increase in divorce levels in most EU countries in recent decades can be regarded as an improvement in the condition of women is debatable. What is remarkable is the great disparity in levels between EU countries. At one extreme, divorce was not permitted in Ireland at all by state law as well as Church dogma until 1995. Italy and Portugal have levels of divorce well below the rest of the EU, while the Nordic countries and the UK have the highest levels. High divorce rates are currently also recorded in most countries of Central Europe and the former USSR (Russia 42 per cent, Estonia 47) as well as in the USA (48 per cent).

Births Outside Marriage

With relaxation of divorce laws and requirements the number of births outside marriage has grown sharply in some EU countries.

Sweden and Denmark lead the way in the EU, while France and the UK have near average levels and Spain, Italy and Greece record the lowest levels. This is another variable measuring female–male relationships that in some respects may be considered to express an achievement in giving women higher status and freedom yet which financially may make single-parent families less secure and well-off on average than two-parent families.

Female Schooling

Among the developed countries of the world the mean years of schooling received by adults over the age of 25 is broadly similar for females (9.6 years) and males (10.4). In contrast, the overall level for females in developing countries is only 2.7 years, whereas for males it is 4.6. Not only are the schooling levels much lower, but the gender gap is much wider. In column (7a), the disparity between average female and average male years of schooling is expressed in relation to the male score, made to equal 100. In six of the EUR 15 countries the level is equal or the female years are longer. Only in Portugal, where women spend 5.2 years compared with men's 6.8 years, is the disparity very marked.

Employment

It has become increasingly accepted in the EU countries in the second half of the twentieth century that employment for women should be normal rather than an exception. At the same time, women are assumed to continue as the main workers in the home (see Box 9.1). Although a full-time or even a part-time job outside the home gives women greater independence and influence, it is debatable whether it is gain for society as a whole. Column (8) shows that in Sweden virtually all women are in employment of some kind, almost equalling men in total numbers. In sharp contrast, in Spain and Ireland, almost three times as many

men work as women. Convergence with respect to the proportion of women in employment throughout the EU is a long way off, and the issue is complicated by the question of unemployment, with women gradually taking jobs from men. Figure 9.1 shows marked contrasts in the participation rate among the countries of the EU.

Wages

Whatever the proportion of females to males in the workforce, the overwhelming evidence is that on average women's wages are well below those of men. Much effort has been put into ensuring that women are paid the same as men for doing the same job. On the other hand, little can be done to change the situation in which women predominate in lower paid jobs, especially in services. Moreover, in the same kind of work, men tend to get promoted more frequently to higher levels of pay. In Sweden and France, the gap between the wages of women and men is at its narrowest, while it is widest in Ireland and Belgium. In both the USA (59 per cent) and Japan (51 per cent) the ratios of women's to men's wages are much lower than in the EU as a whole (unweighted average of 75 per cent).

Science

The variable measuring the position of women in science gives a good idea of the place of women in the professions. For once, Portugal has the highest score in the EU, while the Netherlands has the lowest. In Japan, women hardly have a toe in the scientific establishment.

Parliament

With regard to politics, at least at national level, the gender gap among EU countries is very marked. Greece, France and the UK, accompanied by the USA and Japan, have little

BOX 9.1 A (GERMAN) WOMAN'S WORK IS NEVER DONE

A first-ever large-scale survey, involving 7,200 households, examining time planning in and outside German homes was produced by Germany's Federal Bureau of Statistics (Statistiches Bundesamt) in 1995. In 1992 West Germans did 77 billion hours of unpaid work, about two-thirds of it by women. Paid work done that year by both men and women was only 48 billion hours. Assuming an hourly rate of DM 11.70 (US $8.30) for household workers, a year of unpaid labour would cost $640 billion.

Regardless of whether or not they are employed outside the home, women averaged about twice as much unpaid work in the home as men, about three-quarters of it for housekeeping. Women who worked outside the home and had children still managed to carry out five hours and 20 minutes of household work daily, compared to just an hour and 20 minutes for married working fathers.

> Traditional notions of the division of labour are still strong in Germany ... while both men and women spent about ten hours daily for sleeping, eating and personal hygiene, the similarities ended there. During the remaining 14 hours of the day, men and women performed very different kinds of unpaid labour. Women, for ex-

ample, spent an average of 40 minutes daily doing laundry, compared to three minutes for men. When men perform unpaid labour, they are likely to work on their cars (six minutes per day), do repairs or perform unpaid labour in clubs, associations or political parties.

Some generalisations can be made from this information.

- Broadly similar gender disparities have been noted in other European countries, including Russia.
- In the context of the EU it is a sobering thought that not much more than half of the 'real' production of services (and some goods) is expressed in GDP totals. It would not be in the interest of the national governments of EU Member States to include more than the formal, 'necessary' amount of goods and services produced because their contribution to the EU budget is partly related explicitly to GDP per capita.
- The missing, unpaid work does not include the informal sector, the black economy, since by its nature it does not produce formal account sheets and invoices.

Source: The Week in Germany 1995: 5

more than a token representation of women in their national Parliaments. In the four Nordic countries (counting Norway), a third or more of the representatives are women.

It is only possible to arrive at a rough overall score for each of the EU countries with regard to the position of women, whether based on appropriate variables in Table 9.3 or on some

other set of variables. Even counting divorce and births outside marriage as negative features, Sweden emerges as the country in which the gender gap is narrowest, followed by the other Nordic countries. Apart from some anomalies, the southern countries have the widest gender gap in the EU. Although in principle the EU Treaty contains provisions on gender equality, the application of which is monitored by the European Commission, the moves to change come mainly from within each individual country. It is unlikely that umbrella-waving delegations of Greek women will descend upon Stockholm to battle for a reduction there in the proportion of births outside marriage. Nor did British handbag-wielding women, overtly at least, try to influence the vote on divorce in the neighbouring Irish Republic. At global level, women in the EU at least enjoy higher status and wield greater influence in relation to men than they do in many parts of the developing world.

THE ELDERLY (see Figure 9.2)

While there are elaborate and elegant bodies of theory about social inequalities, gender differences and economic growth, ageing is a simple biological fact and its implications for the population of the EU are clear and stark. Arguably the growth of the elderly population will be the greatest single socio-economic issue and problem in the EU in the twenty-first century. There is a certain irony, therefore, in the almost complete lack of an EU policy on ageing, although it is a trend affecting all 15 EU Member States in a broadly similar way. In contrast,

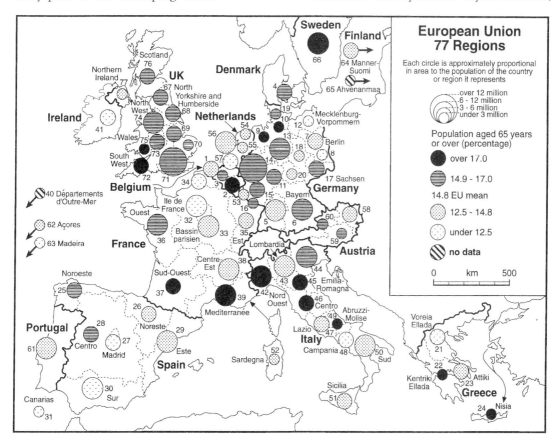

Figure 9.2 Population aged 65 years or over as a percentage of total population. See Table A1 (Appendix) for key to the numbering of unnamed regions
Source: Eurostat 1994a: 36–43

agricultural conditions, issues and problems differ sharply from country to country, yet there is a rigid, draconian Common Agricultural Policy. Although each country confronts its problem of ageing in its own way, each is monitoring the policies and experiences of other Member States.

There is no single precise definition of the age threshold beyond which a person should be considered old or elderly. The most common ages used for statistical and practical purposes are 60 or 65 years, which may also be the age at which retirement from work officially takes place and a state pension becomes available. At the centre of the problem is the fact that many people over retirement age no longer have full or part-time employment, although some (e.g. in the professions, self-employed, in farming) continue to work, while other have private 'unearned' incomes from investments and savings. Whatever the situation of the over 60s, all other things being equal, their need for general care and for healthcare in particular rises sharply with age, as degenerative afflictions such as heart complaints and cancer take hold.

The increasing proportion of elderly among the population of Western Europe in the latter part of the twentieth century results from two main causes: a socio-economic one that has lowered fertility rates dramatically (see Chapter 3) and a medical one that has lowered mortality rates throughout the age spectrum. At times, usually with a limited, sometimes local, impact immigration from outside Western Europe has brought in mainly a young element, slowing slightly the relative growth of the elderly (see Chapter 3). In Eurostat (1995a: 142), the seriousness of ageing in the EU is acknowledged in the familiar bland if not glib language of official EU publications, but no solutions are offered: 'It is the most important social phenomenon of the late twentieth century and poses a number of problems; a drop in the proportion of persons in employment will make it difficult to finance retirement pensions,

stretch social protection budgets and require an ever increasing stock of special housing etc'! In the mid-1990s the European Commission is clearly more involved with if not concerned about such issues as the environment and an integrated transportation system than about ageing.

The ageing of the population of the EU over recent decades can be measured in various ways, while reasonable estimates can be provided for the number and proportion of elderly to be expected in the next few decades, assuming no great demographic upsets. Three indicators of ageing are given in Table 9.4.

Life Expectancy

Average life expectancy at birth has increased greatly throughout Western Europe in the 20th century. Columns (1) and (2) may be compared. They show increases of between about 50 per cent (e.g. UK) and 100 per cent (e.g. Spain).

Most countries of the developing world now have levels of life expectancy well above the lowest values shown in column (1) for Europe a century ago. In the developed world, however, in spite of numerous developments in medical science and the more problematical assertion that in theory people could live far beyond the age of 70 or 80, it is unlikely that a further relative increase in life expectancy of 100 or even 50 per cent could be expected again in Western Europe in the twenty-first century, if only because the cost of providing medical facilities to keep so many elderly alive would be prohibitive.

Percent of Population over 64

A commonly used statistic in demographic sources expresses the proportion of people over age 64 as a percentage of total population. The data in columns (3)–(5) show that in all EU Member States except Ireland the percentage has risen substantially between 1968 and 1993. The trend is expected to continue well

Table 9.4 The growth of the elderly population

	(1) Life expectancy in years	(2)	(3) Percent of population over 64	(4)	(5) Estimate for 2030	(6) No. aged over 64 per 100 aged 15–64	(7)
	c. 1900	early 1990s	1968	1993		1990	est. 2025–30
Belgium	47	77	13	16	21	21	35
Denmark	55	75	12	15	23	23	37
Germany	47	76	13	15	26	23	43
Greece	n.a.	77	10	14	20	19	31
Spain	35	77	9	15	20	19	32
France	47	78	13	15	22	21	36
Ireland	49	75	11	11	15	17	28
Italy	44	77	10	16	22	20	35
Netherlands	52	77	10	13	23	19	38
Austria	40	77	14	15	n.a.	22	38
Portugal	n.a.	75	9	14	18	20	33
Finland	47	76	8	14	n.a.	20	40
Sweden	55	78	13	18	n.a.	27	36
UK	50	76	13	16	19	23	31

Sources: (1) UNDYB 1967 Table 29 Expectations of life; (2), (4) WPDS 1995 (PRB 1995); (3) WPDS 1973 (PRB 1973); (5) Coman 1995; (6), (7) The European 21–4 May 1992, p. 40

into the twenty-first century unless a dramatic change occurs in fertility level or in the number of young non-EU immigrants settling in the EU. The percentage of over 64s varies even more markedly at regional and local level than at national level. For example, according to Eurostat *Regions* (1994, Table 1.3), in 1991 against the French average of 14.1 per cent of over 64s, Limousin (NUTS level 2) had 21 per cent, but Ile de France only 10.9. In Greece (14.2 per cent), predominantly rural Kentriki Ellada (NUTS level 1) had 17.6 per cent, Attiki, with the capital, Athens, only 12.6 per cent. In Spain (13.5 per cent), Galicia (NUTS level 2) with 15.8 per cent contrasted with Canarias at 9.4. Such sharp regional differences within EU Member States reflect internal migration trends as well as differences in natural change, and illustrate the presence of many areas, often with a large agricultural population, with a large and growing element of elderly. More locally, many retirement areas in places perceived as attractive environmentally have even higher proportions of elderly population.

Over 64s per 100 15–64s

Although most children in Western Europe continue their education well beyond the age of 14 it is customary in demographic sources to separate under 15s from people aged 15–64 inclusive, with the young and the elderly regarded as dependent economically. The middle group are potentially in employment, supporting the other two groups as well as themselves. In columns (6) and (7), the ratio of people in the third group to the second group is shown for 1990 and (expected) for 2025–30. Except for Sweden, there were in 1990 about five people in the age group 15–64 for every one over 64. That ratio is expected to change markedly in the next three or four decades, reaching a staggering 1:2 in Switzerland, with Germany expected to have 43 over 64s per 100 15–64s, and, like Switzerland in 2030, about 50 per 100 in 2040. In southern

EU Member States the situation is changing in a similar fashion but with a time-lag of a decade or two.

To appreciate the problem of the elderly in the EU it is necessary to look more closely at the age composition of the over 64s. It is after the age of about 75–80 that general care and healthcare needs begin to increase sharply. Thus in 1995 there were about 4 million people over 75 years of age, while the number expected in 2050 is about 8 million (out of roughly the same size of total population). The growing proportion of over 80s is also being monitored. According to Eurostat (1995a) they accounted for 3.7 per cent of EU population in 1993 compared with 1.6 per cent in 1960. There were a mere 300 people over the age of 100 in the UK in 1951 compared with 4,400 in 1993 (3,900 of them women).

Although most of the over 64s in the EU have little or no source of income beyond their state and in some cases private pensions, they are all members of the electorate. For this reason if for no other, EU politicians at national and at local level need to think and act with care when they begin to address the problem of the elderly. Expenditure on pensions and on other measures to care for and support the elderly has started to increase relative to other sectors of national budgets with, for example, a shift in emphasis from education, as the population of children to total population diminishes in most parts of the EU. Family structure and cohesion vary among countries of the EU (see e.g. Todd (1987)). In England, the nuclear family often breaks up as the children leave home for work elsewhere in the country or to enter further education. In some EU countries it is no longer expected that elderly relatives should be cared for by their children. In terms of sheer cost, as opposed to the advantages of independence, however, it is inefficient for individual elderly people to live alone, as they lose the financial advantages of some degree of economy of scale achieved when living in a full household. In most countries of the EU traditions are changing and increasingly elderly relatives are expected to remain independent. As they get older, and are no longer able to care for themselves, they need care in a residential home, a nursing home or, at worst and most expensively, in a hospital.

According to Buscall (1994):

> The growing cost of the welfare state is threatening the economic competitiveness of countries as both employers and employees are forced to contribute more and more to sustain increasing numbers of pensioners and unemployed.
>
> This is set to worsen as pensions – already the biggest single item of government expenditure in most industrialised countries – grow with ageing populations. It is estimated that the cost of benefits for the elderly will exceed the combined price of healthcare and education in the next 50 years.

Some tinkering with pension systems has already taken place, but more drastic measures will be needed in the future. For example, the age of retirement can be raised, delaying entitlement to a state pension and forcing people to continue longer at work because many people are perfectly capable of continuing in full or part-time employment well over the age of 65. Such a measure would be self-defeating unless jobs are available. A means test could be applied, forcing the better-off (with private pensions) to make their own arrangements for retirement. 'Ageism', the difficulty experienced by older workers in getting employment once out of work, has been the subject of EU recommendations. Prejudice against older job applicants cannot be eliminated by law, but the advantages of the maturity and experience of older people can reduce age discrimination. There remains a long transition period between ages about 50 and 70 in which increasing numbers of people are taking early retirement in their fifties while considerable numbers carry out some kind of paid work (as well as voluntary work) into their seventies. It hardly needs

Plate 9.2 A gipsy settlement in central Greece. Greece has a large number of migrants from neighbouring countries, some permanent, some seasonal. The gipsies in this settlement are mainly from Albania and provide agricultural labour

pointing out that work in the home largely performed by women continues regardless of age, unrecorded, not formally remunerated, but time-consuming and arduous.

The increasing number of elderly people in EU countries is also making an impact on the employment structure of each country and region in the EU. Jenkins (1996) points out that social work is one of the few occupations in the UK in which employment has risen in recent years, together with computer industries. Between 1990 and 1995 the number employed in social work grew from 794,000 to 964,000. A considerable proportion of this work involves community care, the running of residential homes and the provision of home helps, some in the public sector, some in the private sector.

Health activities, of which a considerable proportion is devoted to the elderly members of the population, employed about 1.5 million in 1995. Care of the elderly is clearly a labour intensive activity compared with agriculture, many branches of manufacturing and communications. In the EU as a whole, according to Coman (1995), on average nearly 40 per cent of welfare spending is on pensions and healthcare for the elderly.

As and when the issue of the elderly becomes more central within EU policies greater standardisation of services will be required. In due course it should also be possible for citizens of a given EU country to retire and end their days in any other country. It is not possible to calculate with precision at present to what extent such

BOX 9.2 THE PLIGHT OF THE EU'S IMMIGRANTS: TWO EXAMPLES

Sweden

Like Switzerland, Sweden is among the developed countries of the world with the highest proportion of foreign born to total population, about 860,000, or 10 per cent of the total. Sweden is attractive to migrants from non-EU countries on account of its high standard of living, lavish welfare benefits and general image of relative freedom from nationalist and anti-immigrant attitudes. During 1989–93 alone some 200,000 foreigners entered Sweden, many of them refugees. Like France (see Box 3.1), Sweden has been willing since the Second World War to receive immigrants to work in less well paid, often menial jobs.

The attitude to foreigners is, however, gradually changing, as McIvor (1993) illustrates with reference to the cluster of high-rise flats at Fittja on the southern outskirts of Stockholm. Here about three-quarters of the 70,000 residents in the suburb of Botkyrka are either foreign or of foreign extraction.

> Tainting Sweden's international image as a successful, egalitarian society, the emergence of such ghettos threatens to destroy the country's claims to being a highly homogenised community where newcomers are painlessly assimilated . . . .
>
> Many Swedish parents have moved their children to other schools, claiming that the standard of Swedish language training is poor . . . .

There is discrimination against foreigners on the jobs market and many with high qualifications end up in menial jobs or on the dole:

> Sweden, it is half-jokingly claimed, has the best educated cleaners in the world.

Italy

Italy is among the EU countries with the lowest proportion of foreign-born citizens. Between the two World Wars Italians were discouraged from emigrating and there were few immigrants. After 1945 Italy became one of the chief sources of migrant labour for the countries of north-west Europe, but this flow has now almost ceased. Unlike France and Britain, Italy had few colonies in Africa and Asia and until recently only a trickle of non-Europeans entered the country. In the late 1980s there were about one million foreigners in Italy, less than 2 per cent of the total population. In the mid-1990s, however, about 100,000 legal or illegal immigrants were being added each year, arriving from Africa, Asia and Eastern Europe, including what was briefly an Italian colony, Albania.

Immigrants converge on the larger cities, including not only prosperous industrial centres such as Milan, Turin and Bologna in the north, but also Rome, Naples and Bari, the last two in the more backward regions of the EU. According to Endean (1995), immigration is now the most pressing issue in Italian society: 'At Cagliari, in Sardinia, youths attacked a gipsy camp with Molotov cocktails. In Reggio Emilia, residents have attacked foreign prostitutes with sticks and stones in an attempt to "clean up" the streets.' High on the agenda for Italian politicians now are problems familiar already in some other EU countries: how to cope with the competition between Italians and immigrants for jobs and accommodation; whether to tighten controls on immigration and to increase the grounds on which foreigners can be expelled. Life for Italian politicians is that bit more easy because foreigners do not have the vote. Nevertheless, Italy and to a lesser extent Greece, Spain and Portugal, all once the countries of origin of migrants to the Americas and northern Europe, now face the prospect of a long confrontation with people from the many less affluent countries in the Mediterranean region and further afield.

Source: Endean 1995

intra-EU integration would benefit the country of origin or the destination of elderly migrants but, in general, the flow seems likely to be small.

FURTHER READING

Eurostat (1991) *A Social Portrait of Europe*, Luxembourg: Office for Official Publications of the European Communities, ISBN 92–826–1747–5.

Hadjimichalis, C. and Sadler, D. (1995) *Europe at the Margins. New Mosaics of Inequality*, Chichester: Wiley. Includes issues such as gender, immigrants and capital.

Market Assessment Publications Ltd (1996) *European Lifestyles*, London (4 Crinan St., London N1 9SQ, Price £495).

Olshansky, S. J., Carnes, B. A. and Cassel, C. K. (1993) 'The aging of the human species', *Scientific American*, 268 (4) April: pp. 18–24. The authors discuss the contradiction within medical ethics, which obliges the medical profession to make efforts to postpone death thereby accelerating the ageing of the population.

Wise, M. and Gibb, R. (1993) *Single Market to Social Europe*, Harlow: Longman.

10

THE ENVIRONMENT

'Europe still has important environmental problems which go virtually unnoticed, real time-bombs ticking away . . . such as pollution of acquifers and industrial lands, degradation of forest lands by fires, the decline of urban environments and the loss of biotopes and biodiversitiy.'
Domingo Jiménez-Beltrán (Executive Director, European Environmental Agency) June 1994

- Serious concern over the need to conserve natural resources and to protect the natural and man-made environments from pollution has been expressed in EU countries only in the last two decades.

- There are numerous causes of environmental pollution including the burning of coal, oil and natural gas, emissions from industrial processes and the use of fertilisers in agriculture.

- Pollutants produced locally may be carried over great distances in the atmosphere and by rivers. Many problems are transnational in scale while some, such as the emission of carbon dioxide, are global in effect.

- The reduction and prevention of pollution is desirable and is feasible in theory but detection, the quantitative assessment of the effects and the imposition of sanctions and fines are in practice very difficult.

Environmental policy is relatively recent in Western Europe and it is hardly surprising that the founding treaties of the EC made no reference to pollution or the environment. As the subject became more topical and urgent, attempts were made to lay the foundations for such a policy at EU level, especially given the international nature of many environmental problems. The Single European Act created a special Title in the EC Treaty on environmental protection which set the following objectives:

- to preserve, protect and improve the quality of the environment;
- to contribute towards protecting human health;
- to ensure a prudent and rational utilization of natural resources.

(EC 130r (1))

The increasing importance of the environment as a global issue led to calls for the policy to be developed further through the Treaty of European Union and a further objective was added in the Treaty of Maastricht on

• promoting measures at international level to deal with regional or worldwide environmental problems.

(EC 130r (1))

ENVIRONMENTAL ISSUES AND POLICY

Even the most simple human societies have caused changes in the natural environment. In the last 2000 years Western Europe has experienced increasing pressure on its land, mineral and water resources. For example, in the sixteenth and seventeenth centuries forests were being heavily depleted in many areas through increasing demand for fuelwood and for timber for construction. According to Nef (1977), a shortage of wood in England in the sixteenth and seventeenth centuries led to the growing use of coal as a fuel. Environmental problems are not new. As early as 1877 a directive was established in Osaka, Japan, to control air pollution from blacksmiths' shops and metalcasting factories. By the turn of the century, travel in parts of London was very unpleasant on account of large quantities of horse dung in the streets and smoke from steam locomotives in the cut-and-cover underground system. In the interwar period, excessive ploughing in the Great Plains of the USA caused extensive soil erosion and 'dust bowl' conditions, producing environmental degradation on a massive scale, a process repeated in the 1950s in the 'new' grain lands of the USSR.

In the post-war UK of the 1950s, two separate events, among others, brought home the constant threat of environmental pollution and damage. Dense fog, carrying pollutants, resulted in the widespread presence of 'smog' in some British cities, causing a large number of deaths, especially among older citizens. Subsequent clean air acts introduced 'smoke-free' zones in many urban areas. Early in 1953, many coastal areas of eastern England suffered severe

floods as a result of the combination of very high tides and strong winds from the northeast and east. This event led in due course to the building of the Thames Barrier to protect London from flooding (see Gilbert and Horner 1984), and to concern over the possible impact of even a small rise in sea level on low-lying coastal areas, a concern felt even more strongly in the Netherlands. Los Angeles can take much credit for drawing attention to the influence of motor vehicles as a cause of atmospheric pollution, while Milan and Athens are examples of cities in the EU with such acute problems of pollution from traffic that at times they have been closed to motor vehicles.

In the 1970s and 1980s, thanks particularly to television and instant live viewing through satellites of places anywhere in the world, awareness of a considerable range of environmental problems has grown among politicians and the public alike. A selection of recent environmental disasters is given in Table 10.1. It could correctly be concluded even from this small sample of cases that the accidental release of fuel and toxic chemicals is among the principal causes of environmental disasters.

Man-made disasters exemplified by those in Table 10.1 attract much attention, but gradual processes of man-made environmental damage can be more serious in the longer term. For example, nitrates from fertilisers have for decades accumulated in underground water in many EU agricultural areas in sources of water supply. Some sources of pollution have a local impact only, whereas the pollutants from other sources reach further afield before being deposited, often along rivers, or through the atmosphere in the direction of wind at the time of emission. Still others, including carbon dioxide from the burning of fossil fuels, join the global atmospheric 'pool', in which gases are eventually distributed fairly evenly worldwide. The campaign in the EU to convert cars to lead-free petrol is of immediate benefit because

Table 10.1 Examples of recent energy-related and environmental disasters

Date	Disaster
1960s	Concern over damage to the natural environment in the USSR first received widespread publicity in connection with discharge from factories into Lake Baykal in Siberia. In 1959 there had been contamination from nuclear discharges over an extensive area in the Urals but this did not receive much publicity.
1967	The tanker *Torrey Canyon* spilt 120,000 tonnes of crude oil off the coast of Cornwall, England.
1974	Flixborough, England, an explosion releasing cyclohexane killed 28 people and injured 104.
1976	Seveso, North Italy, an air release of TCCP caused over 200 injuries.
1978	The tanker *Amoco Cadiz* spilt 230,000 tonnes of oil off the coast of Brittany, France.
1979	San Carlos, Spain, propylene transported by road caused 216 deaths and 200 injuries.
1979	Three Mile Island, USA, the failure of a nuclear reactor led to the eviction of 200,000 people.
1984	San Juan Ixhautepec, Mexico, the explosion of a gas storage tank and subsequent fires caused over 500 deaths and 2,500 injuries.
1984	Bhopal, India, leakage of methyl isocyanate caused approximately 2,800 deaths and 50,000 injuries.
1986	Chernobyl, USSR, nuclear reactor explosion, directly causing (only) about 30 deaths and 300 injuries but contaminating many areas of Europe.
1991	Fires at most of the oil wells in Kuwait following the Iraqi withdrawal caused extensive atmospheric pollution until all were finally extinguished in November of that year. The equivalent of roughly 1 per cent of all world energy consumption in a year was burned up.
1996	The tanker *Sea Empress* spilt 70,000 tonnes of oil off Milford Haven, Wales

the lead is diffused over only a very small area before deposition. On the other hand, the drive to stabilise and even reduce carbon dioxide emissions would be largely altruistic if the policy were to be applied only in Western Europe, which accounts for about one-seventh of the world total of carbon dioxide emitted.

The Single European Act had already set out a number of principles, which were built on by the Treaty of European Union, whereby environmental protection should be based on preventive action, should nevertheless 'take into account the diversity of situations in the various regions of the Community', and that 'environmental damage should as a priority be rectified at source and that the polluter should pay' (TEU). EU action on the environment is now concentrated in three main areas: legislation, research, and development and coordination in international fora. Legislative action is through directives on a range of environmental issues such as emission limits to combat air and water pollution, the transport and storage of hazardous waste or the setting of levels for additives in human or animal foodstuffs. R&TD is financed through a number of programmes that are aimed at the improvement of information and knowledge in the field of environmental protection. The EU also participates actively in international conventions for environmental protection in subjects as diverse as the protection of endangered species and long-range trans-boundary air pollution. Environmental protection is also an important component in aid programmes funded by the EU such as PHARE and TACIS for Central and Eastern Europe and the former Soviet Union.

The EU has also established its own European Environment Agency based in Copenhagen, the purpose of which is to provide research and information to EU institutions to promote the development of EU environmental policy and greater public awareness of the need for more protection. Many different national traditions and factors still make a comprehensive EU policy difficult to establish, with Scandinavian countries having higher standards of protection and greater public education on the environment than Mediterranean countries. Lower levels of environmental protection are one of the main concerns over the future accession to the EU of the countries of Central and Eastern Europe. These differences have led to calls for qualified majority voting to be the standard decision-making procedure for environmental policy in the Council and for greater involvement of the European Parliament in this area, since, it is argued, environmental issues transcend national boundaries and traditions. It is, therefore, one of the areas for which reforms were proposed for negotiation in the 1996 Intergovernmental Conference.

A serious problem that is unlikely to disappear quickly is the difficulty both of detecting pollution of the environment and violations, and of enforcing regulations and recommendations. Ziegler (1987) shows that even in the USSR, where before the 1990s central planners had powerful control over the economy, regulations to protect the environment were very difficult to enforce, particularly so because during the Soviet period the state 'owned' virtually all the means of production and so it was meaningless for it to fine its own enterprises. The same principle holds in market economy countries, even though the offenders are mainly privately owned enterprises or autonomous public sector concerns, rather than entities directly under state control. The 'polluter pays principle' sounds admirable, but if a company is fined for a particular environmental infringement, the cost will in due course be passed on to the consumer. As in the Soviet case, fines are usually very small compared with the cost of particular environmental damage done, even if such a cost could meaningfully be calculated with any precision. At present it will almost inevitably be much less trouble for the company causing the environmental damage to pay the fine than to pay the cost of removing the source of the pollution.

Since the late 1980s environmental issues in the EU have been complicated by two new developments, the entry in 1995 of three new Member States, Austria, Finland and Sweden, and a great increase in contacts with Central Europe and the former USSR. EFTA countries had or still have broadly similar environmental policies to those in the EUR 12 countries, with in general rather more exacting standards in some areas. In contrast, the countries of the former Soviet bloc have a range of serious environmental problems of varying scale and complexity, ranging from urban pollution to unsafe nuclear power plants and waste disposal facilities. Environmental policy existed more in theory than in application.

The unification of Germany brought home the enormous extent of pollution in the countries of the former Soviet bloc and the cost of both cleaning up existing affected areas and preventing or limiting future pollution. One of the most sensitive issues in Central Europe and the former USSR is the presence of a large number of nuclear power stations of Soviet design which, according to Rollnick (1995), include six of the most dangerous in the world. In addition to Chernobyl in Ukraine, four others are relatively close to EU Member States: Kola by the White Sea in Russia is not far from Finland; Ignalina in Lithuania is near the Baltic

Sea, Bohunice in Slovakia is close to Austria, and Kozloduy is in Bulgaria, which adjoins Greece. The sixth, at Medzamour in Armenia, has been the cause of greatest concern, particularly to neighbouring Turkey. It was damaged in an earthquake in 1988 and, in spite of acute power shortages in Armenia, was out of use until 1995 when Russian technicians brought it back into commission.

CAUSES AND TYPES OF ENVIRONMENTAL POLLUTION

For the purposes of a more detailed analysis of environmental issues in the EU, it is convenient first to identify and describe various causes and types of pollution and other forms of environmental damage. Table 10.2 shows how the Member States of the EU and four other developed countries score on the 'production' of selected polluting materials and greenhouse gases:

Columns (1)–(4) show the result mainly of the burning of fossil fuels. Apart from Luxembourg, which for its small size has a very large heavy industrial capacity, the USA leads the developed countries in per capita energy consumption and the emission of carbon dioxide. The relatively low levels of carbon dioxide emission from some countries, including France, Sweden and Switzerland, reflect their

Table 10.2 Emissions and waste

Population in millions		(1) (2) (3) (4) Carbon dioxide emissions by source, tonnes per head, 1991				(5) Sulphur and nitrogen emissions, kg per head	(6) Green-house gases per head 1991 (world median = 1)	(7) Percentage of population served by waste water treatment
		Mobile	Energy transforma-tions	Industry	Total			
10.2	Belgium	3.6	2.8	3.5	9.9	71	4.8	23
5.2	Denmark	3.1	6.6	1.3	11.0	89	5.6	98
81.7	Germany	2.2	5.0	2.1	9.1	109	5.5	86
10.5	Greece	2.4	3.4	1.0	6.8	62	3.6	10
39.1	Spain	2.1	2.0	1.2	5.3	56	3.1	53
58.1	France	2.3	1.2	1.7	5.2	46	3.4	68
3.6	Ireland	1.7	3.3	1.7	6.7	88	4.8	11
57.7	Italy	1.9	2.5	1.4	5.8	76	3.5	61
0.4	Luxembourg	9.0	3.5	14.5	27.0	63	11.4	90
15.5	Netherlands	4.3	3.7	1.6	9.6	49	4.7	93
8.1	Austria	2.2	2.1	1.9	6.2	39	3.9	72
9.9	Portugal	1.4	1.8	1.0	4.2	36	n.a	21
5.1	Finland	2.8	3.8	3.0	9.6	106	4.6	76
8.9	Sweden	2.6	1.3	1.4	5.3	59	n.a	95
58.6	UK	4.0	4.3	1.6	9.9	112	4.9	87
4.3	Norway	3.0	1.5	1.9	6.4	66	5.7	57
7.0	Switzerland	3.5	0.2	0.8	4.5	35	3.1	90
263.2	USA	5.7	7.8	3.3	16.8	154	9.0	74
125.2	Japan	2.0	3.3	2.2	7.5	17	4.8	42

Source: UNDP 1995, Table 36 Environment and pollution

extensive use of nuclear and hydro-electric power.

In column (5), sulphur and nitrogen emissions correlate fairly closely with carbon dioxide emission. In column (6), greenhouse gas emissions, measured here in relation to a world average score of 1, show how great the contribution is from the developed countries compared with that from developing ones.

In column (7) the level of treatment of waste water reflects the great need to address the problem in densely populated, highly urbanised countries such as Germany, the UK and the Netherlands. The customary concern over environmental and social problems, typical of Denmark and Sweden, contrasts with the more easy-going life-styles in Greece, Ireland and Portugal.

A number of distinct types of pollution are now considered.

Waste and Chemical Substances

The volume of waste and chemical substances processed and disposed of in the EU amounts to some 2 billion tonnes every year, about 5 tonnes per inhabitant, 30 million tonnes of which are estimated to be dangerous. Relevant legislation is mostly restricted to individual types of waste disposal considered to cause acute problems as, for example, waste from the titanium oxide industry, waste oils, and the use of sewage sludge in agriculture. In the area of waste management, the Union is working to establish guidelines to adopt clean technologies, to reprocess waste, to improve waste disposal, tighten up the transport of dangerous substances, and reclaim contaminated land. Some very dangerous chemical substances, including polychlorinated biphenyls and terphenyls (components of electricity transformers), pesticides and detergents, have been subjected to special legislation, and there are regulations regarding classification, packaging, labelling and shipment, some in force since 1967.

Some of the sectors of economic activity giving rise to the greatest concern have been identified. Seven industrial sectors have been made eligible for EU support for the introduction of clean technologies: surface treatment (e.g. galvanising, cadmium-plating), leather (e.g. chromium salts), textiles (various chemicals), cellulose and paper (e.g. pulp bleaching), mining and quarrying, the chemical industry (e.g. organo-chlorine compounds) and agri-food (e.g. effluents from sugar refineries and oil mills, fertilisers). While specific causes of serious environmental damage can often be pinpointed and dealt with through legislation, the burning of fossil fuels and wood is a problem of much greater magnitude, less amenable to control through legislation.

Water Pollution

Water pollution is of two types: that affecting fresh water on the land surface or in groundwater beneath it, and that in sea water. Both water bodies are the destination of a wide variety of discharges from industrial processes, especially from the processing of raw materials, from sewage removed from centres of population, and from oil released from ships. The issues of water quality include standards for bathing water, the suitability of fresh water for fish life, and the quality of water to be used for drinking. Various quality objectives for water were already established in the 1970s. In 1976, a blacklist of 129 particularly dangerous substances was produced, special targets being discharges of cadmium and mercury.

With regard to marine pollution, three sea areas are of particular concern to the EU: the North Sea, the Mediterranean, especially the western part, and the Baltic, with most of its coastline in the EU now that Finland and Sweden have joined. Several heavily polluted rivers flow into the North Sea and industrial waste and sewage sludge are still dumped there. The Mediterranean has been the subject of EU

[Plate 10.1] The dirtiest fuel of all: shovelling lignite briquettes into a cellar in Zwickau, former East Germany

[Plate 10.2] Twentieth-century pollution: coal-fired thermal electric station at Ratcliffe-on-Soar, England. The chimney on the right emits smoke, but the cooling towers emit only steam. The plant has recently been adapted to burn natural gas

Plates 10.1 and 10.2 Fossil fuel sources of pollution

measures against the dumping of fuel, especially from ships and aircraft. Much of the pollution in the Baltic Sea arrives via rivers from Poland and the former USSR.

Atmospheric Pollution

This pollution includes the presence and effects of the following: sulphur dioxide, nitrogen oxides, and suspended particulates mainly from the combustion of fossil fuels, especially in power stations and industrial plants, lead emissions from motor vehicles, and carbon monoxide from the incomplete burning of fuel. All these sources of atmospheric pollution are potentially harmful to human health, while nitrogen oxides in particular are thought to be the cause of acid rain, widely considered to damage Europe's forests. In 1988 the EC adopted a directive to limit the emissions of sulphur dioxide, nitrogen oxides and dust from power stations above a certain capacity, thus requiring considerable expense to provide desulphurisation and other equipment. Conces-

sions were made to certain plants burning home-produced coal with an excessively large sulphur content, an example of the EC policy of self-sufficiency in energy overriding environmental policy.

The most rigorous environmental protection targets were introduced for the six founding Member States of the EC, together with Denmark, when in 1989 standards were set for the limitation of emissions by motor vehicles, including, for example, tax incentives to reduce lead emissions. Although not strictly pollutants, the emission of carbon dioxide from the combustion of fossil fuels and the use of chlorofluorocarbons (CFCs) in industrial products give concern on account of their contribution to the 'greenhouse effect', thought to cause global atmospheric warming (see Chapter 6). Energy conservation policy in the EU is expected to discourage an increase in the consumption of fossil fuels, while the use of CFCs is to be reduced and phased out.

Sulphur oxides are one of the largest groups of air pollutants. About 100 million tonnes are emitted annually in the world from fuel processing and consumption, of which the EU and EFTA account for about 13 million tonnes. The UK comes highest in kilograms emitted per unit of GDP produced. Central Europe has a much less favourable record in relation to population size than the EU, the former GDR and Czechoslovakia having particularly bad environmental conditions. *HDR 1995* (UNDP 1995, Table 36) gives a figure of 92 million tonnes of sulphur and nitrogen emissions produced yearly by the developed countries (no figure for the rest of the world). Over half came from North America (45.7 million), just under a third from the EU (29.4 million) but only 2.2 million from Japan.

Much background material is available in the OECD publication *The State of the Environment* (1991), the publication of which illustrates growing concern with environmental issues. Europe figures prominently among the more highly industrialised regions of the world which, including the former USSR and Central Europe (neither in the OECD), have less than one-quarter of the world's population, but account for roughly three-quarters of the pollution. As efforts to contain and even reduce environmental pollution in the developed regions of the world take effect, their share of the growing world total should gradually diminish. With population growth expected throughout the developing world and economic growth hoped for in most countries, the use of fuel and materials that cause pollution is likely to increase greatly there, exceeding many times any reduction achieved in the developed countries.

Noise

Not all sources of pollution consist of tangible substances. Noise is not only a cause of discomfort but may also affect health. At EU level, a series of directives establish maximum noise levels from various types of mechanical equipment, thus achieving noise abatement at source. More drastic changes might involve the separation of residential areas from industrial sites and busy traffic routes, but such a scheme would prove extremely costly if applied universally and rigorously.

THE MOVEMENT AND IMPACT OF POLLUTANTS

Pollutants are moved from their sources not only by natural forces, mainly wind and rivers, but also by man-made conveyances. The exact direction and distance they will be carried in the air before being deposited cannot be predicted precisely, but in some areas prevailing wind directions are experienced. In contrast, movement is more predictable along rivers and, unless accidental, when organised by means of waste disposal services such as pipes, lorries and ships. EU policy is aimed to encourage the disposal of waste at the nearest appropriate location, regardless of whether it is in the country in which the waste originates.

The estimated movement of sulphur dioxide in the atmosphere from a selection of three source areas, the UK, Iberia, the Czech Republic and Slovakia (then still one country) is mapped in Figure 10.1 as an example of diffusion through the atmosphere. Sulphur dioxide can be carried hundreds of kilometres, at times even more. Thus, for example, a considerable proportion of the sulphur dioxide originating in Western and Central Europe actually reaches North Africa, itself the occasional source of a reverse flow of harmless particulates from sand storms in the Sahara Desert, reaching far northwards into Europe.

Many rivers in Europe carry large quantities of pollutants, with adverse effects on water quality, fishing and local environments along their courses, and potentially harmful to marine life and to tourist beaches near their estuaries. The courses of selected rivers of Central and

BOX 10.1 THE DOBŘíŠ ASSESSMENT

The first Pan-European Conference of Environment Ministers took place in June 1991 at Dobříš Castle, near Prague. The European Commission accepted a request from the conference to prepare a report on the state of the European environment. The result was a 700-page volume edited by D. Stanners and P. Bourdeau entitled *Europe's Environment: The Dobris Assessment*, produced in 1995 by the European Environment Agency, Copenhagen. The book contains abundant illustrative material, including numerous maps, graphs and pictures, as well as tables. It takes in the whole of Europe as conventionally defined (see Figure 1.1), reaches across the Ural mountains and into Transcaucasia, and it also includes Turkey and Cyprus.

The editors make clear the difficulties of answering the apparently simple question: how healthy is Europe's environment? They state (p. xxiv): 'a full analysis of responses and the state of actions to protect the environment was beyond the scope of the report', and 'no attempt is made at setting priorities or estimating costs or benefits of actions, alternative possibilities, or inaction'.

While economic growth, resulting from increases in production in agriculture, industry and services, is a positive achievement, at least in economic terms, measures to improve the environment and to prevent its further degradation are negative in an economic sense because nothing tangible is produced. The financial outlay and sacrifices needed are not easily quantifiable because, to cite the editors (p. 5): 'when it is stated that the environment is degraded and requires restoration, this involves a quality judge-ment and implies that something is known about what the environment should be like'.

Europe's Environment is clearly structured in six parts, with 40 chapters in total. These are as follows:

- Part I *The context* (three introductory sections).
- Part II *The assessment* (covering both the natural and man-made environments): air, inland waters, seas, soil, landscapes, nature and wildlife, the urban environment and human health.
- Part III *The pressures*: population, natural resources, emissions, waste, noise and radiation, chemicals, technological hazards.
- Part IV *Human activities* (the appropriate chapters in the present book are noted) energy (ch. 5), industry (ch. 7), transport (ch. 4), agriculture, forestry, fishing (ch. 6), tourism (ch. 8), households (ch. 9).
- Part V *Problems*: climatic change, stratospheric ozone depletion, the loss of biodiversity, major accidents, acidification, tropospheric ozone and other petrochemical oxidants in freshwater, forest degradation, coastal zone threats and management, waste reduction and management, urban stress, chemical risks.
- Part VI *Conclusions*.

The twelve environmental problems listed above in V *Problems* are highlighted in General Findings of the Report (p. 599) as being prominent issues of all-European significance, although others, such as soil erosion, which are of regional or local occurrence, should not be overlooked.

Excluding by definition the Asiatic part of the former USSR, Europe occupies less than 7 per cent of the world's land area and about 2 per cent of the total surface of the globe, but it emits into the atmosphere 35–40 per cent of CFCs emitted in the world and about 25 per cent of the sulphur dioxide, nitrogen oxides and carbon dioxide. Apart from some local achievements in the mid-1990s, little was actually being done in Europe either to clean up the environment or to bring in measures to slow down if not prevent further degradation. Even the publication in various languages and the diffusion throughout Europe of this monumental report and, more practically, of a summary, with evidence of past damage to Europe's environment and future threats to it, will not itself ensure action on the part of supranational, national and regional governments. More disasters of the magnitude of Chernobyl are (unfortunately) needed before the unpalatable job of reconciling economic growth and environmental quality and stability is seriously started.

Plate 10.3 Flooding of the River Meuse in Belgium following unusually heavy rain in eastern France, spring 1995. Further downstream large areas of the Netherlands were flooded or threatened by the Meuse, Moselle and Rhine, illustrating the international nature of some environmental problems

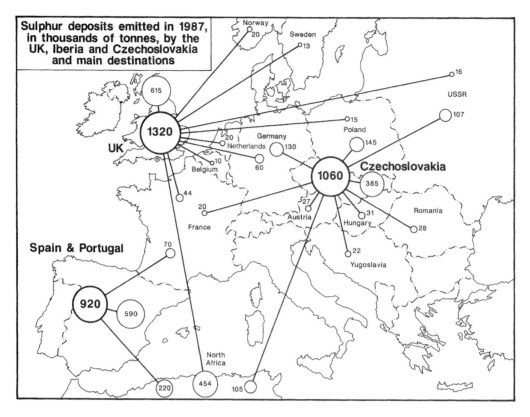

Figure 10.1 Destination of sulphur deposits emitted in the UK, Iberia and Czechoslovakia over Europe and North Africa
Source: European Parliament 1990: 162–3

Western Europe are shown in Figure 10.2, as well as major concentrations of population and industry. Some international boundaries follow the watersheds between the main drainage basins, but many basins are shared by more than one country (e.g. the Rhine, Elbe, Danube), while in places a river actually forms the boundary between two countries (e.g. the Rhine between France and Germany, the Konkamaeno and Tornealven between Sweden and Finland, the Oder between Germany and Poland). Much transfrontier pollution can therefore be expected, a good reason for dealing with pollution problems internationally or under EU legislation.

The quantity and composition of pollutants carried varies greatly from one river to another, and also along different sections of the same river. The nitrate concentrations in a selection of rivers are shown in Table 10.3 to illustrate the contrasts. The Thames has the highest score of all the rivers included in the table, while the rivers of Belgium and the Netherlands, and the Rhine in Germany, are also seriously affected. The Rhine and its tributaries drain the largest river basin in the EU, which contains almost one-fifth of the total population of the EU and an even larger proportion of the metallurgical and chemicals industries of the Union.

The Elbe, which drains a smaller basin than the Rhine, is also heavily polluted. Although its basin is in both the former East and West Germany, it actually starts in the Czech Repub-

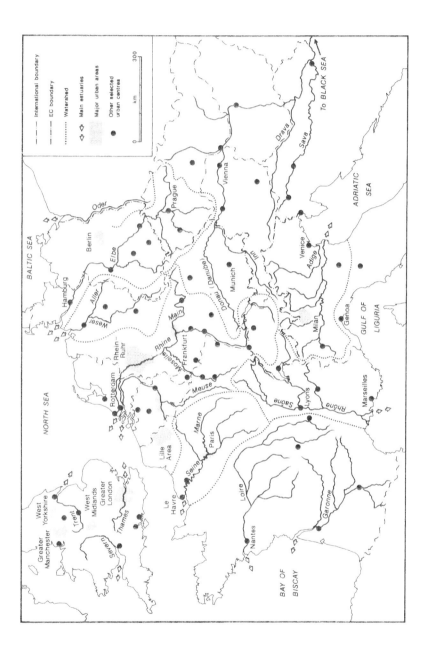

Figure 10.2 Main river basins of parts of Western and Central Europe. No distinction is made between rivers, canalised rivers and canals

Table 10.3 Water quality of selected European rivers: nitrate concentration (mgN/litre)

River	Country	MgN/litre	River	Country	MgN/litre
Thames	UK	7.67	Loire	France	2.53
Escaut-Doel	Belgium	5.06	Danube	Yugoslavia	2.42
Ijssel-Kampen	Netherlands	4.33	Rhine	Switzerland	1.76
Meuse	Netherlands	3.86	Po	Italy	1.68
Rhine	Germany	3.70	Rhône	France	1.38
Guadalquivir	Spain	3.47	Porsuk	Turkey	1.28
Mersey	UK	2.86	Gudenaa	Denmark	1.25

Source: OECD 1991: 61

lic, where it already receives industrial waste and untreated sewage from Prague and smaller cities. As a result, fish taken from the Elbe are mostly inedible, and among various pollutants the river discharges into the North Sea are mercury, copper, phosphates, nitrogen and ammonia. In North Italy, the River Po and its tributaries drain a comparatively small basin which, however, contains much of Italy's industrial capacity, together with the most intensively farmed agricultural land in the country. An excessively large quantity of phosphates originating in the farms has been blamed for the occurrence in recent years of algae on the Adriatic tourist beaches of the region of Emilia-Romagna, just south of the Po delta, giving rise to one of the worst cases of eutrophication in Europe (Marchetti 1985).

While it is obvious that environmental damage is very unevenly distributed in the territory of the EU and along its coasts and offshore seas, the variety of different ingredients that contribute is so great that it is not possible to quantify the overall occurrence accurately. A hypothetical 'density of pollution' map of the EU would show many small areas with very high scores, including some cities, badly polluted mining areas and certain stretches of river and coast, as well as extensive areas affected less seriously in some way or other, and yet other areas virtually free, including the northern parts of Sweden and Finland.

Coasts and Seas

Compared with most other large regions of the world, the countries of the EU share a very long coastline in relation to their total area, but many of the coasts, together with adjoining seas, have become environmental problem areas, and in some cases catastrophes have occurred. The sea receives pollutants carried into it by rivers, sewage and industrial waste directly from the many centres of population along the shores, and waste deliberately dumped there by purpose-built ships or spilt there accidentally.

The deteriorating state of the North Sea has attracted much attention since the 1980s. Eight of the 15 EU Member States have a coastline on the North Sea or lie in the basin of the Rhine, which drains into it: the eastern side of Britain, northeast France, the Benelux, almost all of Germany, as well as Denmark and Sweden, to which the non-EU Switzerland can be added. Indeed, about 40 per cent of the population of the EU lives in areas draining to the North Sea. Virtually every known source of contamination affects the North Sea: atmospheric pollution, especially from the UK, industrial waste and sewage via rivers or from dumping at sea, and radioactive waste. Fishing has been reduced through disease and malformation in the fish, algal growth affects marine life, and many beaches are unsuitable for visitors to use.

At the North Sea Conference in The Hague in 1990 targets were set to reduce pesticides and other hazardous chemicals reaching the North Sea. According to Bond (1995), at a conference for the protection of the North Sea held in Esbjerg, Denmark, in June 1995 it emerged that few goals had been reached. For example, of 18 highly dangerous pesticides identified for elimination, only three had been completely phased out by 1995. Germany, the Netherlands, Norway and Sweden have been more positive and successful in reducing pollution than the UK or Belgium.

In the basins of the rivers entering the Mediterranean, industrial activity is less widespread than around the North Sea, but problems are acute in many localities close to EU ports and industries, for example Marseilles and the Rhône delta area, Genoa on the Gulf of Liguria, and the northern Adriatic. Whereas the North Sea is bordered exclusively by EU Member States, apart from Norway, of the 18 countries bordering the Mediterranean, only four are in the EU – Spain, France, Italy and Greece. Most of the rest are much poorer than the southern countries of the EU.

According to Webster (1995a) most countries bordering the Mediterranean are reluctant to commit themselves to a Spanish proposal that the dumping of toxic waste should cease by the year 2005. The reasons against such a commitment are principally based on the cost to industry in particular and to the economy in general. The Mediterranean is, however, almost entirely shut off from the Atlantic Ocean and therefore particularly liable to accumulate pollutants. Much of the wildlife is at risk, while the number of tourists visiting the coasts is expected to rise from the current figure of 100 million a year to some 260 million by the year 2025.

Some 650,000 tonnes of crude oil is discharged into the Mediterranean annually, while only 20 per cent of coastal towns have plants for treating sewage. Even France, the richest country bordering the Mediterranean, has a poor record. According to Froment and Lerat (1989: 22) Marseilles, Toulon, Nice, Menton, Antibes and Ajaccio, all on the Mediterranean coast, made up half of the principal urban agglomerations of France with no *dépollution* (the application of measures against pollution) in 1980 out of over 100 considered. On the southern side of the Mediterranean population is growing quickly in all of the countries of North Africa, and industrialisation is proceeding apace.

Forests

The world's forests have been shrinking in recent decades, in spite of reforestation in many countries. In the 1980s, a great deal of publicity was given to the felling of the tropical rain forests of the world. At the same time, the effect of acid rain on the coniferous forests of northern and mountain areas of Europe and North America has caused increasing concern. According to the World Watch Institute (*Time*, 9 April 1990) about 70 per cent of the forests of the Czech Republic and Slovakia have been damaged, nearly two-thirds in Britain and around half in Germany, Italy, Norway and Poland. In one influential United Nations report, *The Human Development Report 1990* (UNDP 1990: 7, 62), the following thought is expressed in two places: 'the concept of sustainable development is much broader than the protection of natural resources and the physical environment. After all, it is people, not trees, whose future choices have to be protected.' The area of the EU defined as forest has actually increased in recent decades, but the EU has virtually no influence on the conservation of forests in the tropics other than through the banning of trade in such products as tropical hardwoods and some animal products.

Flora and Fauna

In addition to the problems of agricultural land, forests and fisheries already referred to,

Plate 10.4 Cutting peat in southwest Ireland. Peat is a low calorific fuel burnt locally in homes and also supplying electric power stations. Some is sold in garden centres. An irreplaceable habitat is gradually disappearing

the protection of wildlife, particularly in the form of hundreds of threatened species of plant and animal, receives frequent if sporadic attention, focused usually on only a local or special feature. For example, the wolf population of the EU is now very small, spatially fragmented and under threat of extinction. The population of foxes is much larger, but no more welcome in view of the propensity of foxes to carry rabies, not forgetting their predatory nature and recent tendency to scavenge in urban environments. Many local and migratory species of bird are finding their nesting areas and temporary stop-

ping places increasingly restricted. The killing of migratory birds for sport, widespread in Italy, blood sports in the UK, and even cruelty to animals in general, still common for example in parts of Spain and Portugal, are further examples of national practices regarded as undesirable in other Member States of the Union and therefore possible matters for consideration at EU level.

Agriculture

The intensive use of artificial or chemical fertilisers is causing concern in the EU on account of the pollution caused to both surface and underground water. Between the mid-1960s and the mid-1980s the consumption of nitrogenous fertilisers in the world as a whole increased fourfold. In Europe excluding the USSR, the increase was 2.5 times. Data for selected countries are given in Chapter 6. The wheat yields achieved in Europe were shown in Figure 6.9 to correlate closely with the use of fertilisers. Only in the Netherlands, which has a very high level of consumption already, and in the more environmentally conscious Sweden, has there been virtually no increase in the last 20 years. In Figure 10.3 the large extent of land in northern France affected by nitrogenous fertilisers shown includes much of the total arable area in the country.

Urban Environment

The most widespread and in many cities the most acute environmental problem is pollution from road vehicles. Since the 1970s considerable progress has been made in some EU cities in reducing sulphur dioxide concentrations. The Nordic countries led the way in improving the urban environment. Norway's Prime Minister Ms Brundtland was a particularly enthusiastic and influential advocate of the need for greater concern over the environment. Scandinavia has the advantage (in this respect) of a low

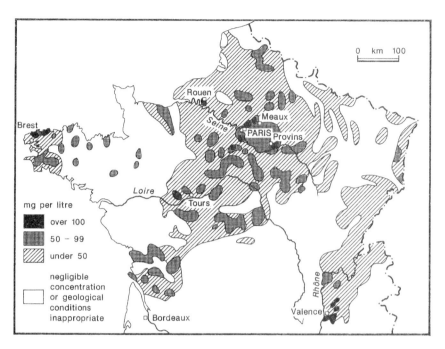

Figure 10.3 Areas of northern France affected by nitrates in subterranean water
Source: Froment and Lerat 1989, Tome 2: 208

density of population and no coal industry, whereas in contrast improvements are bound to take longer in the southern countries (eg Portugal, Greece), where the number of motor vehicles in use is still growing fast and population density is higher.

The European (14–20 April 1995, p. 6) reports that cars, taxis and motor cycles were banned in April 1995 from the historic heart of Athens, an area of 2 sq km, including the most exclusive shopping district. No such solution is possible for Paris where, according to Read (1995), a virtual state of emergency occurred on account of a strike that paralysed public transport. Car owners were warned too late not to use their vehicles to travel to work; massive traffic jams occurred: 'The alert gathered pace with broadcast warnings to the young, the elderly and the thousands of people suffering from respiratory problems not to venture outside, and to avoid physical exertion.'

Even without exceptional weather conditions or traffic hold-ups, there are marked longer term differences in pollution levels in different cities. One might expect that London would be one of the worst cities in the UK if not in Europe for pollution but according to Nuttall (1994b) that is not so. Three other cities, Glasgow, Manchester and Cambridge, exceeded roadside levels of pollution permitted under EU law. Nitrogen dioxide levels were extensively measured and the EU recommended level of 71 parts per billion was exceeded also by Bolton, Oldham, Doncaster and Grimsby, all four, like Cambridge, with populations of less than a quarter of a million.

Low level ozone pollution, which irritates throat and lungs, was extensively monitored in the UK in 1995 (Dawe 1995). During the warm summer of 1990 and on other occasions very high levels of ozone pollution have been found in rural areas, particularly on higher ground, which in several localities coincides with national parks including the Peak District, the Lake District, Dartmoor and the Highlands of Scotland.

PREVENTION AND CURE OF ENVIRONMENTAL DAMAGE

Only in the 1980s did publicity and awareness of environmental problems become widespread in the EU. Politicians have responded in various ways, in particular through the emergence of Green parties and the adoption of green agendas by mainstream parties. Even so, the task of making human activity and environmental stability compatible is only in its infancy and it is therefore difficult to estimate which issues will be foremost in the future, how problems will be tackled, and how much funding will be available.

Davies-Gleizes (1991) notes that with two decades of tough environmental legislation in the USA, US firms are better equipped financially and technologically than EU firms to carry out much of the work on environmental improvement needed in Europe. Nevertheless, a vast industry producing devices to clean the environment has already grown up in the EU, which is ahead of most regions of the world, apart from the USA, in pollution concern and control. The industry has 2.5 million people engaged in making pollution control equipment, and exports from the Union in the late 1980s were already between four and five times the value of its imports of such equipment. The cost of pollution control in the EU was about 30 billion ECU per year in the late 1980s: a necessary expenditure, but still only accounting for less than 1 per cent of the total GDP of the Union.

One regional example of a Union initiative of environmental concern was ENVIREG, which ended in 1993. This initiative addressed the environmental problems of the Mediterranean basin and other Objective 1 regions. Its aim was to demonstrate better methods of dealing with waste water in coastal areas, especially where this threatens the future of tourism, of reducing marine pollution arising from the washing of ships' bilges, and of treating industrial and other toxic wastes properly.

One difficulty of the preventive approach is to ensure that regulations are enforced. Another problem is actually to eliminate or cut the consumption of products that cause damage to the environment. On occasions a policy regarded as favourable and beneficial environmentally has had side effects that are not taken into account when the initiative is started and which may turn out to be negative. A striking example is the protection of deer from natural predators or adequate culling. The herds may grow to such an extent that their presence damages the natural vegetation, including trees, on which they depend, and which itself is valued (see Gill 1991). The widespread use of battery-powered cars (see Casassus 1991) in urban areas would eliminate emissions from the burning of petrol locally, but the increased electricity needed to power the batteries, unless obtained from solar panels in each home running such a car, would have to be generated somewhere specific, a case of transferring or exporting pollution, in this instance from the streets to the vicinity of thermal electric stations. Similarly, the replacement of tractors and other machinery driven by hydrocarbons in the agricultural sector by working animals such as horses would require the provision of large quantities of fodder throughout the year, putting extra pressure on the land and conceivably requiring extra inputs of chemical fertilisers, whereas tractors can be laid up when not in use, or used as snowploughs in Nordic countries.

Some of the measures that would have to be taken to produce a substantial reduction in the consumption of fuel and materials would not be palatable to many people in the EU. The uni-

Plates 10.5 and 10.6 The architectural heritage of the EU

[Plate 10.5] The leaning tower of Pisa in Central Italy is one of the most famous buildings in the EU. Several attempts have been made to prevent it from tilting to such an angle that it would collapse

[Plate 10.6] Italy does not have a monopoly of leaning buildings. This old house in Amsterdam in the Netherlands is one of many in low-lying, artificially drained areas of the Netherlands and northern Germany. The preservation of buildings from both natural and man-made damage and destruction could be a challenge in the EU both financially and technically in the next century. France alone has government support for some 40,000 buildings, but there are hundreds of thousands more needing attention but with no support at all

versal use of smaller, slower passenger cars might, for example, halve the consumption of motor fuel. Organic farming is now practised on about 2 per cent of the farms in Western Europe, the percentage varying from zero in some countries to 5 per cent in Austria. Several countries have targets of 10 per cent organic. If organic farming becomes much more widespread, while it will reduce the consumption of chemical fertilisers, it would also cause a decline in yields and therefore without the extension of cultivation into new areas, a smaller total of agricultural production, eventually making the EU more dependent than at present on agricultural imports. Indeed Hornsby (1991) reports a proposal by a group of English farmers to introduce compulsory nitrogen rationing in the EU with a target of a 20 per cent cut. Although such a measure would reduce overproduction and help to preserve the landscape, it is unlikely to be received enthusiastically by the majority of EU farmers. The generation of electricity from 'clean', renewable sources of energy derived from solar, wind and tidal power is also attractive environmentally (see Box 5.1) but even with the successful development of the technology, and large-scale investment, could by 2010 supply at most about 10 per cent of the quantity of electricity expected to be needed by then in the EU, according to the informed view of Lord Marshall, chairman of the CEGB in the late 1980s.

According to *HDR 1995* (UNDP 1995, Table 35), between 1965 and 1991 there was a great increase in commercial energy efficiency among the industrial countries of the world. In 1965 169 kg of oil equivalent of energy were required to produce 100 dollars of GDP compared with only 25 kg in 1991. Allowance should be made for inflation, the amount of which is not stated in the source. There were, however, big differences in the level of improvement of efficiency, as for example a tenfold reduction of energy used in the UK, Germany and Japan compared with only a threefold reduction in Greece. Rises in oil prices in 1973 and 1979 were a great incentive to importers of oil to use energy more efficiently.

One of the main sources of pollutants and greenhouse gases is the burning of fossil fuels. Out of some 7.5 billion tonnes of oil equivalent of coal, oil and natural gas consumed in the world in 1994, Western Europe accounted for about 1.15 billion tonnes or 16 per cent. The developed regions of the world, Western Europe, Central Europe and the former USSR, North America and Japan, consumed two-thirds of the world total while having little more than one-fifth of the world's population. In Western and Central Europe, and Japan, a very large quantity of fossil fuels are used in a comparatively small area.

As noted above, one way of measuring the efficiency of energy use is to compare energy consumption per inhabitant with GDP per inhabitant (in this case real GDP is used). Figure 10.4 shows this relationship. The countries of the EU and other selected countries are plotted on the graph. The ratio of energy consumption to GDP can be seen with the help of the diagonal lines. The EU average level is about 4,500 dollars of GDP per tonne of oil equivalent of energy used. Italy, Austria and Switzerland have the highest efficiency scores among the richer countries of the world while the countries of the former Soviet bloc are still among the most wasteful in the use of energy.

The amount of energy consumed continues to rise slowly in most EU countries in spite of increases in efficiency, preventing the achievement of goals set in 1992 at the Earth Summit in Rio de Janeiro to stabilise emissions of carbon dioxide. According to Born (1995a), only three of the 15 EU countries, France, Ireland and Spain, are on course to achieve targets. It should be noted, however, that Spain and Ireland (and Germany) were allowed increases of around 25 per cent (Portugal did not set a target) whereas the other countries aimed for stability or a decrease in carbon dioxide emissions. In the longer term it makes only a few years difference one way or the other when a given amount of carbon dioxide is released into the atmosphere. The idea is widespread that once the per capita consumption of energy and raw materials levels out, all the problems of the environment will vanish. For example, if the EU stabilises its annual consumption of energy at the current level of around 1.1 billion tonnes of oil equivalent, it would take 20 years for 22.2 billion tonnes to be released, compared with 17–18 years if the consumption continued

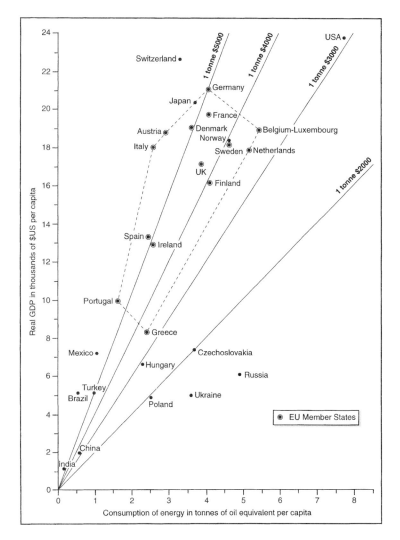

Figure 10.4 The relationship between the consumption of energy per capita and real GDP per capita. Energy in tonnes of oil equivalent, GDP in US dollars

to rise by 1–2 per cent a year, postponing the emission of a given amount of CO_2 by a mere 2–3 years. If global warming is proceeding, as many scientists now assert, and if rising sea level is one consequence, then even 20–30 per cent reductions by industrial countries by 2005 would only put off the time when the 36 members of the Alliance of Small Island States (AOSIS) find their low lying atolls submerged.

CONTINENTAL AND GLOBAL ENVIRONMENTAL ISSUES

As already stressed several times in this chapter, many sources of actual and potential damage to the pollution of the environment affect large areas and even the whole globe. To appreciate adequately the future prospects for the EU it is necessary, therefore, to consider

Plates 10.7 and 10.8 These EU photographs show that there is concern over the disposal of waste and the running of nuclear establishments. Locations are not given. *Source*: European Commission, Direction Générale X, Audiovisuel

environmental issues and prospects beyond Western Europe.

Switzerland is crossed by roads (mainly motorways) heavily used by intra-EU traffic, causing pollution, noise, and congestion. The ultimate solution for Switzerland is to 'ferry' vehicles in transit on trains through specially constructed tunnels, not a new venture since cars are already carried in the Simplon and Lotschberg Tunnels. Like Sweden and Finland, Norway is concerned primarily about the state of its forests and the threat to them from acid rain. With its large oil and natural gas industry, Norway itself is a producer of pollutants in the North Sea, while Swiss industries in the Rhine basin have more than once dispatched unacceptably large quantities of pollutants downstream into the EU.

The former German Democratic Republic deserves special mention because it has been estimated that some 80 billion ECU would be needed to ensure that EU environmental standards are introduced in its territory. Among environmental issues that had to be confronted in the GDR when it was unified with the FRG were the use of lignite as the main source of energy, unsafe nuclear power plants, excessive motor vehicle emissions, the pollution of waterways and the problem of concentrations of livestock in some small areas. In Leipzig in the former GDR, surrounded by lignite quarries, lignite-burning power stations and industries, life expectancy was six years below the average for the remainder of the former GDR. In the industrial centre of Espenhain, near Leipzig, four out of five children suffered from chronic bronchitis. The cost of rendering safe a district near Chemnitz in the extreme south of the GDR, where uranium was extracted for export to the USSR after the Second World War, is expected to be very high.

Poland and, to a lesser extent, Hungary have environmental problems similar to those already taken on board by the EU from the former GDR. The Czech Republic is of particular concern to the EU because much of its mining and heavy industry is concentrated in the extreme northwest (Bohemia), close to Germany. A Report for the European Parliament (EP 1991: 1) referred to catastrophic devastation, with a third of the country affected by sulphur pollution: 'The children of Tiplice are forbidden to play outdoors in winter and for six weeks a year schoolchildren are sent to less polluted areas.' Both lignite-fired and nuclear power stations give concern and bring problems, the former causing atmospheric pollution by burning the fuel without filtering, the latter admitted to have safety problems arising from leaks of contaminated material. Dangerous metals, untreated sewage and damage to agricultural land are other problems in the Czech Republic and Slovakia. The neighbouring Land of Bayern in Germany is adversely affected by pollution from power stations in the Czech Republic (Bohemia), and there is a plan to reduce the generation of electricity there, cutting sulphur emissions, and to top up Czech supplies from Germany itself.

Other countries of Central Europe also have serious environmental problems. Bulgaria's one nuclear power station at Kozloduy is crucial for the supply of electricity to the country but its safety record is poor and its future prospects bleak, while some of the industrial areas in neighbouring Romania have very bad environmental records. The plan to dam and divert the Danube in its middle course is a project of international scope, but one that has been the subject of much controversy. One reason why the EU is likely to be reluctant to admit all the countries of Central Europe to full membership very soon is the great cost that would fall on the richer EU Member States to improve the environment there.

If the environmental problems of Central Europe and the former USSR have only recently become a matter of direct concern to the EU, in the longer term there are likely to be even greater environmental problems elsewhere in the world, especially in many parts of the developing world. Western Europe (or even Europe as a whole) will not be able to isolate itself from global environmental issues. It is therefore appropriate to step back and to look briefly at the EU in a global context.

In the last 200 years, from the time when the Industrial Revolution was already beginning to spread from Britain to other parts of Europe, the population of the world has increased about five times. The use of fuel and raw materials has increased roughly a hundredfold during that time. The use of materials is highly concentrated, with 25 per cent of the world's population using about 75 per cent, and accounting for roughly a similar proportion of all pollution and waste. If the developing countries, almost all with fast-growing populations, are to reach or even approach the levels of consumption of energy and materials currently found in the developed countries, then the pressure on natural resources, including water and agricultural land, would become much greater than now. For example during 1984–94 the production and consumption of coal in China increased by more than 50 per cent, from 730 to 1,110 million tonnes. Droughts have been affecting parts of Africa and the interior of Asia, with implications for food production, while in Brazil and in many other tropical countries, the cutting of the tropical rain forest continues.

Global environmental issues are, however, still of greater concern in the developed than in the developing world, where governments seem less inclined to put environmental protection high on the agenda. The impact of higher environmental standards and measures such as 'green' tax on industry will be a burden on the competitive position of EU industry in the world economy unless comparable measures are applied universally. Nevertheless, the prospect that the earth's atmosphere has been getting warmer for a century or more has prompted debate on the implications of this phenomenon. Assuming that temperatures

have risen significantly and that human activities have been the cause of the rise, what impact if any will the change have on EU countries and elsewhere in the world? Two possible changes have been assumed, a change in world sea level, affecting coastal areas, and a change in temperature and precipitation, affecting the bioclimatic resources of the world. Possible changes to agricultural conditions in the EU are discussed in Chapter 6.

Although the nature and extent of climatic change in the next century is a matter of speculation, in the opinion of Jones and Wigley (1990: 73) 'Although this is obviously an unsatisfactory situation as far as policy implications are concerned, these uncertainties must not be used as excuses to delay formulating and implementing policies to reduce temperature increases caused by greenhouse gases.' In the 1980s a debate began on how to deal with such sources of greenhouse gases as fossil fuels, thus affecting EU policies on various issues (see Table 10.2). With about 7 per cent of the total population of the world, the EU and EFTA account for about 15 per cent of emissions of greenhouse gases, mainly through the burning of fossil fuels. Deforestation and agricultural practices in the developing world also make a substantial contribution. Flooding of low-lying areas by the sea and changes in rainfall and temperature (see Parry 1990) over large areas of the earth's land surface are possible future problems. There is no doubt that the environment, both in the European and global context, will continue to be a very controversial and widely debated issue in the decades ahead.

FURTHER READING

Gattrell, T. and Howes, D. (1993) 'Coming soon to your backyard', *Geographical*, LXV(12), Dec.: 30–3. As the high potential for wind power in parts of Western Europe is developed, areas of great scenic beauty are at risk. A case study from Lancashire.

Haas, P. M. (1990) *Saving the Mediterranean*, New York: Columbia UP.

Malle, K.-G. (1996) 'Cleaning up the River Rhine', *Scientific American*, 274 (1), January: 54–9 on the progress made since the 1960s in international cooperation to reduce pollution in the EU's most densely populated large river basin.

Mnatsakanian, R. A. (1992) *Environmental Legacy of the Former Soviet Republics*, University of Edinburgh, Centre for Human Ecology. An inventory of cases of pollution of various kinds in Russia and all the other 14 Soviet Socialist Republics with data mainly for the late 1980s. Of particular relevance to the EU now that Sweden and Finland are Member States. Russia's North and Northwest regions share a border with Finland. The Baltic Republics and Kaliningrad face Sweden across the Baltic Sea.

Olstead, J. (1993) 'Global warming in the dock', *Geographical*, LXV (9), September: 12–16. Is global warming occurring? If it is, does the release of carbon dioxide into the atmosphere really cause global warming?

Thornes, J. (1993) 'Last resort for the Mediterranean', *Geographical*, LXV (4), April: 24–8. The Mediterranean is under pressure from agricultural, industrial and recreational uses. Urgent action is needed to tackle the problem.

Wright, M. (1992) 'Funds in fragile places', *Geographical*, LXIV (1), January: 34–8. The environmental impact of EC funding on less developed regions can have adverse effects on ecologies.

11

REGIONAL POLICIES AND ISSUES

'The most pertinent observation that can be made with reference to the employment situation in the European Union is the high level of unemployment across the Member States, particularly when compared with the Union's principal competitors.'

(Commission of the EC 1995a: 173)

- How to narrow the gap in material levels between the more affluent and lagging regions has been a matter of unending debate since the EEC was founded.

- Differences in rates of unemployment and GDP per capita are not only great between Member States but also between individual regions in the EU.

- EU policy to reduce regional disparities is limited in impact through low funding, although mechanisms do exist for assistance.

At the beginning of this book it was emphasised that one of the policy aims of the EU is to narrow the existing gap in economic performance and in the production of goods and services between poorer and richer regions. Disparities have been shown to exist between EU Member States and also between regions at levels below state in every aspect of EU life examined so far. As implied in Chapter 1, to some extent the actual layout of the regions used to subdivide EU Member States influences the perception and interpretation of disparities. In this chapter, various aspects of the region are brought together. Two indicators of economic development, GDP per capita and level of unemployment, are examined at regional level. The causes of regional disparities in the EU are then considered in Chapter 12, measures to reduce disparities are discussed, and the influence of non-economic regional features is assessed.

REGIONAL POLICY

The EC Treaty contains a section, Title XIV, entitled 'Economic and social cohesion'. The objectives of the EU in regional policy terms are made clear:

In order to promote its overall harmonious development, the Community shall develop and pursue its actions leading to the strengthening of its economic and social cohesion.

In particular, the Community shall aim at reducing disparities between levels of development of the various regions and the backwardness of the least-favoured regions, including rural areas.

(Art. 130a EC)

In the context of regional policy, economic growth is considered to be beneficial for two reasons. First, it raises living standards in general and, second, it has been associated with economic and social convergence between Member States and between regions, an equalising process that is reduced or ceases during economic recessions. With or without concurrent economic growth, a deliberate policy to reduce regional disparities can be achieved in various ways. People can move or be moved from more backward to more advanced regions, a process that has generally been slow in the EU apart from special occasions such as the exodus of East Germans to the West in 1989–90. Alternatively, financial assistance can be provided by more advanced to more backward regions through the EU budget. Such a transfer can be made either sectorally as, for example, under the Common Agricultural Policy, or spatially, as with the transfer of resources through the Cassa per il Mezzogiorno from North to South in Italy, especially in the 1950s and 1960s, in this case a national rather than an EU initiative.

The EU Structural Funds are both regionally and sectorally based. They include the European Regional Development Fund (ERDF), the European Social Fund (ESF) and the European Agricultural Guidance section of the EAGGF. As was shown in Chapter 2, the EU budget is modest in proportion to the total GDP of the 15 Member States, and inter-regional transfers of resources are correspondingly small.

Two different ways can be used to raise consumption and living standards in disadvantaged regions through the transfer of resources: first to assist in public housing, unemployment benefits and similar ways of improving living standards or, second, to improve infrastructures. The second approach is considered preferable to subsidies because it helps lagging regions to attract capital and jobs, and enables them to increase their productivity and output of goods and services, and therefore the per capita GDP. This important concept is discussed as follows in COM 90–609 (1991): 8–4 to 8–5.

> What then will be the real effects of Community resources on recipients' production, income and employment levels? If these resources are used for consumption instead of investment in human and physical capital, barely any lasting effects on production potential, output growth and income levels can be anticipated. If instead these resources are used for additional investment in raising labour force qualifications, infrastructures and the real capital stock of firms (actions which are 'eligible' in Community terms), substantial lasting effects should materialize. It is of course for this reason that the maintenance of additionality is of crucial importance. While the direct and indirect dynamic effects of using transfers to enhance economic capacity cannot be quantified at present for data reasons, it can be taken for granted that the increase in regional GDP will exceed substantially the value of the transfer itself over the medium and longer term and help to set the weaker regions on a path to faster growth, consistent with the aim of converging economically on the stronger regions. Of course, in the light of the size of disparities described earlier, and of the time required to reduce these disparities . . . a marked relative improvement in the situation of the weaker regions remains a long term challenge, even after the doubling of the structural funds.

Various theories have been developed and policies adopted to ensure that the best use is made of resources transferred from richer to poorer regions, whether internally in Member States, or at EU level. One of the criteria taken into consideration in the allocation of resources is based on the assumption that each region should be encouraged and if necessary assisted to specialise in what it is best suited to perform (COM 90–609, 1991: 79–82), known as the concept of comparative advantage. Unfortunately some regions are better endowed overall than others in their attributes and their locations. In some cases, the best that poorly endowed regions can hope for is to be encouraged to concentrate on what they do least badly.

Plate 11.1 Abandoned farm buildings in the heart of one of Spain's extensive poor regions in the north-centre of the country. The reinforced concrete posts are close to a main road to the right of the picture

One notable characteristic of the EU has been its readiness to accept new members on several occasions since the EEC came into existence in 1958. The list and distribution of advanced and backward or problem regions in the EU has therefore changed. Before 1973, the South and Islands of Italy were the main problem regions of the EC, although there were other smaller regions with difficulties, such as Corse in France, areas in West Germany adjoining the Iron Curtain, and parts of southern Belgium. More emphasis was given then to poor agricultural regions than to industrial regions in decline.

The entry of the UK, Ireland and Denmark in 1973 brought into the EC industrial regions already in decline in the UK, together with the comparatively backward, still very agricultural,

Irish Republic. The entry of Greece (1981) and then of Spain and Portugal (1986) greatly increased the number of people in backward agricultural regions and added some stagnating industrial regions in Spain. The inclusion of the former GDR added a substantial new industrial problem region. Although they have local problem regions, Austria, Finland and Sweden in general were comparable with the more affluent existing EU Member States when they joined in 1995. With the exception of Denmark, before 1995 all the additions to the original six of the EEC have been countries with GDP levels per capita below the EC average at the time of their entry. In anticipation of Chapter 13, in which further possible new members are considered, it should be appreciated that like the former

GDR, Poland, Hungary, the Czech Republic and Slovakia are a long way below the EU level, whatever criteria are used to determine their economic conditions.

In the late 1980s the EC was already having to devise new regional policies to take into account the three new, comparatively backward, members. The entry since then of the GDR has had repercussions on the Community and serious negative effects, initially at least, on the FRG. Given the state of uncertainty and rapid change in the EU in the early 1990s, various aspects of regional features and problems may be drastically modified at short notice. Nevertheless, basic distributions and regional characteristics will change only slowly. Stuttgart, Birmingham and Milan will long remain far more highly industrialised than the Canarias, Crete or Calabria.

In order to implement policy decisions about regional inequalities it is necessary to have appropriate information about them, preferably standardised to allow comparisons between the Member States of the EU. A number of periodic reports have been published by the European Commission to describe the economic performance of the regions of the EU down to NUTS level 2. By the time material is gathered, processed and duly made available, it is usually at least two or three years out of date. In Cole and Cole (1993) extensive use was made of COM 90–609 (1991) (see pp. 230–8). For this second edition use is made of *Regional Profiles*, Sept. 1995, DG XVI-A4. Where available, data are published in that source for 17 variables at national level and at NUTS levels 1 and 2. Data for Austria, Finland and Sweden are available for only some of the variables. Data from *Regional Profiles* (1995) are reproduced in Table A1 (Appendix). They include NUTS level 1 data for GDP per capita and unemployment, the two variables on which attention is focused in this chapter (see Box 11.1 for greater detail).

BOX 11.1 VARIABLES INCLUDED IN *REGIONAL PROFILES* (1995)

1 Five variables on aspects of area and population.
2 Five variables on the overall employment.
3 Three variables on the main sectors of employment: agriculture, industry and services.
4 Four variables on GDP: Purchasing Power Standards at current prices, per total population and per person employed.

For the purposes of this chapter and the following chapter data for national and NUTS level 1 regions are given in Table A1 for five variables selected from the seventeen referred to:

1 Average GDP per inhabitant for the years 1990–1991–1992 in Purchasing Power Standards in ECUs.
2 Industrial employment as a percentage of total employment.
3 Rate of annual average change in employment in industry.
4 Participation rate (total): labour force as a percentage of total population.
5 Average of the harmonised unemployment rates for the years 1992–1993–1994.

Of the five variables, the first and the last are the focus of attention in this chapter, since they are frequently used both in EU reports and in the press to measure and determine the strength and weakness of regions. GDP and unemployment are considered in separate sections and the two variables are then compared.

GDP PER CAPITA IN THE REGIONS OF THE EU

Gross Domestic Product (GDP) measures the income generated in the Member States and regions by the resident producer units and counts the total production of goods and services in a given period. GDP is the most comprehensive single index of the economic success or lack of success (by EU standards) of a region and is used as a guide to the eligibility of the region for assistance. When GDP is used to compare regions that differ in population size, as indeed all the EU regions do, it should be

expressed in terms of the average GDP per inhabitant, whether of the whole population or of some subset of it, such as per person employed. The participation rate (percentage of population actually economically active) varies greatly between regions in the EU, so measurement of GDP against total population and against economically active population does not bear a constant relationship. Another problem in the interpretation of GDP data, especially for some lower-level regions in the hierarchy, is that people may work in one region but reside in an adjoining one (e.g. there is a net flow of commuters from Kent to Greater London and

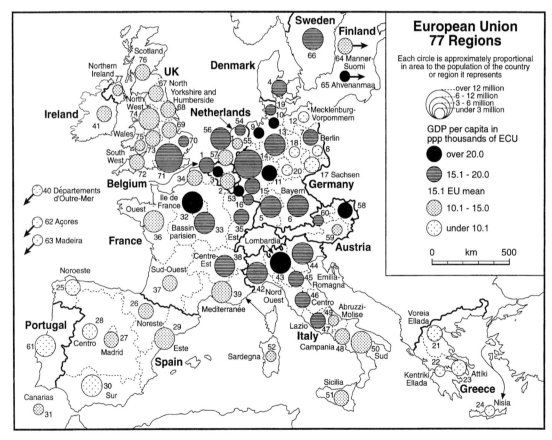

Figure 11.1 GDP per capita at NUTS level 1 in thousands of ppp ECUs. See Table A1 (Appendix) for key to the numbering of unnamed regions
Source: Regional Profiles 1995

Table 11.1 GDP per inhabitant* in four selected NUTS level 1 regions of the EU and at NUTS level 2 within them, average 1990–2

Bayern (Germany)	18.8	Southeast (UK)	17.7
Oberbayern	23.4	Greater London	22.0
Mittelfranken	19.2	Berks, Bucks, Oxon	17.5
Schwaben	16.9	Bedford, Herts	15.5
Oberfranken	16.0	Hampshire, Isle of Wight	15.2
Unterfranken	15.9	Surrey, Sussex	14.6
Oberpfalz	14.9	Kent	13.9
Niederbayern	14.5	Essex	12.6
Noord Nederland	**15.6**	**Sud (Italy)**	**10.6**
Groningen	19.9	Puglia	11.5
Drenthe	13.6	Basilicata	10.0
Friesland	13.1	Calabria	9.2

Source: Regional profiles 1995
*Note: *thousands of ECU at parity purchasing power

back). Unless steps are taken to allow for such anomalies they will contribute to the GDP of the region in which they work but will be counted against the GDP of the region in which they reside. Figure 11.1 shows the GDP per capita of the NUTS level 1 regions of the EU based on the data in Table A1.

The effects of the aggregation of population and of production into regions must now be noted. First, there is a general tendency for the extremes in GDP per capita between regions to increase in accordance with the number of regions into which the EU (or any other major region) is subdivided. This effect is enhanced by the lack of consistency in the size of the regions at each level in the hierarchy. Second, in each region there will be disparities between the subregions of which it is composed. This effect is illustrated in Table 11.1 in which GDP per inhabitant is given for four NUTS level 1 regions, Bayern (Germany), Southeast (UK), Noord-Nederland and the Sud (Italy), and for regions within them at NUTS level 2. The scores for the subregions of each NUTS level 1 region vary markedly. Further disparities exist within the NUTS level 2 regions when they in turn are subdivided.

The regions of the EU at NUTS levels 1 and 2 with high GDP per inhabitant are ordinarily net providers of assistance through the EU budget. Therefore their features and problems are of less immediate concern than those of regions with low GDP levels per inhabitant. The very 'success' of the top regions should, however, be appreciated and accounted for, since they are at levels to which more backward regions should in theory be aspiring. Some of the top regions do also have serious problems, such as the great size and congestion of London and Paris, and the heavy dependence on natural gas in Groningen (the Netherlands). The top 17 regions of the EU in GDP per inhabitant at NUTS level 2 are listed in Table 11.2 and brief comments about them are included. One group comprises large cities with sophisticated services, a considerable concentration of industry and a comparatively small surrounding area: Hamburg, Ile de France, Greater London, Bremen. Another group consists of sophisticated, highly industrialised regions with industries developed mainly in the twentieth century: Darmstadt, Stuttgart, Lombardia. Groningen and Grampian have a natural resource of exceptional importance and a small population. The Valle d'Aosta is so small (115,000 inhabitants) that it would not merit its NUTS level 1 status but for

Table 11.2 The seventeen NUTS level 2 regions with the highest GDP per capita*

	GDP per capita	Region (NUTS level 2)	Country	Features
1	29.1	Hamburg	Germany	Major seaport, city only
2	26.2	Wien	Austria	Vienna and suburbs
3	25.9	Darmstadt	Germany	Frankfurt, industries
4	25.7	Ile de France	France	Paris and suburbs
5	25.7	Bruxelles	Belgium	Brussels and suburbs
6	23.4	Oberbayern	Germany	Munich, industries
7	23.3	Bremen	Germany	Major seaport, city only
8	23.2	Luxembourg	G.D.	Services, industry
9	22.0	Stuttgart	Germany	EU's most highly industrialised region
10	22.0	Greater London	UK	London and suburbs
11	20.7	Ahvenanmaa	Finland	Islands in Baltic Sea
12	20.3	Lombardia	Italy	Milan, industries
13	20.3	Stockholm	Sweden	Stockholm and suburbs
14	20.0	Salzburg	Austria	Services
15	19.9	Groningen	Netherlands	Natural gas
16	19.8	Grampian	UK	Oil and gas
17	19.6	Valle d'Aosta	Italy	Alpine valley

Source: Regional Profiles 1995
Note: *thousands of ECU at parity purchasing power

a degree of autonomy granted to it on account of its cultural uniqueness in Italy, while Ahvenanmaa in Finland has only 25,000 inhabitants and is an archipelago of small islands.

Thus the group of regions at the top of the GDP 'league table' is rather mixed. Together the 17 regions contain 12.7 per cent of the total population of the EU. When broken down into smaller areas and then into families and individuals, it can be seen that in spite of their general affluence they contain some extremely poor people such as the homeless who sleep on the streets in London and Paris. Indeed, unemployment levels are high in some regions at the 'top', such as Ile de France (10 per cent) and Groningen (9 per cent). By definition the unemployed have no earned income and unemployment benefit alone leaves them low on the financial scale. Concentrations of immigrants from developing and Central European countries are also growing in and around several EU capital cities and ports including London, Paris, Marseilles, Naples and Berlin.

The lower end of the GDP spectrum looks very different. At NUTS level 2, 45 regions have a GDP per inhabitant of less than 75 per cent of the EU average of 15,100 ECU per capita. They are distributed by countries as follows:

Greece	all 13
Portugal	6 out of 7
Spain	10 out of 18
Ireland	1 out of 1
Germany	9 out of 40
Italy	4 out of 20
UK	1 out of 35
France	1 out of 23

None: Belgium, Denmark, Luxembourg, Netherlands, Austria, Finland, Sweden

Note: the 45 include Ceuta and Melilla (Spain), the four Départements d'Outre Mer (France) and Madeira and Açores (Portugal)

The above 45 regions have approximately 19 per cent of the total population of the EU. Most of the regions are in Greece, Portugal, southern Spain, southern Italy and former East Germany. They are more dependent on agriculture than most other regions of the EU and are generally more rural.

Just as the regions of the EU with the highest GDP per capita contain relatively poor people, although by class rather than by areal concentration, so the regions of the EU with the lowest GDP per capita contain affluent people. What is also noticeable is the low level of unemployment in most of the Greek and Portuguese regions, a result presumably (at least partly) of the recording and definition of the unemployed, which apparently differs from that elsewhere in the EU and is less rigorous in the registration of people out of work.

In Box 11.2 a simple method for measuring disparities, the Lorenz curve and gini coefficient is explained and the results of its application are discussed. Disparities in GDP per capita in various EU and other regions are compared.

BOX 11.2 MEASURING SPATIAL IRREGULARITIES

For simplicity and to facilitate comparisons, the data used in this Box are restricted entirely to GDP or GRP (Gross Regional Product). Through a series of steps and with the help of an appropriate computer program it is possible to calculate and plot a curve representing the degree of concentration of one distribution in relation to another, the Lorenz curve. The degree of concentration in relation to a perfectly even distribution can then be calculated. The index of this measure, the gini coefficient, can fall anywhere between 0, a perfectly even distribution and 1, a completely concentrated distribution. It must be appreciated that gini coefficients themselves are more useful as relative values that allow comparisons than as absolute values.

In Figure 11.2, diagram a, two Lorenz curves of fictitious distributions are compared. The whole area of the square of the graph is 25 units. The diagonal from lower left to upper right subdivides the square into two triangles each 12.5 units in area. The area between the diagonal and 'curve' A is 3.5 units of area, that between the diagonal and 'curve' B is 6 units of area. The gini coefficient of concentration is the ratio of the area between the diagonal and each curve to the area of the whole triangle, the latter expressed as 1. Thus 3.5 is 28 per cent of 12.5 so the gini coefficient of concentration for A is 0.28 while that for B is 0.48 (6 is 48 per cent of 12.5). Curve B therefore represents a situation with a much higher degree of concentration than curve A.

Before various regions in the real world are examined it must be appreciated that when spatial data are being studied, the aggregation of individuals or small groups of people such as settlements can depend upon the administrative areas used. In diagrams b and c, for example, the same population of rich, medium and poor people is subdivided into two different sets of administrative areas. In Equalia, exactly the same proportions of rich, medium and poor people are found in each unit. As a result, the gini coefficient for Equalia is 0 (see diagram e). On the other hand, in Irregula the administrative units are formed in such a way that there is a concentration of rich in one area and of poor in another. Irregula has a gini coefficient of 0.231.

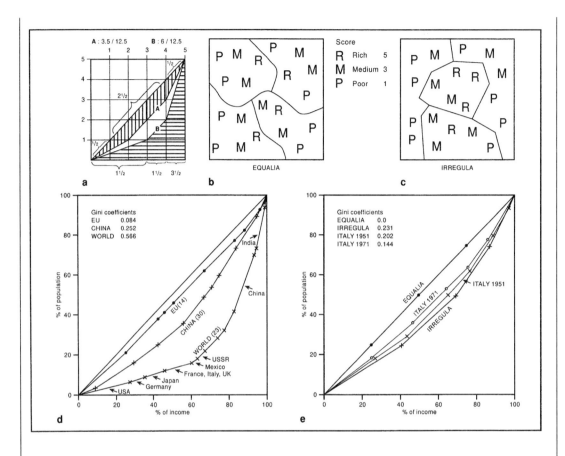

Figure 11.2 Quirks of regional inequalities resulting from the aggregation of data by areas and the gini coefficients of selected distributions from Table 11.3

Table 11.3 shows gini coefficients for the world, a selection of countries and a selection of EU units. In diagrams d and e, some of the Lorenz curves are shown graphically. Many are similar and it is therefore not possible to draw more than a few on a single graph. Table 11.4 is subdivided into five subsections,

1   As might be expected, the disparities in GDP per capita between the 23 largest countries of the world in population are very marked, and the gini coefficient is 0.566 (see WORLD (23)

in diagram d). At province level in China (also diagram d), in spite of more than four decades of Communist rule, the per capita distribution of GDP varies markedly, giving a gini coefficient of 0.252. By comparison, regional disparities in the EU (diagram d) and the USA are comparatively low.

2   Equalia and Irregula are included for comparison.

3   Situations in the EU
    At NUTS level 1, the effect of disag-

Table 11.3 Regional inequalities in GDP per capita according to gini coefficients

1	*Global and international comparisons, early 1990s data*		
1.1	World	23 largest countries in population	0.566
1.2	China	30 province level units	0.252
1.3	EU	14 at national level (Belgium includes Luxembourg)	0.084
1.4	USA	9 major regions	0.047
2	*Effects of varying spatial aggregation*		
2.1	Equalia	4 units	0.000
2.2	Irregula	4 units	0.231
3	*Effects on EU data of differing numbers of units, early 1990s data*		
3.1	EU	NUTS level 1	0.168
3.2	EU	National level plus 4 original Visegrad	0.166
3.3	EU	National level, 12 countries before 3 newest additions	0.087
3.4	EU	National level, 14 countries (Belgium includes Luxembourg)	0.084
3.5	EU	7 groups of countries	0.075
3.6	EU	3 groups of countries, regionally grouped	0.073
3.7	EU	Original 6 countries	0.031
3.8	EU	3 groups of countries, biased to give low coefficient	0.023
4	*Contrasts within largest EU countries at NUTS level 1*		
4.1	Italy	11 units	0.169
4.2	Germany	17 units	0.162
4.3	France	8 units	0.110
4.4	Spain	7 units	0.101
4.5	UK	11 units	0.076
4.6	West Germany	11 units	0.063
5	*Italy through time*		
5.1	Italy	NUTS level 2, 1951 20 units	0.202
5.2	Italy	NUTS level 2, 1971 20 units	0.144
5.3	Italy	NUTS level 2, 1990 20 units	0.171

gregation is evident (see 3.1). The coefficient of 0.168 shows greater concentration when 72 NUTS level 1 regions are considered than the coefficient of 0.084 when the 14 countries (3.4) (Belgium includes Luxembourg) are considered. The effect of adding Poland, the Czech Republic, Slovakia and Hungary (see 3.2) to the 14 EU countries referred to above is also to increase the degree of concentration to 0.166. On the other hand, the addition of the three newest members to the EU has made little difference (0.087 before, 0.084 after). 3.5–3.8 show the way different groupings of the countries of the EU give different coefficients.

4 Disparities within individual EU countries

Italy and post-unification Germany have the greatest internal diversity among the larger EU states, France and Spain are more uniform, while the smallest degree of concentration is found in the UK and in West Germany *before* unification.

5   The comparison of distributions through time is illustrated with the example of Italy at NUTS level 2. Between 1951 and 1971 there was a sharp drop in the level of regional inequality from 0.202 to 0.144, to some extent, at least, the result of the efforts to improve conditions in the South and Islands through the Cassa per il Mezzogiorno. Between 1971 and 1990, however, the level of regional equality in Italy has increased considerably, with stagnation in the industrial sector of the South and Islands but continuing industrial growth in the North and most of the Centre. The accession of Greece, Spain and Portugal has diverted EU attention and funding from the poorer parts of Italy, before 1981 the main backward region of the whole Community.

UNEMPLOYMENT

One of the main aspects of regional policy in the EU has been to reduce unemployment levels overall and to reduce regional disparities. Until recently it has been difficult to compare unemployment levels between different countries of the EU because, as argued by the authors elsewhere in this book, definitions have not been consistent. In this section the data used to measure unemployment are from *Regional Profiles* (1995), in which 'Harmonised unemployment rates, comparable between Member States' are given.

The average level of unemployment for EUR 15 countries during 1992–4 was 10.4 per cent. This compares with an unweighted mean for December 1995 of 10.0 per cent. Four non-EU developed countries with market economies had much lower levels than the EU average for that month: USA 5.6 per cent, Norway 4.1, Switzerland 4.0 and Japan 3.2. In the EU, only Greece and Austria, both with 4.7 per cent unemployment, fell within that range, while in particular Finland at 16.5, Spain 15.4 and Belgium 14.4 greatly exceeded the levels in the four non-EU countries given above.

Figure 11.3 shows differences in the EUR 15 countries at NUTS level 1. Regional disparities in the level of unemployment can be appreciated better at NUTS level 2 than at NUTS level 1. For EU policy-makers, with convergence and cohesion a high priority, it is bad news that at NUTS level 2 the disparity in unemployment levels is between extremes of 2.5 per cent for Vorarlberg in western Austria and 31.2 per cent for Andalucía in southern Spain. The extremes are thus considerably more marked at regional than at national level (Austria 3.9 per cent, Spain 21.5 per cent). Table 11.4 shows the NUTS level 2 regions with the highest and lowest levels of unemployment in 12 of the EUR 15 countries; the remaining three are not subdivided. This table may be compared with Table 10.5 in Cole and Cole (1993: 237). Within countries, particularly the five largest in population, there are marked differences in level. Even over a few years, unemployment scores in NUTS level 2 regions can change considerably, as for example with Berkshire, Buckinghamshire and Oxfordshire, which in 1990 had a level of 2.2 per cent, the lowest in the UK in that year, yet in 1992–4 had an average of 6.4 per cent. Even so, the broad contrasts in the EU as a whole change only gradually.

[Plate 11.2] A demonstration about unemployment in Ireland

Plates 11.2 and 11.3 Unemployment in the EU illustrated by EU pictures in Ireland and Italy
Source: European Commission, Direction Générale X, Audiovisuel

[Plate 11.3] Less demonstrative unemployed in Italy, the SA on the car number plate indicating Salerno in Campania, one of the regions in the EU with the highest unemployment among young people

The general area of the EU with the lowest unemployment levels consists of most NUTS level 2 regions in southern Germany, Austria and northern Italy, to which Switzerland can be added. Before unification, the regions of East Germany officially also had very low unemployment levels, but in the process of transformation to a market economy, levels have risen dramatically. The low unemployment heart of the EU extends into the Czech Republic, which had only 2.8 per cent in 1995, but levels are much higher in Poland, Slovakia and Hungary. Elsewhere in the EU, unemployment levels are below the EU average in the rest of former West Germany, the Netherlands, Sweden, much of the UK and also Greece and Portugal.

In contrast, almost all the regions of Spain, southern Italy, the former East Germany and Finland have higher than average levels of unemployment, as do some regions in the

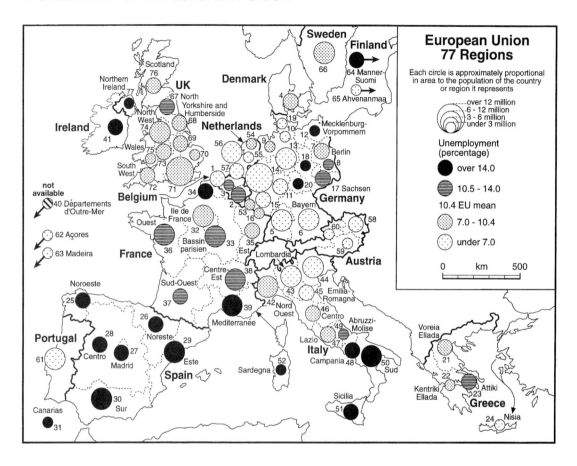

Figure 11.3 Unemployment as a percentage of total economically active population at NUTS level 1. See Table A1 (Appendix) for key to the numbering of unnamed regions
Source: *Regional Profiles* 1995

UK, Ireland and France. In general, unemployment tends to be highest either in areas in which agriculture is relatively prominent (e.g. Ireland, southern Italy) or in industrial areas in which mining, heavy industry and textiles predominated in the past if not now. At a more local and detailed level, it is not uncommon to find adjoining regions with markedly different levels. Thus for example in Belgium, West-Vlaanderen has 4.8 per cent unemployment but neighbouring Hainaut, a mining and heavy industrial area, has 13.5 per cent. In Italy, Lazio, which contains Rome, has 10.2 per

cent unemployment, whereas adjoining Campania, with Naples, has 20.2 per cent. Similarly, Unterfranken, with 4.5 per cent, faces Thüringen, with 15.0 per cent, across the former Iron Curtain.

It is not possible to generalise rigidly about the relationship between unemployment and GDP per capita levels, prevailing types of economic activity or the influence of large cities. Thus for example, in the case of capital cities, Madrid and Berlin have lower unemployment levels than the regions adjoining them whereas Wien (Vienna) and Attiki (Athens) have the

Table 11.4 Unemployment (percentage) at national level and in regions with the lowest and highest scores in each country

	(1) under 25s	(2) All	(3) Lowest		(4) Highest	
Austria	n.a.	3.9	Vorarlberg	2.5	Wien	5.1
Portugal	9.9	5.7	Centro	3.4	Alentejo	9.2
Netherlands	13.8	6.5	Utrecht	5.3	Friesland	8.1
Germany	n.a.	7.5	Oberbayern	3.3	Mecklenburg-V.	16.9
Greece	n.a.	7.5	Ionia Nisia	3.6	Attiki	10.4
Sweden	n.a.	7.9	Smaland M.O.	6.0	Mellersta N.	8.9
Belgium	16.9	8.2	West-Vlaanderen	4.8	Hainaut	13.5
UK	15.5	10.0	Grampian	5.6	N. Ireland	15.0
Italy	31.0	10.4	Trentino-A.A.	3.6	Campania	20.2
France	20.9	11.1	Alsace	6.9	Languedoc-R.	14.8
Finland	n.a.	16.4	Ahvenanmaa	5.1	Pohjois-Suomi	19.1
Spain	37.1	21.5	Navarra	12.9	Andalucía	31.2

Source: Regional Profiles 1995; Eurostat 1994a, Table 11.4 for column (1)
Note: Not subdivided: Denmark 10.3 (under 25, 11.5), Luxembourg (G.D.) 10.4 (5.1), Ireland 15.4 (27.8). Contrast the level of 2.6 per cent given for Luxembourg (G.D.) for Dec. 1995 in *The European*, 14–20 Dec. 1995, p. 20.
n.a. – no data available

highest levels in Austria and Greece respectively. The regions containing the major ports of Germany, the Netherlands and Belgium have comparatively low levels of unemployment, Merseyside (Liverpool) the highest level in the UK apart from Northern Ireland.

The reasons for the differing unemployment levels in the EU can to some extent be explained by job losses in different sectors of the economy. Well before the EEC came into existence, almost everywhere in Western Europe employment in the agricultural sector (including forestry and fishing) had been in decline (see Chapter 6). Where employment in agriculture is still at a high level (by EU standards), notably in most of the regions of Spain, Portugal, southern Italy, Greece and Ireland, further job losses can be expected in the decades to come. Since the Second World War and particularly since the early 1970s there has also been a net loss of jobs in mining (especially of coal) and some branches of manufacturing. In particular, the older industrial regions of Western Europe, especially those originally located on or near coalfields (see Chapter 5), have

experienced heavy job losses. Until the 1980s most regions of the EU enjoyed an increase in employment in the service sector but the increasing use of machines in offices and in retail outlets has led to job losses also in the service sector. Even so, the number of economically active people in the EU has increased steadily over the last four decades, particularly thanks to the increasing number of women in full- or part-time jobs.

In Eurostat *Regions* (1994) the level of unemployment of persons under the age of 25 is given for 1993. Column (1) in Table 11.4 shows the national average for this variable for countries with data available. The EU average is 19.1 per cent, almost twice the level of 10.4 per cent for all age groups, while the EU level for those of 25 years and over is 8.7 per cent. From the data in columns (1) and (2) of Table 11.4 it can be seen that the ratio of unemployment among under 25s to total unemployment differs markedly from country to country. In Italy, for example, it is about three times as high, in the UK only 50 per cent higher. The absolute value of 37.1 per cent for Spain points to an acute shortage of new jobs. In some countries the level of unemployment among under 25s at NUTS level 2 is very high in some regions compared with national levels. In Belgium, for example, the figure for Hainaut is 30.4 per cent, compared with only 8.9 per cent in adjoining West-Vlaanderen. In Spain, approaching half of under 25s were unemployed in 1993 in some regions: Andalucía and País Vasco both 45.8, Asturias 45.5, Extremadura 43.2. Even in Spain's main industrial region, Catalunya, the level was 32.5 per cent. In southern Italy the level is even higher, exceeding 50 per cent in Campania (58.4), Sicilia (56.9), Calabria (55.2) and Basilicata (53.4).

Although data were not available for the new Länder of Germany and therefore not available for Germany as a whole, they are available for the Länder of former West Germany. In complete contrast to Spain and Italy, the level of unemployment of under 25s was close to the overall level of unemployment, mostly between 3 and 6 per cent. Insofar as unemployment among under 25s reflects prospects for the economy in the future, Germany is apparently better placed than the rest of the EU, while the UK, the Netherlands and Portugal emerge relatively favourably. The youngest cohort of job seekers is not the only part of the workforce to have difficulty in finding employment in many regions of the EU. For older people seeking work, especially men, there is some degree of discrimination, with age a drawback (see Chapter 9).

In comparison with the USA and Japan, most

Plate 11.4 An ERDF scheme for retraining in Ireland
Source: European Commission, Direction Générale X, Audiovisual

of the EU countries perform very badly with regard to unemployment. The constant drive to increase efficiency in many sectors of the economy means that jobs are shed, a procedure now taking place in both the private sector and the public sector. It is difficult to compensate for such job losses without defeating the purpose of achieving greater efficiency. A cosmetic 'remedy' is possible through a redefinition of unemployment or a narrowing of the age limits between which people are 'eligible' to be counted as unemployed, lowering for example the age of retirement. To introduce measures to restrict the employment of women would be unacceptable. In general the northern countries of the EU have higher participation rates than the southern countries, indicating a greater proportion of women in work (see Chapter 9) but, paradoxically, most EU regions with high unemployment levels actually have low participation rates already.

Given the frequent steep 'gradients' over short distances in the EU between unemployment levels in different regions, there are many situations in which in theory people could be expected to migrate from regions with high unemployment to regions with low unemployment, as happened after the Second World War, when many southern Italians moved to northern Italy and beyond. Unless more jobs materialise, however, the reduction in the disparity in unemployment levels between different regions would not alter the total number of jobless in the EU. Although various measures have been introduced to enable people to seek employment anywhere in the EU, for many the prospect of living in a country with a different language and culture is a disincentive, while the loss of accommodation and the sheer stress and expense of moving are serious obstacles.

New jobs are, of course, being created all the time in the EU, but if they result from private investment, whether from sources in the EU itself or from non-EU countries such as the USA and Japan, there is no guarantee that regions with high unemployment (and/or low GDP per capita) will be preferred and chosen. Only through the public sector in each country and through the EU budget can jobs be created in regions with high unemployment.

GDP AND UNEMPLOYMENT COMPARED

In order to represent the complex economic situation in the EU visually, GDP per capita and percentage unemployed have been plotted together graphically in Figure 11.4 (p. 286). The vertical axis measures the GDP per capita of each country and NUTS level 1 region, the horizontal axis their unemployment level. Thus for example region 1, Vlaams Gewest in Belgium, has scores of 16.3 (16,300 ECU) on the vertical axis and 6.1 on the horizontal axis. It is shown by a B and a 1 on the graph. Region 30, the Sur of Spain, has scores of 9.2 and 30.1 respectively, and appears at the extreme lower right of the graph. The position of each of the 15 Member States is shown by a circle with a cross inside. Denmark, Ireland, Luxembourg, Portugal, Finland and Sweden are not subdivided at NUTS level 1 in the data source used and each, therefore, is a single NUTS level 1 region in its own right. The remaining nine countries are subdivided and a different letter is used for the regions of each country. Regions are numbered on the graph as in Table A1 (Appendix), but the overseas regions of France and Portugal, included in Table A1, are not shown on the graph. All countries are named either on the main graph, or on the inset graph, upper right, which enlarges the most crowded parts of the main graph, and selected regions are also named where space allows.

In order to show the position of selected countries in what might be referred to as economic space, polygons have been drawn on the graph joining the outer regions of the relevant countries. Germany has been represented by

BOX 11.3 THE GREAT UNEMPLOYMENT DEBATE OF 1995–6

As noted in the text, before the 1980s unemployment was so low in most parts of the EEC except southern Italy that it was not a major issue or policy area. Levels subsequently rose sharply in various countries, although not simultaneously everywhere in the EU. The entry of Spain, of the GDR, with heavy job losses following unification, and of Finland, have added countries that had very high levels in the mid-1990s, and have thus caused increasing concern about unemployment in national governments and EU institutions alike.

In 1990 unemployment in EUR 12 was 8.3 per cent (Com 90–609, Table A). It had risen in 1993 in EUR 15 (including therefore the three members to be) to 10.6 per cent (*Regional Profiles* 1995), at which level it stood also in 1995. A slight drop to 10.1 per cent was recorded for the period around January 1996 (*The European*, 11–17 April 1996, p. 21). No less disturbing, in 1995 youth unemployment was 20 per cent in EUR 15, with over 40 per cent in Spain, but only 8 per cent in Germany (*EP News*, 10–14 July 1995).

The big countries of the EU tend to carry more weight and receive greatest publicity when things go wrong. Thus in both Germany (4 million unemployed, 11 per cent, and with little change during 1995–6) and France (3 million, 12 per cent, and a serious strike) political leaders committed themselves to creating more jobs. Unemployment was just as high in Italy, but more a fact of life, while in the UK it was near 8 per cent. The fact that within the EU, Austria, Portugal, Sweden and Luxembourg all have levels appreciably lower than those in four of the five largest countries, is a matter for thought.

Similarly Norway (4.3), Switzerland (4.6), the USA (5.5) and Japan (3.3) are all well below the German and French levels. For what it is worth, it should be recalled that between the two world wars much higher levels of unemployment were experienced in the industrial countries. During 1931–3, for example, the level in the UK exceeded 20 per cent.

Early in 1996 Chancellor Kohl called for unions, employers and the government to work out incentives to businessmen to help to create 2 million new jobs in Germany by the year 2000. As elsewhere in the EU, however, in Germany various branches of both industry (e.g. mechanical engineering, construction, motor vehicles, electronics) and services (e.g. retailing) have actually recorded heavy job losses in 1995, while in most cases production has actually risen. A less ambitious but still substantial proposal by President Chirac of France was put forward in 1995 during his election campaign to create 250,000 new jobs in 1996 for the young.

How, then, can unemployment be cut? Examples earlier this century in Europe include the drafting of millions of men into the armed forces in Italy and Germany in the 1930s, and the management of employment by the state in the USSR from the early 1930s, with state planning achieving full employment for men and about 90 per cent employment for women, whatever the cost: inefficiency, overmanning, the 'key lady' on each floor in hotels syndrome. Less drastic measures include keeping more young people in education, lowering the retirement age, and excluding one of two married (or unmarried) partners from employment.

But creating new jobs can be costly, as for example the £1.7 billion to be invested in Newport, South Wales, by the LG company of South Korea to produce consumer electricals. About 6,000 new jobs will be created at an average cost of £280,000 per job. In addition, up to £30,000 is needed from British government support for each job. One figure for the EU average is much less, 60,000 ECU per new job. Are there to be a few highly paid new jobs or more numerous low paid jobs? Public works such as the creation of the trans-European networks (TENs), or collecting litter along the highways and in cities? Should new jobs be created in regions with the highest levels of unemployment, even if they are relatively prosperous, or in the poorest regions?

In the context of the policy of ensuring fair competition without subsidies from national governments, job creation schemes could violate such a policy if implemented at national level. Job creation could be overseen by the European Investment Fund, but the resources of this fund are very limited. It has been proposed either that the fund should be greatly enlarged or that somehow part of the 350 billion ECU a year paid out in unemployment and related benefits in the EU could be used to create jobs.

A further twist in the job creation debate is the prospect that the raising of funds from taxes and other sources would jeopardise the progress required in most EU Member States to allow them to satisfy the criteria for joining the single currency. What is worse, it has been estimated that up to 10 million jobs could be lost through the introduction of the single currency, raising the unemployment level in the EU to about 15 per cent. The economic cake is expected to grow gradually in the rest of the 1990s, but if unemployment is reduced quickly, there will be more slices of the cake but many of them will be thinner. In *The Merchant of Venice*, Shylock's fortunes foundered over the problem of making the correct cut.

two distinct polygons (region 7, Berlin, has not been included), while the regions of Italy are scattered over such a large swathe of the graph that their polygon would be too confusing to draw. Without Ile de France, France occupies a small space around the means of the two axes, as (see inset graph) does Great Britain.

The closer on the graph the members of any pair or larger cluster of countries or regions are, the more similar they are on the basis of the two variables. Of the 15 countries, the UK and Italy are situated very close to the mean centres of the two variables, while six other countries are fairly close. Of the original six members of the EEC, only Luxembourg is not in this 'central area' of the economic space. The remaining five are accompanied by the UK, Denmark and Sweden. The other seven countries of the EU are situated well away from the centre. Whereas most of the regions of France and the UK are near to the centre of the graph, those of Germany and Italy fall into two distinct subsets, with only Berlin (7) in Germany and Abruzzi-Molise (49) in Italy close to the centre of the whole graph. The very high level of unemployment in Spain pulls all its regions far to the right of the mean percentage of unemployment of 10.4 per cent for the whole of the EU.

Intuitively one might expect any particular region to be similar to its neighbouring regions and to nearby ones in geographical space, but there are many exceptions to this tendency. For example, Ireland and Northern Ireland (UK) are very similar, but they both also closely

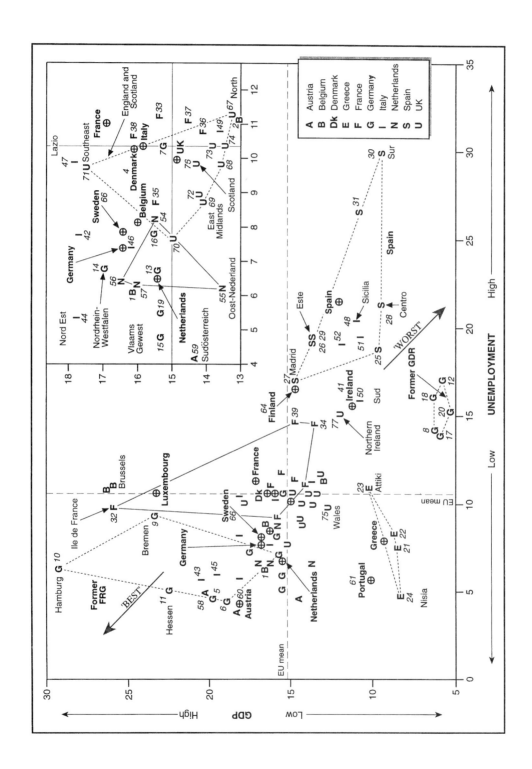

Figure 11.4 The relationship of percentage unemployed to GDP per capita at NUTS level 1. Inset map shows central part of main graph on a larger scale and gives numbering of regions. See Table A1 (Appendix) for key to the numbering of regions not named on the graph

resemble the Sud region of Italy. The five regions of the former GDR are remarkably similar, forming a ring in the lowest part of the GDP axis. On the other hand, countries and regions far apart in geographical space may be very similar. Thus Finland and Madrid are almost identical, as are Lazio (47 in Italy) and the Southeast (71) in the UK. At NUTS level 2 a still more complex picture would emerge, with some subdivisions of NUTS level 1 regions scattering quite widely in economic space, a feature of NUTS level 1 regions already noted in relation to whole countries. Two

further aspects of the distribution of the EU NUTS level 1 regions on the graph deserve comment, their distribution in relation to the mean centre and their location in relation to 'good' and 'bad' scores.

With the help of a number of circles of increasing radius centred on the central point of the graph (where 15.1 GDP and 10.4 unemployment intersect) it is possible to subdivide the 72 regions in Figure 11.4 into four groups according to their distance from that centre. The resulting scores are mapped in Figure 11.5. The closer a region is to the centre, the

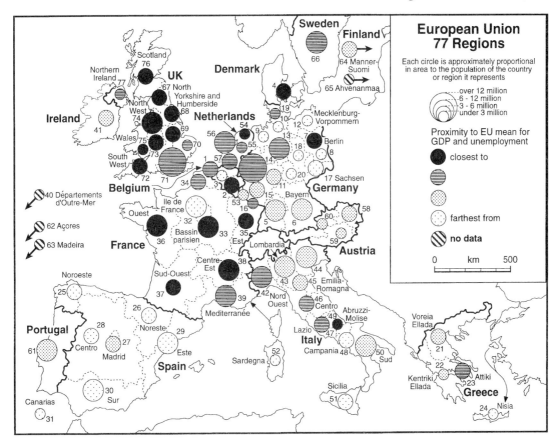

Figure 11.5 Proximity of NUTS level 1 regions to the EU average for a combination of GDP and unemployment levels. The regions have been divided into four classes according to their distance from the centre of the graph in Figure 11.4. See Table A1 (Appendix) for key to the numbering of unnamed regions

closer it is to the EU average for the two variables combined. Of the 18 regions closest to the centre, 8 are in England and Scotland, 6 in France and neighbouring Wallonie (Belgium), and the remaining four are Noord-Nederland, Denmark, Berlin and Abruzzi-Molise (Italy). At the other extreme, the 18 regions farthest from the centre of the graph consist of two distinct subsets at opposite corners of the graph. Ile de France (ie Greater Paris), Brussels, Bremen and Hamburg are at the 'most favourable' upper left corner of the graph and stand out primarily thanks to their very high levels of GDP per

inhabitant. The remaining 14 are in three distinct parts of the EU, five in Spain, four in the southern part of Italy and Greece, and five in the former GDR. All are characterised by below average GDP per inhabitant and all but Nisia (Greece) by above average unemployment.

The economic distances between regions on the graph extend over a large range for GDP between Thüringen (20), which scores 5.2 per capita, and Hamburg (10) which scores 29.1, and also for unemployment between Westösterreich (60) at 3.4 per cent and the Sur of Spain (30) scoring 30.1. The ranges on the two vari-

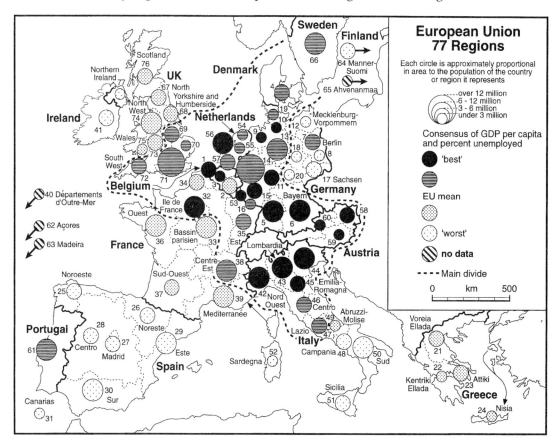

Figure 11.6 Favourable and unfavourable regions with regard to GDP (high level = 'good') and unemployment (low level = 'good'). The NUTS level 1 regions have been divided into four classes according to their distance along a line on the graph in Figure 11.4 from upper left (best) to lower right (worst). See Table A1 (Appendix) for key to the numbering of unnamed regions

ables are very large in economic terms. Cohesion seems a remote prospect in the short term, and indeed the accession of Greece, Spain, Portugal and the GDR since 1980 has made 'economic distances' in the EU much greater than they would be without these countries. The accession to the EU of Poland, Hungary, the Czech Republic and Slovakia would add the equivalent of about four GDRs roughly in the same area of the graph occupied now by East Germany and their addition would, of course, move the GDP and unemployment averages.

The second aspect of the distribution of regions on the graph shows another feature of the EU. Roughly equal weight is given to each of the two variables, because fortuitously the ranges of the score on the two variables are almost identical. Since high is 'good' on GDP and low is 'good' on unemployment, the 'best' place to be in economic space is therefore at the upper left of the graph, the 'worst' at the lower right. The regions are divided into four quartiles according to their economic level on the consensus of the two variables with the help of a series of diagonal lines on the graph (not drawn) running from lower left to upper right. The resulting scores are mapped in Figure 11.6.

The 'best' 18 regions fall into two groups, first city and service dominated regions that are small in area: Ile de France (Paris), Brussels, Vlaams Gewest, West Nederland, Bremen and Hamburg, along the northwest 'face' of the original EEC, and second the 'newer' (i.e. not coalfield based) industrial regions of southern Germany and northern Italy, together with Austria, whose low unemployment level contributes to its favourable economic position. At the other extreme, the least favourable 18 regions are in four groups, all peripheral, southern Italy, Ireland, Spain and the former GDR, together with Finland, unfavourably placed compared with its fellow Nordic countries Denmark and Sweden on account of its high level of unemployment. The picture for the whole of Western and Central Europe can be completed by adding Switzerland to the 'best' 18 and visualising an economic abyss (see Figure 13.1) to the northeast, east and southeast of Austria, with various countries and their numerous regions all joining the 18 least favourable of the EU.

Much of the material in this chapter has shown that economic differences between the regions of the EU are very great and that the accession of new countries in the 1980s has indeed increased distances in economic space. In Chapter 12 the causes of regional inequalities will be examined. Attention will then focus on the measures being taken to reduce these inequalities.

FURTHER READING

Commission of the EC (1984) *The Regions of Europe* (Second Periodic Report . . .) Luxembourg: Office for Official Publications of the European Communities.

Commission of EC (1991) *Employment in Europe 1991*, Com (91), 248 final, Directorate-General Employment, Industrial Relations and Social Affairs, Luxembourg, ISBN 92–826–2916–3.

12

NARROWING THE GAPS IN THE EUROPEAN UNION

'The Commission insists that for every 100 ECU invested through the structural funds, between 22 and 33 ECU flows back to the more advanced regions in payment for the know-how and equipment they provide.'

(Mann 1995a)

- Cohesion and convergence are objectives of the EU regional policy. The degree to which these have so far been achieved is difficult to assess objectively. Convergence is more likely to be achieved in periods of economic growth than in times of recession.

- Existing regional disparities result from long-term differences between regions in their natural resource bases, the availability of capital, the level of education and skill of the population and, more spatially, on location within the EU, particularly whether 'central' or 'peripheral'.

- Obstacles to progress towards cohesion in the EU are both economic and non-economic.

- A considerable part of the EU budget is devoted through the Structural and other Funds to assisting lagging regions.

Two basic aims of political leaders in both the EU as a whole and its individual Member States are to make the economic 'cake' bigger and to slice it up more equably. Ensuring that the economy of the EU continues to grow is related both to internal policy and to external influence. What is happening in the rest of the world, discussed in Chapters 13 and 14, is of increasing relevance to the future of the economy of the EU. On the other hand, ensuring that the economic cake is sliced up more equably *within* the EU depends largely on internal situations, policies and decisions. The second policy aim will be the theme of this chapter.

In the first section attention will be drawn to the difficulty of determining the extent to which regional differences in economic and social conditions are changing. The following section contains a number of estimates of convergence or lack of it in the EU. Subsequent sections are concerned with the following areas:

- Causes of regional inequality.
- Obstacles to the attainment of greater regional equality.
- Measures that are being taken or are planned to reduce regional inequalities.
- Regional differences at NUTS level 2 for the

five most recent joiners of the EU, the former GDR, Austria, Portugal, Finland and Sweden, in Box 12.1.

COHESION AND CONVERGENCE

The terms cohesion and convergence are widely found in both technical and more popular EU publications. They have been used and defined in earlier chapters of the present book. For the purposes of this chapter cohesion refers to a broad concept related to the goal of narrowing the gap between the more developed and the more backward regions of the EU. The achievement of cohesion implies the creation of institutions and systems that facilitate such features as greater integration in the EU, closer links through the removal of barriers at national boundaries, greater standardisation of products, and the creation of comparable if not identical living standards in all the Member States. On the other hand, convergence is more a means of progressing towards cohesion, the use of appropriate measures to reduce regional disparities through time.

In the view of the authors there is no entirely objective way of measuring regional inequality and of determining with great precision the intensity of change in levels. In the quagmire of socio-economic data sets, inconsistent definitions and subjective weightings of different factors, it is difficult to produce a single definitive measure of convergence. In some circumstances it is possible to select and justify the use of given sets of criteria either to demonstrate convergence in the EU or in the same situation to show no change or even divergence. Some of the pitfalls will now be described. For convenience they are numbered, but the list is neither exhaustive nor is it made up of perfectly mutually exclusive factors.

1 Convergence for which regions? Convergence at national level of Member States of the EU can be measured either for all EUR 15 countries or for particular subsets of the 15. Only the original EUR 6 founder members have been directly affected by EU membership throughout its existence. The last six Member States to join, together with the former GDR, have been in the EU for relatively short periods of time. If, however, convergence among the original EUR 6 is measured from 1957 to the mid-1990s, then the impact of subsequent joiners on their fortunes cannot be ignored. If convergence among all EUR 15 countries over the last four decades is studied then Spain, Portugal, the former GDR, Austria, Finland and Sweden have been directly affected for less than a quarter of the time. When convergence is assessed at a more detailed spatial level than that of the 15 sovereign states, similar problems arise. However, the measurement and comparison of regional disparities through time *within* sovereign states can be carried out in a more straightforward way.

2 The scope of any study of regional disparities depends on the data available and on their accuracy. It usually takes a few years for regional data about newly joined Member States to be integrated into EU data sets. Data for variables of importance are not available below NUTS level 1 for some countries. Other crucial data such as employment in industry and unemployment vary from one source to another in spite of attempts to standardise definitions, as for example in the case of unemployment (see Chapter 11) or in the application of parity purchasing power standards to compare the GDP per capita in different countries. The comparison of GDP data at different times is particularly difficult on account of different rates of inflation (or deflation) and changes in exchange rates.

Variables measuring more specific phenomena than GDP can be used to demonstrate change over time, but some are more robust than others. For example, infant mortality rates (provided they are measured accurately) mean the same in 1950 as in 1990. On the other hand, energy consumption, however measured, as for example in terms of tonnes of coal equivalent, differs markedly between 1950 and 1990 because each tonne used has been used more efficiently as time has passed.

3 There is frequent confusion between relative and absolute change. As will be shown below in the case of energy consumption, the level in country A can increase at a faster rate than the level in country B while at the same time the absolute gap widens. A less common but less forgivable practice is to compare total production or consumption of some item in regions with different populations. Data should be expressed in terms of per capita.

4 When convergence (or divergence) between countries or regions is being traced through time a set of data may be taken to illustrate some feature without thought for the elements in the set. For example, at NUTS level 2 it can be demonstrated that between 1980 and 1992 the highest and lowest regions in GDP per capita (PPS) has narrowed. In 1980, Groningen (in Noord-Nederland) had a score of 207 in relation to the EUR 12 level of 100, while in 1992 its score had dropped to 132. Groningen is a NUTS level 2 region with a small total population

(550,000) but a huge production of natural gas. Its apparent decline from 207 to 132 was largely the result of a re-assessment of its GDP. Hamburg's per capita GDP was similarly tampered with.

5 The effect of aggregation must always be borne in mind whether for regions or for classes (such as income groups). Ideally individuals, or at most households, should be considered. At a much more general level, the inconsistencies of the systems of regional divisions at NUTS levels 1 and 2, due to the isolation of certain large cities from their surrounding areas, produce regions at both levels with very high GDP per capita scores. Table 12.1 illustrates the problem with the example of Hamburg, which has a GDP per capita almost twice the average for EUR 15 and in 1992 had the highest score of all NUTS levels 1 and 2 units in the EU. If, however, like some other cities of comparable size in the EU, Hamburg did not have special status as a German Land, but was embedded in a larger unit made up also of Schleswig-Holstein (NUTS levels 1 and 2) to the north and Lüneburg (NUTS level 2) to the south, the combination of the three would have only 18,700 ECU per capita, only 124 compared with the EU average of 100 (see Table 12.1 for the calculations). Hamburg, at 196 when by itself, would be replaced at the top by Brussels, at 174, to

Table 12.1 The Hamburger illusion

	Population million	GDP billion ECU	GDP per capita ECU
Hamburg	1.7	49.5	29,100
Schleswig-Holstein	2.7	41.9	15,500
Lüneburg	1.55	20.0	12,900
All three above	5.95	111.4	18,700
Germany			16,600
EUR 15			15,100

which a similar procedure could justifiably be applied by merging it with surrounding Brabant, with a score of 122, thus lowering its score in a similar way.

At the other end of the scale, it is questionable whether it is realistic or appropriate to include in studies of convergence the seven extra-European regions of the EU, those of France (four Départements d'Outre-Mer), Portugal (Açores and Madeira) and Spain (Ceuta y Melilla). Similarly, since comparable data for the former GDR before German unification are either not available or reflect such a different political and economic system, it may be argued that the new regions of Germany should likewise be excluded in convergence calculations.

CONVERGENCE IN THE EU: THE OFFICIAL VIEW

The following three sections contain some examples of studies of convergence in the EU to illustrate different approaches to the question and different interpretations of what has happened. Three sources are used, official publications of the EU, the findings of other authors and case studies by the authors of the present book. The more technical publications of the EU cover at great length and with commendable impartiality the successes and failures of cohesion and convergence. On the other hand, the more 'glossy' EU publications tend either to give oversimplified, sometimes glib accounts of progress, or to include very little of substance on the subject.

In COM 87–230 final (Third Periodic Report), published in May 1987 (therefore just after the entry of Spain and Portugal, and before German unification), problems of convergence and cohesion are covered at considerable length. It is noted that convergence took place both between and within Member States during most years up to 1977 but that after 1977 little

change occurred. According to COM 87–230 (pp. 52–3):

> Real convergence is one of the Community's fundamental objectives and is essential for its cohesion. As a result of the first oil shock and the major worldwide disequilibria of the last fifteen years, the process of real convergence was interrupted and partly reversed. It now [1987] needs to be set in motion again. To achieve convergence in living standards, the countries and regions lagging behind need to record above-average growth rates of income generation, i.e. of employment and productivity. The number of jobs in the weak areas must also grow at a much faster rate than elsewhere, because present unemployment, structural underemployment and demographically induced future growth in the labour force all tend to be highest in the weak regions. But real convergence is a process that can produce results that will become discernible only gradually. For this reason, regional policy must take a long view, short-term successes being no measure of its effectiveness.

and COM 87–230 (pp. 58–60):

> One essential difference between the two main periods must not, however, be overlooked: regional convergence within the individual Member States up to 1973 was partly attributable to migration from the weaker to the stronger regions and only partly to comparatively stronger growth of production in the weaker regions. Convergence prior to 1973 was not all positive therefore. In numerous cases there seemed to be no sufficiently powerful regional policy to supplement general economic policy. In the period since 1973, net regional migration has declined significantly, and this, together with the generally lower rate of growth, helps to explain why regional income disparities too displayed the trends described above [i.e. little convergence occurred].

The general message of the above extracts in particular and of the relevant sections in the Third Periodic Report in total was that in the first 30 years of its existence convergence in the EU was limited in scope and was disappointing, although of course the total production of goods and services increased greatly during that period. In other words the EC cake got much bigger, but the shareout changed little. Convergence in the EU has at best been two steps forward and one step or even two steps backward as external influences (especially oil prices) and the entry of new, poorer than average, countries have interfered with whatever policies were being implemented during particular periods.

In Commission of the EC (1994a: 33–4) (the Fifth Periodic Report) economic growth is also related to the narrowing of regional disparities:

> After a prolonged period of slow growth in the first half of the 1980s, the Community economy picked up significantly in the second half of the decade. Growth at the Community level is an important pre-condition for narrowing regional disparities in output and income . . . .
>
> These general trends were accompanied by considerable variations in the performance of Member States and regions. At Member State level, the net effect of the differences in performance over the past decade can be summarised in terms of a period of slight widening in disparities in GDP, in per capita terms, between 1980 and 1984, followed by a steady narrowing (real convergence).

Reference is then made to the good performance relative to the Community average of the four weakest Member States, Greece, Spain, Ireland and Portugal. In 1986 the average GDP per head of the four was 64 per cent of the EU average but by 1993 it had reached 70. It should be pointed out, however, that progress has been slow in Greece, while Spain after the death of Franco in the mid-1970s and Portugal once it eventually gave up trying to keep its African colonies, were both already experiencing rapid economic growth *before* they joined the EU in 1986.

CONVERGENCE IN THE EU: UNOFFICIAL VIEWS

In a paper published in *West European Politics*, Leonardi (1993) distinguishes two opposing views of regional economic development theory as relevant to the question of convergence in the EU. One view is that without the application of policies to improve conditions in existing backward/less developed regions, divergence is likely to occur because core areas (not necessarily located centrally from a geographical point of view) will prosper and develop at the expense of lagging regions. The opposite view is that developments in favoured regions will in due course spread outwards to less favoured regions.

Leonardi argues that the 'core-dominance thesis' has been widely accepted as the explanation of regional trends in the EU. In other words, EU policy to achieve greater cohesion is going against the 'natural' economic trend. He divides the history of development in the EU into two periods, 1950–70 and 1970–90, and focuses on the EUR 9 countries. He finds evidence in the work of Molle, van Holst and Smit (1980) that for the first period: 'the gap between the rich and poor regions – as well as the gap between rich and poor Member States – has decreased . . . [they] unequivocally concluded that the core dominance thesis was not at all supported by the data'. For the second period, Leonardi himself uses EU sources to analyse convergence trends at national and regional level. He concludes:

> The conclusions that can be drawn from the above analysis are quite clear. During the last 40 years, economic and social cohesion has taken place in the European Community. Levels of productivity and social well-being have increased, and the gap between the poorer and richer regions and countries of the Community has decreased. Convergence theories perform much better in predicting the course of European development than do divergence theories in their emphasis on backwash effects and the importance of economies of scale. In addition, the assumptions adopted by the neofunctionalists on the economic pillar and the role of economic expansion as the driving force for political integration are supported by the data analysis.
>
> (Leonardi 1993: 512)

His expectation of further convergence in the future is stated thus (p. 513):

> To a certain extent, the expectations on cohesion written into the preamble of the 1957 Treaty and emphasised in the Single European Act and Maastricht Treaty are being promoted by economic and political integration, and when further integration takes place through the dynamics of the Single Market and Monetary Union the prospects for the weaker regions in the Community will improve rather than deteriorate. These results help to explain why peripheral regions and countries are so supportive of the European Community, the integration process, and the pursuit of active regional development policies at the European level. Peripheral, underdeveloped economies are net gainers from market integration due to spread effects. The incentives for domestic as well as foreign investors to enter the markets of peripheral areas are not only supported by middle to long-term calculations of profitability but also due to immediate returns to investments made possible by the Community Support Frameworks.

The reader who feels the need for a refresher course on regional development theories of the second half of the twentieth century should read the first part of Leonardi's paper. The empirical evidence used to support Leonardi's interpretation of convergence is to be found in the latter part of the paper. As pointed out earlier and to be illustrated below, it is not difficult to pick out the experience of particular countries or of NUTS level 1 or 2 regions to demonstrate (if not prove) either that convergence has taken place or that divergence has taken place.

Leonardi assumes that to achieve convergence, a considerable public sector is needed, combined with power in national governments or the EU institutions to direct assistance and investment to less favoured areas. The world now knows how it was possible in the former USSR, in which the State controlled over 90 per cent of the means of production, to achieve rapid convergence by industrialising backward regions and providing healthcare, educational

and other services throughout the country. The cake grew and the slicing was fairly equable by the 1980s. One problem was that the ingredients of the cake were unbalanced and the final product contained dollops of poison.

If Leonardi appears to have set out to demonstrate that convergence has occurred in the EU between 1950 and 1990, Helgadottir (1994) used the latest report of the European Commission to question the effectiveness of efforts to help the less successful EU regions. Her conclusions will please those who question the need for structural funds and indeed for the EU itself. Data are shown for the richest and poorest regions in nine of the EUR 12 countries in 1980 and 1991. Values are for GDP per capita and they are expressed as percentages of the EU average (= 100). Thus, for example, in Belgium (itself at 108 in relation to the EU average), the scores of the richest and poorest regions, Brussels (a NUTS level 1 and 2 region) and Hainaut (a NUTS level 2 region) changed as follows between 1980 and 1991: Brussels increased from 166 to 171, Hainaut dropped from 84 to 80.

In this example, the gap in Belgium itself therefore widened. Helgadottir's use of the richest and poorest regions within the nine EUR 12 that have appropriate subdivisions was unfortunate, if not undiscerning, because in four of the nine countries there were no data for 1980 for one of the two regions compared. Within the five countries for which the data could actually indicate the change between 1980 and 1991, in Belgium, Spain and Portugal the gap did widen considerably, in Italy the change was minimal, while in Greece it actually narrowed. A flaw in her choice of regions may also be noted. The top and bottom regions in 1991 were not necessarily the same as the top and bottom regions in 1980.

Watson (1995a) cites 'recent' figures to show how EU polices have helped certain countries to 'bridge the economic gap':

The EU's economy grew by 1.7 per cent between 1989 and 1993, but the economies of its four least well-off members – Greece, Portugal, Spain and Ireland – grew by 2.2 per cent. Without the benefit of EU funds, it is estimated that figure would have been 1.6 per cent.

When Ireland joined the EU in 1973, its gross domestic product (GDP) was 57 per cent of the Union average (measured on the basis of the 15 existing members). This year it is projected to rise to 85 per cent. Similar but less dramatic progress has been registered in Portugal and Spain. The GDP of the first has risen from 54 per cent in 1986 to 68 per cent and the second from 70 per cent to 79 per cent.

The purpose of referring to the very different approaches and conclusions of Leonardi, Helgadottir and Watson at some length is to make the following point: Leonardi's paper was carefully researched but is likely to be read by only a very limited, generally professional, readership, while the articles of Helgadottir and Watson will have reached a much wider readership. The present authors do not claim to be able to assess the extent of convergence or divergence in the EU up to the present with accuracy. The aim of the next section is to show three data sets that in different ways contribute both to highlight the difficulties of measuring convergence and to provide examples of evidence that can be used.

CONVERGENCE IN THE EU: THE APPROACH OF THE AUTHORS

Many studies of convergence in the EU use data for GDP, appropriately processed to show parity purchasing power or 'real' GDP. GDP has the advantage of recording the total production of goods and services, but it is very abstract, since no account is taken of the various elements that contribute to the total. The authors will argue below that even if in theory the ultimate goal of convergence is that every country and even every region should have the same GDP per capita (in real terms), some parts of the EU may not in reality need such high levels as others. In the three brief studies that follow GDP per capita is considered first. There follow studies of infant mortality and consumption of energy.

The six highest and lowest ranked NUTS level 2 regions in GDP per capita in the EU are listed in Table 12.2. Neither the new regions of Germany (no data for 1980) nor the overseas regions are considered. The GDP per capita of each region is expressed as a percentage of the EU average of the time, namely 1980 = 100, and 1992 = 100, although the 100 for 1992 represents a larger *total* GDP than the 100 for 1980. The membership of each pair of sets is different in the two years, as is the ranking.

In order to spread the representation of the highest and lowest regions, rather than simply taking the highest and lowest of all (as in Helgadottir (1994)), the average GDP per capita score for six regions has been calculated. In 1980 the ratio of the average for the highest six, 171, to the average for the lowest six, 46, was 3.7 to 1. The corresponding ratio in 1992, 171 to 48, was 3.6 to 1. On this evidence the amount of apparent convergence was negligible. At the rate of 2 percentage points in 12 years, to 'climb' from 46 to 48 of the EU average, would need about 300 years to reach the EU average, which itself would be moving upwards anyway!

As noted earlier in this chapter, Groningen was subjected to drastic statistical surgery in the late 1980s so it 'unfairly' raises the average for the six highest in 1980 (or lowers it in 1992). If Groningen is replaced by the seventh ranked region in 1980, Oberbayern, with a score of 141, then the average for the highest six in that year is only 160. The ratio of the highest to the lowest six drops from 3.7 to 1 to below 3.5 to 1.

The brief interpretation of the set of data in Table 12.2 shows some of the problems of using GDP to identify convergence among the

Table 12.2 GDP per capita in NUTS level 2 regions (EU average = 100)

Highest 6 in 1980		Highest 6 in 1992	
Groningen	207	Hamburg	196
Hamburg	186	Brussels	174
Brussels	167	Darmstadt	174
Ile de France	161	Ile de France (Paris)	169
Bremen	157	Oberbayern (Munich)	157
Darmstadt	148	Luxembourg	156
Mean of highest 6	171	Mean of highest 6	171
Lowest 6 in 1980		Lowest 6 in 1992	
Anatoliki Makedonia G	49	Ionia Nisia G	53
Dytiki Ellada G	49	Dytiki Ellada G	52
Centro (Portugal)	47	Centro (Portugal)	48
Extremadura (Spain)	45	Ipeiros G	47
Ipeiros G	45	Voreio Aigaio G	45
Voreio Aigaio G	43	Alentejo (Portugal)	41
Mean of lowest 6	46	Mean of lowest 6	48

Source: Eurostat 1995b
Note: G – Greece

regions of the EU. Most of the regions at the highest and lowest extremes of the scale are small in population, and as noted earlier in this chapter, among the highest there are several virtual city regions (Hamburg, Bremen, Brussels, Luxembourg). Changes in the fortunes and the ranking of at least some regions not at the extremes should also be taken into account. In Table 12.3 two more specific and restricted measures of convergence among many possible ones are illustrated with data sets chosen in each case for three different years.

Although infant mortality levels reflect the availability and quality of only a small part of all healthcare services, they serve as a broad measure of the state of healthcare. At national level, convergence in the EUR 15 countries has been remarkable (see Chapter 8 for NUTS level 1 data for infant mortality). As always, it must be remembered that only six of the countries have been in the EU all the time between 1963 and 1993. In 1963 there was a *relative* gap in infant mortality rates between Portugal (73) and Sweden (15) of almost 5 to 1. By 1993

the gap between highest and lowest, Portugal and now Finland, not Sweden, was less than 2 to 1 (8.6 to 4.4). Even more striking was the drop in the *absolute* gap, the difference between the scores of the two countries dropping from 58 to 4.2. In the case of infant mortality and of other economic and social features (e.g. total fertility rate), in which the indices are declining because low is 'good', there is no problem in identifying a limit. It is impossible to have an infant mortality rate of below zero, while the practical limit seems to be around 4.

In the case of energy consumption per capita, the level has increased in all the EUR 15 countries during most years from 1959 to 1994. In contrast to infant mortality, there is no theoretical final *upper* limit to which per capita energy consumption could rise, although limits related to supplies and to pollution make it unlikely that the process of growth of energy consumption will go on everywhere indefinitely. To the most immediate question, whether per capita energy consumption in the EU converged between the 1950s and the 1990s or not, the

Table 12.3 Infant mortality and energy consumption in EUR 15 countries in selected years

	(1)	(2)	(3)	(4)	(5)	(6)
	Infant mortality, deaths per 1,000 live births			*Energy consumption per capita in tonnes of coal equivalent*		
	1963	*1978*	*1993*	*1959*	*1977*	*1994*[1]
Belgium	27	12	7.6	3.9[3]	5.8	7.8
Denmark	19	9	5.7	2.5	5.3	5.9
Germany	28[2]	15[2]	5.8	3.4[4]	5.6	6.1
Greece	39	19	8.3	0.4	2.0	3.7
Spain	41	16	7.6	0.8	2.3	3.6
France	25	11	6.1	2.3	4.2	6.0
Ireland	27	16	6.0	2.0	2.8	3.9
Italy	40	18	7.4	1.0	3.0	3.9
Luxembourg	29	11	6.0	n.a	14.2	n.a.
Netherlands	16	10	5.9	2.7	5.6	7.8
Austria	31	15	6.2	1.9	3.7	4.2
Portugal	73	39	8.6	0.4	1.0	2.4
Finland	18	9	4.4	1.4	5.0	6.7
Sweden	15	8	4.8	3.0	5.7	7.4
UK	22	14	6.6	4.6	4.9	5.6

					Difference	
					Relative	*Absolute*
Infant mortality						
1963	Portugal	73	Sweden	15	4.9:1	58
1978	Portugal	39	Sweden	8	4.9:1	31
1993	Portugal	8.6	Finland	4.4	2.0:1	4.2
Energy consumption						
1959	UK	4.6	Portugal, Greece	0.4	11.5:1	4.2
1977	Belgium	5.8	Portugal	1.0	5.8:1	4.8
1994	Netherlands	7.8	Portugal	2.4	3.3:1	5.4

Sources: UNDYB 1967, Table 12, pp. 272–3 for 1963; *WPDS 1980* (PRB 1980) for 1978; *UNSYB 1963*, Table 129; *UNSYB 1979/8* Table 189; BP 1995: 34
Notes: n.a. not available
    1 1994 in tonnes of oil equivalent × 1.5
    2 West and East before unification
    3 Includes Luxembourg
    4 FRG only

answer is yes and no. Relatively the ratio of the highest level in 1959 (that of the UK) was 11.5 times higher than the lowest level (Portugal and Greece) whereas in 1994 it was only 3.3 times higher in the Netherlands and in Belgium (the highest) than in Portugal (still the lowest). Absolutely, however, the gap has actually widened considerably, from 4.2 tonnes per capita (between UK and Portugal) in 1959 to 5.4 per capita (between the Netherlands and Portugal) in 1994. It is worth noting that a similar situation occurs when relative measures of GDP change have been used in studies to trace convergence when the EU average of 100, against which scores are calculated, has itself represented a greater real level of GDP per inhabitant as economic growth has proceeded over the years and decades.

Two further by-products of the energy consumption data set may be noted here. First, it is evident that there is no reason to argue that levels of energy consumption in Portugal or Greece need ever be as high as they are in the Netherlands or Sweden. In this sector of the economy energy makes a larger relative contribution to GDP in the two northern countries than in the two southern countries. Only if some other item of production needed to be correspondingly larger in the southern countries than in the northern ones, could it be justifiable to argue that GDP per capita should converge. Second, in Table 12.3, columns (4)–(6), Luxembourg is included with Belgium in the calculations. The individual value for Luxembourg is also shown for 1977. If the score of Luxembourg, as a sovereign Member State, is used to represent the highest value in that year and Portugal the lowest, the ratio is 14.2 to 1 instead of 5.8 to 1 when Belgium (including Luxembourg) is used, a further example of the problem of having a set of political and administrative areal units that vary greatly in population size at all levels in the EU.

It may be asked, then, if convergence has occurred in the EU since its creation. Before a conclusion is reached, it should be made clear what has converged, and whether the convergence is absolute or relative (or both). If convergence has occurred it may not necessarily be the result of the existence of the EU. It might have happened without the EU as, for example, the rapid economic growth of Spain and Portugal (noted above) that took place before their membership of the EU and, also during the first years of membership, before the EU effect could be credited with much influence. For the record, the Czech Republic has been experiencing faster economic growth than most EU countries in recent years, thereby converging without being a Member State. The reverse of this theme is the question of whether, for example, Scotland or Catalunya would be richer or poorer if they had been completely sovereign, as the Republic of Ireland has been from the UK for over seven decades. Similarly, would the gap between North Italy and South Italy be greater or smaller than it is now if unification in the 1860s had not followed the path it did and each had remained independent? One can only speculate. History cannot be re-run on different lines and economists cannot experiment on whole countries or even large regions.

No doubt the debate will continue well into the next century both on how to evaluate the impact of regional policies and how to reduce regional disparities. A report (European Parliament (1996)) summarises the issue of evaluation of the regional impact of community policies. Among other aspects of the subject it stresses the need to reconcile top-down and bottom-up approaches to impact evaluation.

THE CAUSES OF REGIONAL DISPARITIES

It is of little more than academic interest for the inhabitants of various regions of the EU to know that according to one set of theories on regional development their regional location or their natural resource endowment predetermines their economic position in the EU league, while according to another set of theories their position can be changed given suitable conditions in the EU. In addition to considering the macro-economic view of the EU situation it is important to look closely at the features and problems of each region. The fortunes of some regions in Western Europe have fluctuated over the centuries. Other regions,

[Plate 12.1] Recognition of ERDF funding amid nostalgic posters at the North of England Open Air Museum, Beamish, County Durham. The site includes a colliery, tramway and nineteenth-century buildings

Plates 12.1 and 12.2 European Regional Development Fund helping English projects

[Plate 12.2] Central Nottingham gets assistance to redevelop a run-down area of warehouses and other premises

such as North and Central Italy and the Low Countries, have, however, generally prospered, while still others, such as the Massif Central in France and South Italy, have long been relatively poor. In order to provide a framework for studying the causes of regional disparities in the EU, two of many possible aspects will now be discussed. First, the attributes of regions will be assessed and, second, a summary will be given of the findings of a survey by the European Commission covering some 900 companies.

The geographical literature of some decades ago contained much evidence intended to support the view that the natural environment affects human activities profoundly. A more controversial explanation of differences between societies, now discredited, was related to the actual mental capabilities and cultural attitudes of various groups of people. Much has also been said about the importance of location on the success (or otherwise) of economic activities. For various conceptual and methodological reasons, the influence of environmental and locational differences on regional development has been played down in recent decades, if not ignored altogether. Instead, overwhelming importance is attached to economic organisation (e.g. the movement of capital, the cost of labour) and the effect of political and economic decisions. The present authors are of the view that even within the comparatively small EU, regions differ so profoundly in their attributes and locations that the study of their past performance and future prospects must take into account these differences, notwithstanding the great influence of organisational decisions. For simplicity, two influences have been distinguished: first, the various elements that make up the region and, second, its location in a given context relative to other regions. The two influences are together comprehensive enough to cover most aspects of regional disparities and to answer most questions about them.

Each region contains a particular combination of natural resources, people (often referred to as human resources) and means of production (capital resources). Natural resources include water, bioclimatic resources (the quality of the land and soil), fossil fuel and non-fuel mineral reserves. Among the attributes of people, educational levels and skills in new and growing activities are important, as is also financial organisation. Means of production, for example the quantity, quality and efficiency of factories, farms and hospitals, and the level of mechanisation, are guides to the state of the means of production of goods and services in sectors that are expanding, stagnating or declining. Location in the EU, however it is defined, is broadly related to the supposed advantage of being central rather than peripheral in a spatial sense (see Chapters 3 and 4).

It is not possible in the scope of the present book to provide an exhaustive account of all the regions of the EU. In order therefore to exemplify the possible effects of the *attributes* and the *locations* on regional disparities in the EU, four NUTS level 2 regions are compared in Table 12.4, two from Italy, two from the UK. At present levels of performance, Lombardia in Italy and South Central England are roughly similar, and each is well ahead of arguably the most backward region in each of the two countries, Calabria and Northern Ireland. On the scale used, consisting of points subjectively awarded by the authors in order to illustrate the way the situation can be viewed, no EU region could score 75 out of 75 or 0 out of 75. Some regions of southern Germany might exceed Lombardia by a few points, while some in Greece and Portugal are somewhat below Calabria. The scores of the four regions in Table 12.4 show that there is little scope for profound changes to be made over a short period. Calabria and Lombardia have been vastly different for centuries, as have South Central England and Northern Ireland. Indeed, the creation of the EU may actually have slightly enhanced the locational advantages of Lombardia and South Central England over their national positions in a pre-EU context.

Location cannot be changed, although places can be brought closer together in time distance by the improvement of communications. On the other hand, a firm looking for the optimum location in the EU for a given function has locational freedom, and the decision about location would presumably take into account the various relevant advantages and disadvantages of a number of possible regions. The accessibility scores for representative places in the EU discussed in Chapters 3 and 4 serve as a rough guide for identifying favourable locations. Some of the features, such as improved educational levels, and the introduction of new industries, can produce changes gradually.

An extensive survey of the views of 900 EU Companies (see COM 90–609 (1991)) on regional issues was directed at firms in three types

Table 12.4 A comparison of the attributes of four NUTS level 2 regions of the EU (Notional scores: maximum 5 most favourable, 0 least favourable)

	Lombardia	Calabria	Berkshire Buckinghamshire Oxfordshire	Northern Ireland
Water	4	2	3	3
Quality of agricultural land	4	1	3	3
Forests	3	1	1	1
Hydro-electricity	3	1	0	0
Fossil fuels	0	0	0	0
Non-fuel minerals	0	0	0	0
Higher education	4	0	4	2
Financial	4	0	4	1
Agricultural sector	4	1	3	3
Industrial sector	4	1	4	2
Service sector	4	1	4	2
Research	4	0	5	1
Tradition and skills	4	0	4	3
EU location	3	1	5	1
National location	3	1	5	1
Local variations	great	great	small	small
Total score	48/75	10/75	43/75	23/75

Note: The scores for each attribute range between 5 (for a position among the top few in the whole EU) to 0 (either absence of the attribute or a position among the bottom few in the EU). The judgements are subjective, and although made with considerable thought, are presented primarily to draw attention to reasons why regional differences exist in the EU rather than, for example, as a possible basis for regional policy-making.

of regions: backward agricultural regions, declining industrial ones, and a control group. A feature of the findings is a lack of concern about location in the EU. If a location is unsatisfactory for a particular industrial or service activity, then that activity can be moved to a place regarded as more favourable, but the locational features of its initial position can only be changed slightly by an improvement in transport and communications links with other regions. Since improvements in the transport sector are numerous, ongoing, and widely distributed in all kinds of regions, the relative improvement of a particular link may be matched and in effect cancelled out by improvements in others.

Many problems connected with non-locational features were universally regarded as serious in the survey.

- Many of the companies questioned hoped that income and corporate taxes could be lowered, although this was somewhat unrealistic.

- The quality and suitability of the skills of the labour force was a consideration.

- Economic growth was regarded as essential for progress.

- In the lagging regions, the cost of credit caused concern, together with the attitude of bureaucracy. In Italy, for example, interest rates tend to be higher in the South than in the North.

- Business services were lacking and inadequate.

- In Portugal and Ireland the provision of the energy supply was regarded as a serious issue.

- In Ireland, southern Spain, Sicilia and Sardegna, transport (including presumably the negative effects of a peripheral location) was referred to as a problem and suitable industrial sites were not available in southern Italy.

- Educational provision is inadequate at all levels in Spain.
- In declining industrial regions, a need to retrain unemployed people and to improve the qualifications of the workforce was referred to, while at the same time labour costs have to be reduced. The former GDR, not included in the survey, on joining the EU immediately became its most critical declining industrial region.

According to the survey, even the most successful regions had problems, with labour shortages a common feature. Indeed, vocational training and professional expertise are perhaps the greatest single need for the 1990s. According to COM 90–609 (1991: 39): 'in an increasingly competitive world in all regions both process and product innovations are constantly needed'. In the EU a related problem is the great concentration of research and development in a few regions, as in Madrid for the whole of Spain and in northwest Italy for that country. Due to a long-standing policy of decentralisation in France, Paris is less dominant than it was some decades ago. The great concentration of research contracts in R&TD in a small number of centres is emphasised in Commission of the EC (1994a: 100–1):

> A limited number of such islands in the Community stand out from the rest: Greater London, Rotterdam/Amsterdam, Ile de France, the Ruhr area, Frankfurt, Stuttgart, Munich, Lyon/Grenoble, Turin and Milan. Up to three quarters of all public research contracts, including those funded by the Community, are estimated to be concentrated in these few places. They also tend to work closely together as part of a highly exclusive network.

A century ago, Glasgow, Manchester and Birmingham might have been included in such a group.

In spite of the dramatic recent reduction in fertility rates, the population of many regions of the EU continues to grow and immigrants add significantly to the number of young adults in some regions, so there will be more young people seeking jobs, as the extremely high unemployment levels among under-25s show. How many young people in EU countries will be able and willing to continue in full-time education well into their twenties is likely to be a prominent issue in the years to come.

Although it is convenient to consider each region as homogenous, taking average values for variables, in reality there are great disparities, both environmental and human, within regions, however the regions are delimited. To be sure, the more finely a part of the world is subdivided, the smaller should be the differences within each region. In the EU, only at NUTS level 3 are regions small enough in some cases to have a considerable degree of homogeneity. Two examples will serve to illustrate the diversity found in the EU regions at NUTS level 2.

1 *Lombardia* (Italy), which is an entity at both NUTS levels 1 and 2, consists of nine provinces. Three of these are entirely or partly in the Alps, where a particular set of problems is to be found. A zone of highly industrialised centres of various sizes, including Milan, extends over all or part of several more provinces. Finally, the southern part of Lombardia, in the North Italian Lowland, contains some of the most productive farmland in the EU.
2 *Derbyshire and Nottinghamshire* (NUTS level 2) include the Peak National Park, with upland farming, many coalfield settlements, most now with their collieries closed, two large manufacturing centres specialising in engineering and textiles, and a zone of mainly agricultural land of reasonably good quality.

In both the regions described above, it is unrealistic to generalise about conditions and to pretend that development assistance, if forthcoming, should be distributed evenly over the region. A comparable degree of diversity can be found in many other NUTS level 2 regions of the EU, and usually even greater diversity at NUTS level 1.

Another consideration in the planning of the best allocation of resources to regions regarded as backward is to ensure that resources actually help the more needy. In the United Nations Development Programme, this issue is given prominence:

> A major conclusion from all the evidence is that not all government spending works in the interest of the poor and that great care must be taken in structuring social spending to ensure that benefits also flow to them. The very rationale for government intervention crumbles if social expenditures, far from improving the existing income distribution, aggravate it further.
>
> (UNDP 1990: 33)

For example, a road improvement programme in a relatively backward agricultural region of southern Spain or Italy might improve the conditions for public passenger and goods traffic but would also benefit car owners, by definition already the more affluent members of the community. Ideally, every family, household or even individual should be considered if the most appropriate benefactors are to be targeted. In practice, however, with about 370 million individuals in the EU, generalisations have to be made and decisions based on these.

ECONOMIC OBSTACLES TO COHESION IN THE EU

The reduction of regional disparities in GDP per capita and in material conditions in the EU will no doubt be a matter of concern as much in the future as it has been in the first 40 years of the existence of the EEC. While economic features of the EU can be changed gradually to the advantage of the poorest regions if appropriate measures are taken, it can be more difficult to solve problems based on cultural and social conditions. For example, the boundaries of most of the sovereign states of the EU do not coincide precisely with linguistic or religious divisions. Greater autonomy, if not complete independence, is sought by appreciable sized minorities, if not majorities, in a number of 'suppressed' nations or ethnic groups. Cross-border allegiances cause problems, as between the Unionist and Nationalist groups in Northern Ireland, with Great Britain and the Irish Republic respectively, and through the links between Spain and France in the land of the Basques. In this section economic reasons for regional disparities are summarised. Non-economic features that can present obstacles to cohesion and convergence are then discussed.

Geographical Location

In both EU and non-EU publications in which regional disparities are discussed the existence of a core and a periphery is widely assumed to influence the economic performance and living conditions in the regions of the EU. The potential impact of location on different economic activities is not usually quantified. There is, however, agreement that to be in a 'core' coinciding roughly with an area enclosed by lines joining consecutively Amsterdam, Munich, Milan, Paris, London and Amsterdam is advantageous for some activities. It is self-evident, also, that on the extreme periphery, places such as the smaller Greek islands in the Aegean Sea, the Canarias of Spain, the islands off Western and northern Scotland, and the extreme north of Sweden and Finland are disadvantaged in a locational sense although, as with the Greek islands and Canarias, climatic and scenic conditions, combined with modern air travel, have favoured the development of large-scale tourism. How much weight is given to location in the development and prospects of different economic activities depends among other factors on the extent to which the activity serves all or a large part of the EU and, if it produces goods rather than services, on how heavy/bulky/

[Plate 12.3] A sign announcing the provision in Portugal of 155,000 new telephones near Bragança in the extreme northeast of the country

Plates 12.3 and 12.4 ERDF funding for infrastructure on the periphery of the EU

[Plate 12.4] Sewerage project supported by the ERDF in the far west of Ireland. Achill Island's population depends heavily on pastoral farming

perishable the products are and therefore the time and cost of reaching EU markets.

Natural Resources

Although bioclimatic products (food, timber), fossil fuels, non-fuel minerals and even fresh water can in theory be imported into any region, the local/indigenous natural resources are generally significant in determining the types of activities that predominate in the region, and the economic progress, or lack of it. Bioclimatic resources for cultivation, pastoral and forest activities and fishing, under modern conditions, vary greatly from region to region in the EU according to water and thermal resources, slope and soil quality. The presence of coalfields formed the basis for modern industrialisation in many regions of the present EU in the nineteenth and first half of the twentieth centuries but with the sharp decline in coal production since the 1950s, this resource is now of limited importance. The organisational environment in each region is also a crucial influence. In particular the educational level, pool of technical skills and access to research and innovation affect the attractiveness of the region to local, national, EU and non-EU capital.

Gross Domestic Product

Sharp variations in GDP per capita, one of the principal measures of development or lack of it, have been illustrated in Chapter 11 and already in this chapter. To a large extent, the level of GDP per capita reflects the particular combination of different types of economic activities in each region. Where unemployment is high, and low paid jobs predominate, GDP per capita will tend to be low. Thus, for example, in the NUTS level 2 region of Andalucía in southern Spain, unemployment is very high, productivity per worker in the agricultural sector is much lower than in the most favoured agricultural areas of the EU, and many people have low paid jobs in tourist centres along the coast. Although tourism brings in a large amount of revenue, airlines serving the resorts and hotels are generally owned by outsiders, public or private sector. In contrast, in Brussels and Luxembourg, for example, the direct and indirect influence of the presence of EU institutions on GDP has contributed to the high per capita levels, as has the presence of some of the most successful industrial enterprises in the EU in the NUTS level 2 regions of Stuttgart and Oberbayern (Munich) in southern Germany.

An example of the way the choice of location of new enterprises may affect the fortunes of particular regions is provided by the implementation in 1993 of the decision at the Edinburgh summit to establish new European agencies. Instead of concentrating all or most of the agencies in one centre, a future supranational 'capital' of the EU, sixteen agencies were shared among 12 cities in 11 countries (none went to France). In addition to five which went to Luxembourg, the remainder, almost without exception, were located within each country in cities in regions with the highest or a very high per capita level of GDP relative to the national level. It is not argued that the presence of these agencies alone will greatly affect per capita GDP levels in the places in which they are located, but what happened serves to illustrate how certain regions may benefit in terms of job creation by particular decisions, while others are ignored even though economically their claims may be much stronger.

NON-ECONOMIC OBSTACLES TO COHESION IN THE EU

Language

Over 100 different languages are spoken in Europe. In Figure 12.1 the main language divides are shown, those between the Germanic, Romanic and Slavic groups, which are all subdivided. Finnish, Hungarian and Greek do not belong to these groups. Some of the implications of the presence of many languages in the EU have been described in Chapter 2. Here it may be noted that in several of the Member States of the EU, with minority languages having varying numbers of speakers, a vociferous minority if not a majority of citizens seek greater autonomy if not complete sovereignty within a national framework. Examples are the País Vasco region in Spain, Wales in the UK and the German speaking population in Bolzano-Bozen province in northeast Italy. When this suppressed linguistic nationalism results at times in violence, the effect is to risk sacrificing a favourable regional image and to lose private investment, if not assistance from the national government or the EU budget.

Figure 12.1 The distribution of languages in Western and Central Europe. Belgium has no language of its own. Dutch (Flemish) is spoken in the north, French in the south. The language of administration in Luxembourg is French. German is also used. In the former Yugoslavia, the subdivisions and newly separate States are shown, but Serbo-Croat is spoken by about 17 million people in Croatia, Serbia and Bosnia–Herzegovina

Sources: based on Rand McNally 1987: 152–3 for language areas and on Gunnemark and Kenrick 1985 for numbers speaking main languages

Religion

In the sixteenth century virtually universal Christianity in Western Europe ended with the emergence mainly in northwest Europe of Protestantism. Most of the EUR 15 countries are either predominantly Roman Catholic (e.g. Italy, Spain), Protestant (e.g. Denmark, UK), or mixed (e.g. Germany, Netherlands). All branches of Christianity as well as non-Christian reli-

gions are now free to practise throughout the EU. Nevertheless, the ostensibly religious conflict continues to damage the image of Northern Ireland to such an extent that before the cease-fire there in 1994, Watt (1993) reported that Catholic families were considering leaving Ulster altogether. In due course, future expansion of the EU is likely to bring in Orthodox Christian countries and, if Bosnia, Albania and Turkey eventually join, also a large number of

Muslim regions. The scope for further religious tension and conflicts at both international and internal levels is considerable.

Other Aspects

Other features may make the image of some EU regions seem unfavourable. Sicilia and more recently Calabria have been poisoned by the negative activities of the Mafia, Sardegna by the presence of bandits and kidnappers, Corse (France) by a strong nationalist movement. In the former GDR, the more industrialised regions have legacies of pollution from the burning of lignite and from nuclear waste. Even when linguistic or religious minorities do not exist in a country, national feeling may be strong, as in the case of Scotland, whose geographically distinct location in relation to England and long tradition of independent culture until early in the eighteenth century give it a high national profile.

Changes in Central Europe and the former USSR have shown that one of the most centralised and strictly controlled large regions of the world, CMEA, could suddenly disintegrate. Czechoslovakia, Yugoslavia and, above all, the USSR, were countries in which two or more national groups were merged into single sovereign states, yet after decades of centralised rule they are now fragmented. There are at least two reasons why new strength may be given to movements within the EU towards more autonomy for suppressed or subnations: first, the example of the Republics of the former USSR and, second, the growing strength of the supranational nature of the European Union. The second influence impinges on the role of the national parliaments of the Member States. As shown earlier, within the NUTS level 1 regional framework, areas such as Wales, the País Vasco, Sicilia and the two halves of Belgium, Vlaanderen and Wallonie, are respected in the EU regional system and kept intact. Economically, for example, it might have been

more realistic to join North Wales to northwest England and South Wales to southwest England, while industrially Wallonie in Belgium and Nord-Pas-de-Calais in France have very close links.

Regional Initiatives Enterprise

Until a decade ago there was little direct contact between Brussels, as representing the EU, and regions within Member States. In the 1990s, according to Watson (1995a),

> Regions are pressing to bolster their position within the EU structure. The Committee of the Regions is less than two years old, but is already lobbying for new powers . . . .
>
> Regions continue to open up offices in Brussels – well over 100 are represented in the Belgian capital – in order to be closer to EU decision-making.

Ritchie (1995) summarises the new situation:

> The EU's more centralist governments – the UK, for example – can stop Brussels dealing directly with the regions, but they cannot stop the regions increasingly dealing directly with Brussels.
>
> Scotland is a classic case. While UK Prime Minister John Major insists subsidiarity stops at Member State level, the Scots (and the Welsh and Northern Irish) are increasingly taking what has been known as the 'Westminster Bypass' and setting up their own missions in Brussels.

Cross-border Contacts (See Figure 12.2)

Although some parts of the national boundaries between EU Member States date from the nineteenth or twentieth century, the influence of such boundaries has been strong even after frequent moves to allow free movement across them. Until recently, neighbouring regions in different countries, separated by a national boundary, have tended to have little contact. According to Watson (1995b), the Interreg programme of the EU is designed to encourage

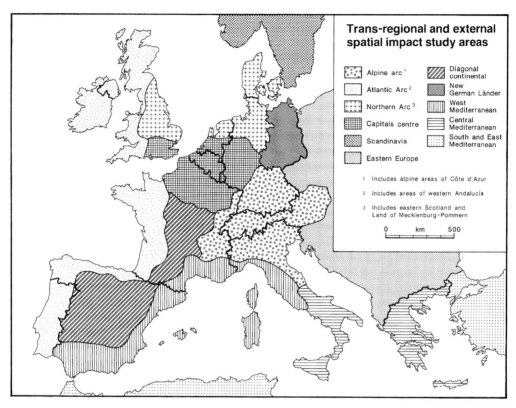

Figure 12.2 Trans-regional and external spatial impact study areas (see text for explanation). Note that Atlantic Andalucía is not shown and Pommern is Vorpommern
Source: Commission of the EC 1991: 23

cross-border cooperation not only between regions in the EU, but between EU regions and the contiguous regions on borders between EU and non-EU countries. Cooperation may be, for example, in education and training, environment and tourism projects, energy, communications and transport, and to halt smuggling.

Examples of cross-border links *within* the EU are numerous and include, for example, cooperation between Salzburg in Austria and Bayern in Germany, between the south of Corse in France and the north of Sardegna in Italy, and between Karlsruhe in Germany and northern Alsace in France. Mann (1995b) describes 'cross border cooperation' on a much larger scale in the case of the Atlantic Arc Commission, a

group of 32 regions extending from northern Scotland to southern Portugal, truly transnational in the EU context, with 12 UK regions, Ireland, 6 French, 5 Spanish, 5 Portuguese and 3 more associated. 'Their association shows that solidarity would be crucial in fighting perceived disadvantages they faced compared to other, more favoured areas of the Union.' There is generally less enthusiasm over contacts and cooperation between neighbouring EU and non-EU countries. The East Germans have become cool towards their neighbours in Poland, Greece is not on good terms with adjoining countries and Italy rejects the young Slovenia's attitude towards the long-standing Italy–Yugoslav border dispute.

THE REDUCTION OF REGIONAL DISPARITIES IN THE EU

Given the irregular progress of economic growth in the EU since the late 1970s, the accession of new members, and the great diversity within the Community in spite of its inclusion under the umbrella of the 'developed world', it is a daunting task for the Commission to decide where regional assistance should be directed. According to COM 90–609 (1991): 'the ten least developed regions, located mainly in Greece and Portugal, presently have average incomes per head which are one third of the average of the ten most advanced regions'. Similarly, disparities in unemployment are

noted: 'the regional differences remain substantial, and in 1990, in the 10 regions with the lowest unemployment, the rate averaged just over 2.5 per cent while in the ten regions with the highest rate it averaged 22 per cent'. Cole and Cole (1993, Chapter 10) may be referred to for details of the regional situation in the early 1990s. In this section attention focuses on the later part of the 1990s.

First, what are the objectives of the Structural Funds of the EU – the allocation of funds? Figure 12.3 shows that about 85 per cent of the area of the EUR 12 countries is eligible for some kind of assistance, although the population affected is only about 70 per cent because it is mainly the most densely populated regions

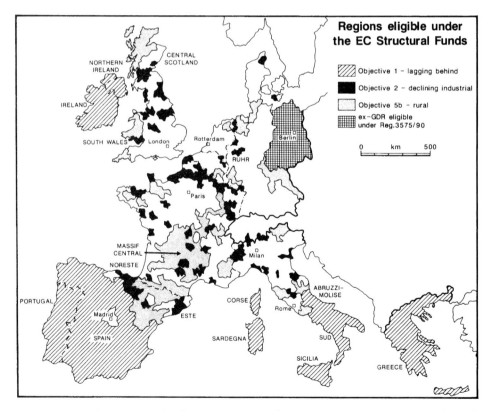

Figure 12.3 Regions of the EU eligible for certain classes of assistance through EU Structural Funds. Classes (Objectives) 3, 4 and 5a have less direct significance and are not shown. At the time of writing, there was no up-date of this map, which is reproduced in Eurostat 1995a

Table 12.5 Appropriations for the Structural Funds by Member States for the period 1994–9 in millions of ECU at 1994 prices

	(1) Total	(2) Per capita	(3) Objective 1	(4) 2	(5) 3 & 4	(6) 5a	(7) 5b	(8) 6	(9) Community initiatives
Belgium	1,859	182	730	160	465	195	77	—	233
Denmark	767	148	—	56	301	267	54	—	89
Germany	20,586	252	13,640	733	1,942	1,143	1,227	—	1,902
Greece	15,066	1,435	13,980	—	—	—	—	—	1,086
Spain	32,810	839	26,300	1,130	1,843	446	664	—	2,428
France	12,750	219	2,190	1,765	3,203	1,932	2,238	—	1,422
Ireland	6,004	1,668	5,620	—	—	—	—	—	384
Italy	20,679	358	14,860	684	1,715	814	901	—	1,705
Luxembourg	89	223	—	7	23	40	6	—	13
Netherlands	2,084	134	150	300	1,079	165	150	—	241
Austria	1,623	200	184	n.d.	n.d.	n.d.	n.d.	—	—
Portugal	15,396	1,555	13,980	—	—	—	—	—	1,416
Finland	1,704	334	—	n.d.	n.d.	n.d.	n.d.	511	—
Sweden	1,420	160	—	n.d.	n.d.	n.d.	n.d.	230	—
UK	10,265	175	2,360	2,142	3,377	450	817	—	1,119
EU Total	143,102	384	93,994	6,977	13,948	5,452	6,134	741	12,038
EU percentage	100[1]		65.7	4.9	9.7	3.8	4.3	0.5	8.4

Source: Eurostat 1995a: 66–7
Notes: n.d. – not determined; — negligible, none, or does not apply
1 EU totals of individual columns add to 97.3 because the allocation of some funds has not been determined yet.

that are not eligible. Table 12.5 shows the allocation of Structural Funds by country.

The application of the Structural Funds of the EU is based on four principles:

- To fulfil six priority objectives (discussed below).
- To ensure partnership and cooperation between the Commission of the EU, the Member State concerned and appropriate authorities within that State.
- The principle of additionality implies that the use of funds should have real economic impact.
- Programming of an overall strategy to overcome problems of the weakest areas.

The priority Objectives (see columns (3)–(8) in Table 12.5) are as follows:

- *Objective 1* To promote the development and structural adjustment of underdeveloped regions (allocated 65.7 per cent of the 1994–9 Structural Funds).
- *Objective 2* To redevelop regions seriously affected by industrial decline (4.9 per cent).
- *Objectives 3 and 4* To combat long-term unemployment and to assist workers to adapt to industrial change (9.7 per cent). These Objectives are not explicitly restricted to particular parts of the EU but by their nature they apply most extensively to regions of industrial decline.
- *Objective 5a* To modernise production,

Plate 12.5 Reindeer herding is the traditional economic activity of the Lapp population of the extreme north of Finland (here) and Sweden. The area has been recognised as eligible for special EU assistance by its designation as the only Objective 6 region in the Union

BOX 12.1 THE LATEST COUNTRIES TO JOIN THE EU

At the time of writing it was still difficult to obtain data for the newest countries to enter the EU, the former GDR, Austria, Finland and Sweden. Moreover, in *Regional Profiles* (1995) Portugal, Finland and Sweden are not subdivided at NUTS level 1 apart from the island regions of Portugal and Finland. Table 12.6 therefore contains NUTS level 2 data for unemployment and for GDP per capita for the countries referred to above. Together they contain only 12.3 per cent of the total population of the EU but 19 per cent of the NUTS level 2 regions (39 out of 206). The former GDR excluding Berlin has five regions at NUTS level 1 (see Table A1 – Appendix), but nine regions at NUTS level 2, although the subdivisions of Sachsen and Sachsen-Anhalt do not have separate data. Austria's three NUTS level 1 regions are subdivided into nine. The former GDR is included to introduce the regions and to show their location (see Figure 12.4). Points about the remaining four countries will now be discussed.

Although Portugal is much poorer in terms of GDP per capita than the three newest Member States of the EU, in all four the region containing the capital city has a much higher GDP per capita than any of the other regions (except Ahvenanmaa in Finland). In Portugal there are marked regional differences in the level of unemployment (Centro 3.4 per cent, Alentejo 9.2 per cent) and also in GDP per capita, the Centro and Alentejo being the lowest on the mainland. By EU standards the Norte is highly industrialised (42 per cent of employment), whereas the Centro still depends heavily on agriculture (22 per cent of employment),

Lisboa and the Algarve on services (70 and 71 per cent respectively). Attention has been drawn to the sharp regional contrasts at NUTS level 2 in mainland Portugal to remind the reader of the great diversity in the EU and the problem of directing EU Structural Funds to the various backward parts of the country, the whole of which is designated as Objective 1.

Austria's GDP per inhabitant is almost twice as high as that of Portugal but even so its eastern extremity, Burgenland, qualifies for Objective 1 status. In the west, Austria adjoins Europe's most prosperous country, Switzerland, whereas in the east it borders the Czech Republic, Hungary and the former Yugoslavia. There is a very sharp GDP per capita gap between Switzerland and Central Europe. The gradient runs 'down' Austria from west to east, with the southeastern part of the country lagging in comparison with the rest. Austria is likely to be a net loser in the allocation of funds from the EU budget, but it could be compensated to some extent by exploiting its central position in Europe as a whole. Already it is on busy routes between Italy and Germany, environmentally a liability but economically a positive feature. Routes between northwest Europe and the Balkans, and between Italy and the Czech Republic and Slovakia, are likely to grow in importance in the next century.

Finland and Sweden, like Austria, will be net losers in the allocation of EU budget funds. Gothenburg in Sweden is the main centre of iron and steel production and transport engineering (ships, motor vehicles) and the only major centre in the Nordic countries that resembles the

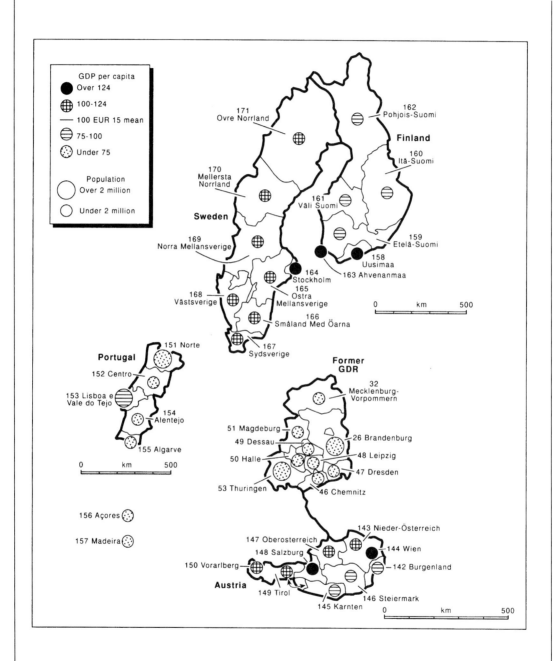

Figure 12.4 Regional variations in GDP per capita in the latest EU joiners, related also to the EU average. Note that the former GDR and Austria are on a larger scale than Portugal, Finland and Sweden

Source: Regional Profiles 1995

Table 12.6 Unemployment and GDP NUTS level 2 data for five latest EU joiners

		(1)	(2)	(3)	(4) GDP per capita Country = 100	(5) EUR 15 = 100
		Population in millions	Unemployment (%)	Thousand ppp ECU per capita		
	Former GDR (excluding Berlin)	14.2	15.0	5.8	100	38
26	Brandenburg	2.5	14.0	6.2	107	41
32	Mecklenburg-Vorpommern	1.8	16.9	5.6	97	37
46	Chemmitz					
47	Dresden	1.7	13.7	5.8	100	38
48	Leipzig	1.8	13.7	5.8	100	38
49	Dessau	1.1	13.7	5.8	100	38
50	Halle	0.6	15.9	6.0	103	40
51	Magdeburg	1.0	15.9	6.0	103	40
53	Thüringen	1.2	15.9	6.0	103	40
		2.5	15.0	5.2	90	34
	Österreich (Austria)	7.7	3.9	18.1	100	120
142	Burgenland	0.3	4.8	11.5	64	76
143	Niederösterreich	1.5	3.3	15.3	85	101
144	Wien	1.5	5.1	26.2	145	174
145	Kärnten	0.5	3.9	14.3	79	95
146	Steiermark	1.2	4.1	14.4	80	95
147	Oberösterreich	1.3	4.1	17.1	94	113
148	Salzburg	0.5	2.9	20.0	110	132
149	Tirol	0.6	2.9	18.1	100	120
150	Vorarlberg	0.3	2.5	18.5	102	123
	Portugal	9.9	5.2	9.9	100	66
151	Norte	3.5	4.5	8.8	89	58
152	Centro	1.7	3.4	7.0	71	46
153	Lisboa e vale do Tejo	3.3	6.5	13.7	138	91
154	Alentejo	0.5	9.2	5.9	60	39
155	Algarve	0.3	4.7	8.5	86	56
156	Açores	0.2	5.0	6.2	63	42
157	Madeira	0.3	3.9	6.6	67	44
	Suomi (Finland)	5.1	16.4	14.6	100	97
158	Uusimaa	1.3	13.3	18.8	129	125
159	Etelä-Suomi	1.8	16.8	13.9	95	92
160	Itä-Suomi	0.7	18.7	11.7	80	77
161	Väli-Suomi	0.7	16.2	12.6	86	83
162	Pohjois-Suomi	0.6	19.1	12.7	86	84
163	Ahvenanmaa	0.025	5.1	20.7	142	137
	Sverige (Sweden)	8.7	7.9	16.6	100	110
164	Stockholm	1.7	6.1	20.3	122	134
165	Östra Mellansverige	1.5	7.1	15.1	91	100
166	Småland Med Öarna	0.8	6.0	15.8	95	105
167	Sydsverige	1.2	6.9	15.6	94	103
168	Västsverige	1.7	7.1	16.1	97	107
169	Norra Mellansverige	0.9	8.3	15.2	92	101
170	Mellersta Norrland	0.4	8.9	16.8	101	111
171	Övre Norrland	0.5	8.7	16.9	102	112
	EUR 15	369.7	10.4	15.1	—	100

Objective 2 regions of older EU Member States. The proportion of population engaged in agriculture, forestry and fishing is small (6.9 in Finland, only 3.3 in Sweden). Living conditions in rural areas in the two countries are satisfactory with a good provision of services. The very low density of population in central and northern Finland, and comparable areas in the north of Sweden, have been designated 'special' under the new Objective 6 of the EU Structural Funds. As in Norway, one of the problems in these most northern areas of Europe is to make conditions attractive enough to prevent or minimise depopulation through migration to cities further south.

processing and marketing structures in the agricultural, forestry and fishing sectors, again without explicit application in particular parts of the EU (3.8 per cent).

- *Objective 5b* To promote the development of rural areas (4.3 per cent), the location of which is shown in Figure 12.3 and which cover mainly thinly populated areas not included in Objective 1 regions.
- *Objective 6* since the beginning of 1995, to assist in the development of thinly populated, rural, areas in Sweden and Finland north of 60°N.

The data in column (2) of Table 12.5 show that (as has been the case earlier), the allocation of Structural Funds differs greatly among Member States. The Netherlands receives about one-third of the EU average of 384 ECU per inhabitant, Ireland more than four times the EU average. It is interesting at this point to speculate how the allocation would change should, for example, Poland, the Czech Republic, Slovakia, Hungary and Slovenia join. Would the grand total be increased, or would the current proportion allocated to the present weakest Member States be cut?

There follows a summary of the appropriations of the Structural Funds for 1994–9 by Objective and by country.

Objective 1 Regions

These regions receive the largest share of the allocation. Four Member States, Denmark, Luxembourg, Finland and Sweden, do not have any areas defined as lagging in development. At the other extreme, the whole of Greece, Ireland and Portugal are eligible. In between, Germany is subdivided along the divide between former West and East, with the whole of the GDR except Berlin qualifying. Most of Spain and roughly the southern half of Italy also qualify. As if to emphasise the dynamic and flexible nature of Objective 1 funding, the Abruzzi region of Italy will hold Objective 1 status only during 1994–6. In the remaining five countries, only relatively small parts of the total population are within Objective 1 zones:

- In France, Corse, part of the former coal-mining area of Nord-Pas-de-Calais, and the four Départments d'Outre-Mer.
- In the UK, the NUTS level 2 region of Merseyside (with Liverpool), Highland and Islands (northwest Scotland), and Northern Ireland.
- In Belgium the heavy industrial region of the Borinage (Hainaut).
- In the Netherlands, Flevoland, the newly reclaimed, predominantly agricultural region by the Zuider Zee.

- In Austria, Burgenland, which lies along the border with Hungary.

The smaller areas designated for Objective 1 funds listed under the last five countries are lagging for specific reasons. On the other hand, in former East Germany, Greece, Spain, Ireland, Italy and Portugal, entire countries or large parts of them have blanket coverage, which includes both agricultural and industrial areas, with both predominantly rural settlements and large cities. In the smaller areas, special surgery is required to solve local problems of backwardness. In the large areas the whole infrastructure needs to be upgraded if convergence is to be achieved. Portugal, in particular, has economic and social indices that place it among the most developed countries of the developing world (see e.g. *HDR 1995*, UNDP 1995, Table 1), similar to Argentina, Chile and Uruguay in Latin America and also to the Czech Republic and Slovakia in Central Europe. It will be of more than academic interest to follow the hoped for convergence of Portugal, Greece and former East Germany towards the EU mean in the decades to come. It has been pointed out, however, that the mean itself will move upwards, making the laggards chase a receding target. If the present 'development gap' in the EU continues in its present form, then there is little chance that in the world as a whole the much wider gap, involving many times more people, can be narrowed.

Objective 2 Regions

These regions are mostly small in area but are numerous and are scattered widely over areas not designated as Objective 1. They include almost all of the coalfield areas of the UK, France and Germany as well as centres of the metallurgical industries, shipbuilding and textile manufacturing. The newer manufacturing areas established away from the main coalfields,

characteristically in southern Germany and Lombardia, are not defined as 'declining industrial' whereas the Turin (northwest Italy) and Barcelona (Spain) areas are.

Objective 5b Regions

These regions, like Objective 1 and 2 regions, are located on the map in Figure 12.3. They are rural areas in need of development. Although extensive in the area they cover, they are limited in total population and receive only 4.3 per cent of the total Structural Funds. They are widespread in mountain areas (e.g. the Pyrenees, Alps, Apennines, Wales), the regions in former West Germany adjoining the former Iron Curtain, and in areas with poor natural conditions for agriculture (e.g. parts of northern Germany, Bretagne in northwest France, the Ebro valley in Spain). The need to ensure the adequate provision of services is one of the problems in Objective 5b regions. The dispersed nature of rural population may result in greater average transport costs required to ensure access to services.

The allocation of resources explicitly to raise levels in backward or declining regions of the EU, whether provided at national or at EU level, results in a modest transfer from richer to poorer areas. Figure 12.5 shows areas that have received the most assistance over a long period. Much of the investment is placed to improve infrastructure in poorer areas, building roads, providing telecommunications and upgrading water supply. A much greater amount of investment in the EU comes from private sources, but such investment is placed where conditions are (or are perceived to be) the most favourable to ensure returns on capital. Whether richer or poorer areas receive the investment is incidental. For example, the US motor vehicle companies General Motors and Ford have had investments in Western Europe since before the Second World War. They are

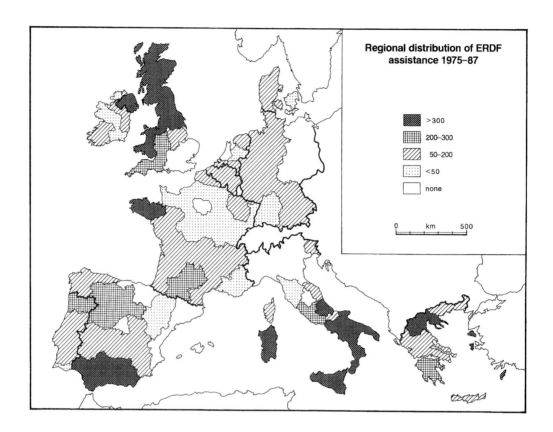

Figure 12.5 Regional distribution of ERDF assistance 1975–87 in millions of ECU at NUTS level 2 (UK at NUTS level 1)
Source: Commission of the EC 1989: 65, Table 20

mainly in the London area (Luton, Dagenham), Belgium (Genk) and the Ruhr area of Germany, among the most affluent regions of the EU.

Most of the private investment in the EU originates in Member States. Increasingly companies are spreading their investment beyond the confines of the Member States in which they originated and are based. Capital from EU companies is also invested outside the EU in developed countries, especially the USA, and also in developing countries, especially South and Southeast Asia. This 'loss' of investment is roughly balanced by investment in the EU from non-EU countries, especially the USA

and Japan. Such investment is placed within the EU to reach the very large Single Market without having to penetrate its common tariff. This investment will be examined more closely in Chapter 14, but it may be noted here that certain countries, particularly the UK, Ireland and the Benelux countries, are preferred for investment by many companies and that a number of regions in the UK with declining industrial sectors have been chosen. The influence of the English language on choices of location is difficult to quantify but seems to be considerable.

FURTHER READING

Cole, J. P and Cole, F. J. (1991) 'The New Germany', *Focus*, 41 (3), Fall: 1–6.

Commission of the European Communities (1991) *Europe 2000 Outlook for the Development of the Community's Territory*, Office for Official Publications of the European Communities.

Ellger, C. (1992) 'Berlin: legacies of division and problems of unification', *Geographical Journal*, 158, March: 40–6.

European Communities (1988) *Disadvantaged Island Regions*, Economic and Social Consultative Assembly, Brussels, July.

European Parliament (1996) *The Regional Impact of Community Policies Executive Summary*, Regional Policy series, W-16 External Study, European Parliament. Directorate General for Research. A full version is available in French.

European Voice (1995) 'Survey: regions', 23–9 Nov., pp. 13–18, covers several aspects of new trends in the role of EU regions.

'Finland' (1995) *The Geographical Magazine*, Supplement, LXVII (10), October.

'Germany' (1996) *The Economist*, 339, (7964), 4 May: 21–3.

Grimes, S. (1992) 'Ireland: the challenge of development in the European periphery', *Geography*, 77 (1), January: 22–32.

Jones, B. and Keating, M. (1995) *The European Union and the Regions*, Oxford: Clarendon Press.

Keir, M. (1992) 'The strange survival of nationalism', *Geographical*, LXIV (7), July: 25–9. Nationalism as a force in European politics has not been consigned to the dustbin of history.

King, R. (1992) 'Italy: from sick man to rich man of Europe', *Geography,* 77 (2), April: 153–69.

King, R. and McGrath, F. (1993) 'The immigration isle', *Geographical*, LXV (11), November: 22–5. The Western extremity of Ireland, problems of a rural community on the EU periphery: Achill Island.

Mead, W. R. (1991) 'Finland in a changing Europe', *The Geographical Journal*, 157, Part 3, November: 207–315.

Naylon, J. (1992) 'Ascent and decline in the Spanish regional system', *Geography*, 77(1), January: 46–62.

Neven, D. and Gouyette, C. (1995) 'Regional Convergence in the European Community', *Journal of Common Market Studies*, 33 (1), March: 47–65.

Scargill, D. I. (1991) 'Regional inequalities in France, persistence and change', *Geography*, 76 (4), October: 343–57.

Spencer, K. (1994) 'Investing in Northern Ireland', *Geographical*, LXVI (9), 43–5. Northern Ireland is having greater success attracting foreign investors than British ones.

Wild, T. (1992) 'From division to unification: regional dimensions of economic change in Germany', *Geography*, 77 (3), July: 244–60.

Wild, T. and Jones, P. (1993) 'From peripherality to new centrality? Transformation of Germany's *Zonenrandgebiet*', *Geography*, 78 (3), July: 281–94.

Wild, T. and Jones, P. N. (1994) 'Spatial impacts of German unification', *The Geographical Journal*, 160, Part 1, March: 1–16.

Williams, R. H. (1996) *European Union Spatial Policy and Planning*, London: Paul Chapman. Deals with the influence of EU policies and programmes on spatial planning in a supranational context, at both EU and overall European levels.

Winchester, H. P. M. (1993) *Contemporary France*, Harlow: Longman.

13

ENLARGEMENT OF THE EUROPEAN UNION

'In the longer term, after 1990, given current and prospective development efforts, Soviet natural resources could conceivably be more important in the world economy – or less'.

(Jensen *et al*. 1983)

- The break-up of the Soviet bloc has freed a large number of countries, including former Soviet Republics, to establish stronger links with the EU. Many of these countries are seeking membership of the EU.

- The EU is increasingly involved with four main groups of countries, several not technically in Europe: the remaining countries of EFTA; the countries of Central Europe; Russia and other former Soviet Republics; and the countries of the Mediterranean basin.

- Apart from the EFTA countries, together with Malta and Cyprus, all prospective new members of the EU are much poorer that the EU average and in addition have social and environmental problems that could cause complications if they join.

According to the general spirit and principles of the European Union, any country in Europe should be able to join the EU, providing that it satisfies certain conditions, already discussed in Chapter 1 (see also Figure 1.1 on the traditional eastern limit of Europe). When the Treaty of Rome was signed and indeed until the late 1980s the theoretical readiness to consider countries in Eastern Europe was a hollow, academic statement of intent since very few imagined that the Soviet bloc would disintegrate. When the Soviet Union relaxed its hold on Central Europe and then in 1991 itself broke up, it became possible both for 'neutral' countries, notably Austria, Finland and Sweden, to join the EU without Soviet opposition, and for

former CMEA partners quickly to increase ties with Western Europe. It should be appreciated that both Russia and Turkey are only partly in Europe, but they do qualify for EU membership with respect to being part of Europe. Most of Russia's territory is in Asia but not its population, and most of Turkey's territory and population are in Asia.

EXTENDING THE EU

Three of the former EFTA countries are full members of the EU and two of the remainder, Norway and Iceland, are in the European Economic Area (EEA). Future enlargement of the

Plate 13.1 The Berlin Wall, the removal of which at the end of 1989 started the disintegration of the Soviet bloc, with far-reaching consequences for the EU

EU therefore involves non-EFTA countries of Europe, as many as 12 of which are considered to be possible candidates for membership. In this chapter the prospects for membership of a large number of European countries will be examined, many of them emerging and gaining international recognition only in the 1990s. The countries are grouped for convenience and are discussed in descending order of the likelihood that they could join the EU. In passing, it may be noted that although further integration and cohesion can be continued as the EU enlarges, the diversity and proliferation of backward economies, the addition of new languages, and other features of prospective new candidates, make simultaneous enlargement and

greater integration a difficult process. Why enlarge at all?

One of the main economic reasons for creating and subsequently enlarging the EEC was to produce a large market in which, thanks to economies of (large) scale, pooled capital resources and collaborative research, the manifest economic strength of USA, and more recently that of Japan, could be matched and rivalled. In the event, the total GDP of EUR 15 is only slightly less than that of the North American Free Trade Agreement area (the USA, Canada and Mexico) and almost three times as large as that of Japan. The only other 'developed' bloc, that of the USSR and its CMEA partners, no longer exists. There are now other competitors, however, in particular the emerging economies in the Far East, against which the strength of the Single Market still needs to be concentrated and for which the economies of scale and benefits of cooperation within the EU still apply.

Until the early 1990s there was a strategic reason for enlarging the EC, namely to weaken the Soviet bloc, in reality the Russian Federation, by collecting any fruits that fell from the Soviet tree. Thus East Germany was absorbed instantly, once the opportunity came. The high cost of achieving convergence, to West Germany in particular and to the EU in general, was known, but was regarded as a worthwhile price. At first it was thought that the USSR itself would remain intact, or at most give up control of the Baltic Republics. In effect, as well as the Russian Federation itself, nine other former Soviet Socialist Republics *in Europe* have now become independent countries. Although Russia is sensitive about the prospect of losing control entirely over its former colonies not only politically but also strategically and economically, there is nothing to prevent, for example, the Baltic Republics, Ukraine or Georgia from applying for EU membership and indeed the Baltic States have already applied. The main argument used to justify enlargement to take in new Member States from Central and Eastern Europe is political, since the fragile democracies recently established in these countries need membership of the EU to foster and strengthen their political and social development. Although there is still a lack of security, with instability in the region, in particular in the Russian Federation, which remains a nuclear power, the strategic argument for enlargement is less significant than before.

Before 1990, the foreign trade of the countries of CMEA with Western Europe was already increasing, especially with West Germany. About half of the total value of Soviet foreign trade was with its CMEA partners in Central Europe and about a sixth with Western Europe, particularly with West Germany. In the 1990s, investment from Western Europe in Central Europe and Russia has increased appreciably. Earlier in the life of the EEC, the increase of trade between the countries of Western Europe was used as a strong argument for the establishment of the customs union and the Single Market. As the trade of Central European countries with the former USSR diminishes and their trade with the EU increases, the same argument may be applied.

On the other hand, the apparent advantage of adding new countries to make even larger the already large 'home' market of the EU could bring problems. Among the non-EU countries in Europe, only Norway, Iceland and Switzerland match up to the present economic level of the EU, while Malta and Cyprus are very small and are similar in development to southern Italy and Greece. Although economic conditions and GDP levels per capita differ enormously among the countries of Central Europe and the former USSR, all are much poorer that the EU average and most are much poorer even than Greece or Portugal. In addition to the restructuring of their economies and the need for greatly improved infrastructure, much needs to be spent on cleaning up present pollution and preventing further pollution in the future. Once in the EU they would become the source of legal migrants moving up the GDP ladder into the more affluent EU Member States. Since they also have a surplus of cheap agricultural products, these would be a problem for the EU's Common Agricultural Policy. They could themselves absorb much of the EU budget for many years to come, thus depriving the poorest regions of the present EU of some if not all of the future assistance they might justifiably otherwise expect.

In addition to the economic disadvantages to the present EU of accepting a number of new members, there are organisational problems. Most countries are small by EU standards but almost all would bring a new language, while each would expect its allocation of MEPs and at least one commissioner. In the next section some of the criteria relevant to the enlargement of the EU are examined in general terms. In the remaining sections groups of non-EU countries in or close to Europe are described and their positive and negative attributes in terms of EU membership are assessed. A possible 'new map' of Europe, with strengthened links between the EU and Central Europe, and within Central Europe, is envisaged.

WHAT IS NEEDED FOR A COUNTRY TO JOIN THE EU?

Future applicants for EU membership will have to satisfy many conditions before their applications are likely to gain the unanimous acceptance of current members, at least unless majority voting is used to determine future membership. In this section a number of important considerations are noted, some of them not, however, pertinent for all countries. Some of the points are related to numerical data in Table 13.1.

Location

The location of the potential applicant in relation to the existing limit of the EU. Is it contiguous (e.g. Poland) or is it separated by non-EU territory (e.g. Romania), as Greece is at present? How far is the country from the centre of population of EUR 15 (near Strasbourg), from the centre of area (near Namur, Belgium), or from the centre of economic production (near Karlsruhe, Germany)? Is the country an island? Does the country share a boundary with the Russian Federation (e.g. Belarus)? Is it landlocked?

Demographic Features

Two considerations are of outstanding importance. First, (column (1) in Table 13.1) a large population size (e.g. Poland is nearly as large as Spain) implies a large impact on EUR 15, while a small population has an organisational nuisance value (the Luxembourg syndrome). Since almost every country in Europe now experiences little if any population growth and many countries can be expected to decline in population (see column (11)), population change is a major consideration only in Turkey and the Maghreb countries.

Political Aspects

Applicants should have (and keep) in place a democratic structure, with elections by universal adult franchise, and a multi-party system. There should be respect for human rights such as gender and ethnic equality as well as freedom of association and of the press. The call to 'bring back the communists' and the election of former party members of the communist period in various countries of the former Soviet bloc is a matter of concern in the EU.

Economic Level

Applicants for membership of the EU should have economic levels and conditions compatible with those of the market economies of the EU countries. Depending on the measure of production/product used, there is an average gap of somewhere between 3:1 and 5:1 between the EUR 15 countries and those of Central and Eastern Europe. The annual World Bank Report recognises four income classes in the world. Out of 134 countries considered, 24 are high-income economies, 22 upper middle income and 39 lower middle income (the rest are low income). Thirteen of the EUR 15 countries are high-income economies, while Greece and Portugal, together with Slovenia, Hungary, Estonia and Belarus, are upper middle income. All the remaining countries eligible for EU membership or in the Maghreb are in the third category, lower middle income.

Table 13.1 Data on possible new members of the EU and on other selected countries

	(1) Population in millions 1995	(2) GNP per capita 1993	(3) Real GDP per capita 1992	(4) Infant mortality per 1,000	(5) Life expectancy in years	(6) Education index	(7) Human devt index	(8) % in agriculture 1994	(9) Has cultivated per 100 population	(10) Cereal yields kg per ha 1992-4	(11) Total fertility rate	(12) Privatisation % 1994
Selected EU and EFTA countries												
1 Denmark	5.2	26,730	19,080	5.7	75.3	94	920	4	49	5,190	1.8	n.r.
2 Greece	10.5	7,390	8,310	8.3	77.6	88	907	22	34	3,650	1.4	n.r.
3 Switzerland	7.0	35,760	22,580	5.6	78.0	91	925	3	7	6,110	1.5	n.r.
4 Norway	4.3	25,970	18,580	5.8	76.9	95	932	4	21	3,490	1.9	n.r.
PHARE and associated countries												
5 Hungary	10.2	3,350	6,580	11.6	69.0	88	856	10	48	3,620	1.7	55
6 Czech Rep	10.4	2,710	7,690	8.5	71.3	89	872	9^2	32	4,120	1.7	65
7 Poland	38.6	2,260	4,830	13.7	71.1	91	855	18	38	2,570	1.8	55
8 Slovakia	5.4	1,950	6,690	15.6	70.9	90	872	9^2	30	4,110	1.9	55
9 Slovenia	2.0	6,490	n.a.	7	73	n.a.	n.a.	22^1	15	3,640	1.3	30
10 Romania	22.7	1,140	2,840	23	70	85	703	17	44	2,440	1.4	35
11 Bulgaria	8.5	1,140	4,250	16	71	84	796	11	49	2,770	1.4	40
12 Estonia	1.5	3,080	6,690	16	69	89	862	n.a.	74	1,760	1.3	55
13 Latvia	2.5	2,010	6,060	16	69	89	857	n.a.	66	1,790	1.5	55
14 Lithuania	3.7	1,320	3,700	16	70	88	769	n.a.	61	2,030	1.7	50
15 Croatia	4.5	n.a.	n.a.	12	70	n.a.	n.a.	22^1	29	4,120	1.4	40
16 Macedonia	2.1	820	n.a.	24	72	n.a.	n.a.	22^1	31	2,460	2.2	35
17 Albania	3.5	n.a.	3,500	33	72	80	739	46	21	2,510	2.9	50
Selected MEDITERRANEAN countries												
18 Turkey	61.4	2,970	5,230	53	67	74	792	46	45	2,110	2.7	n.r.
19 Tunisia	8.9	1,720	5,160	43	68	63	763	21	57	1,210	3.4	n.r.
20 Algeria	28.4	1,780	4,870	55	67	60	732	22	29	840	4.4	n.r.
21 Morocco	29.2	1,040	3,370	57	63	41	554	33	37	930	4.0	n.r.
Selected TACIS countries												
22 Ukraine	52.0	2,210	5,010	15	69	87	842	n.a.	67	2,950	1.6	30
23 Belarus	10.3	2,870	6,440	13	70	90	866	n.a.	61	2,600	1.5	15
24 Russia	147.5	2,340	6,140	19	68	89	849	n.a.	88	1,610	1.4	50
Malta and Cyprus												
25 Malta	0.4	n.a.	8,280	9	76	83	880	3	negl.	n.r.	2.0	n.r.
26 Cyprus	0.7	10,380	15,050	9	77	88	906	19	23	2,430	2.3	n.r.

Sources: (1), (4), (11) WPDS 1995 (PRB 1995); (2) World Bank 1995, Table 1; (3), (5)–(7) UNDP 1995, Table 1; (8)–(10) FAOPY 1994, (FAO 1995, Tables 3, 1, 15); (12) Naudin 1994a: 18

Notes n.r. not relevant; n.a. not available; negl. negligible
1 For the whole of Yugoslavia, 1990
2 For the whole of Czechoslovakia, 1992

Fuller descriptions of the variables
(2) Gross National Product per capita in US dollars, 1993.
(3) Real Gross Domestic Product per capita at parity purchasing power (ppp) in US dollars, 1992.
(4) Deaths of infants under one year of age per 1,000 live births, early to mid-1990s.
(5) Life expectancy at birth in years, early to mid-1990s.
(6) Education index, a composite index on a possible scale from 0, no educational facilities, to 100, the highest obtainable at the time of calculation, 99 achieved by Canada. The elements in the index are adult literacy (more discriminating among developing countries) and first-, second-, and third-level gross enrolment ratio.
(7) The human development index combines life expectancy (to represent health, the body), educational level (the mind) and basic income, cut off above a ceiling of 5,374 US dollars per capita, thus discriminating among developing countries while differentiating only slightly among more affluent countries. Theoretical maximum score 1,000, minimum score near zero. Canada with 950 is the highest scorer, Niger with 207, is the lowest, ranked 174. The world average is 759, while the 'industrial' countries average 916.
(8) Economically active population in agriculture as a percentage of total economically active population 1994.
(9) The cultivated area of each country (1993) consists of arable land under field crops, fallow and land under permanent crops such as vines and olive trees. Its availability is shown here in relation to population, the area being expressed in hectares per 100 inhabitants. As a result of a number of factors, the productivity of a given area of cultivated land varies from region to region on account of both inherent natural conditions (moisture and thermal resources, slope, quality of soil), and the intensity of application of means of production such as fertilisers. In addition, some countries are much more generously endowed per capita than others with natural pastures or forests.
(10) Cereal yields in kilograms per hectare, average for 1992–4.
(11) Total fertility rate (TFR), the average number of children a woman will have assuming that current age-specific birth rates will remain constant throughout her child-bearing years (usually considered to be ages 15–49), early to mid-1990s.
(12) The percentage share of the private sector in GDP, mid-1994.

The World Bank uses GNP calculated according to current exchange rates. These inflate the GNP of countries with strong currencies (e.g. Switzerland, Norway, Japan) and reduce the GNP of those with weak or artificially fixed currencies (e.g. China, India). In contrast, the UN Human Development Programme uses 'real' GDP, or parity purchasing power US dollars, thus removing the stretching effect of exchange rates (see column (3) in Table 13.1). Thus the Central European countries compare very unfavourably with the EU average GNP according to World Bank data, roughly at 5:1, but less unfavourably, at about 3:1, according to real GDP, column (3). Progress towards privatisation differs greatly among the countries of Central Europe and of the former USSR, column (12).

Agriculture

In terms of EU budget expenditure, agriculture is the most 'expensive' sector, involving the Common Agricultural Policy plus funds spent on backward, predominantly rural, areas. The data in columns (8)–(10) draw attention to the large proportion of economically active population still in agriculture in Central and Eastern Europe, to the comparatively large area of cultivated land per inhabitant, and also to the low cereal yields compared with those in most of the EU. Thus jobs will be lost from the land, while a more efficient and productive system of organisation could greatly increase total agricultural output through higher yields.

Natural Resources

Apart from Russia in general and individual countries for specific products (e.g. Poland coal, Algeria oil and gas, Morocco phosphates), by world standards the mineral resources per capita of all the countries are modest.

Quality of Life

Infant mortality (column (4)) may seem an esoteric aspect of the quality of life to measure but it serves as a clearer shorthand indicator of health and hygiene conditions than life expectancy (column (5)). Education (column (6)) is calculated on level of literacy and time spent in various levels rather than on the nature and content of what is taught. The HDI index (column (7)) reveals a considerable gap between all EUR 15 and EFTA residue countries, except Portugal, and all the other countries considered.

Cultural/Human Environment

These aspects are even more difficult to quantify than quality of life. The EUR 15 might be reluctant to take on board new members with ethnic, nationalistic, linguistic and ideological problems. As shown in Chapter 12, within the EU itself, such problems are considerable in some countries. Almost all the countries outside the EFTA residue would bring in Slavonic or other languages new to the EU. Romania, Russia, Turkey and Morocco all have ethnic minorities, while all former Soviet Republics contain a considerable minority of Russians. Islam is the main religion in several countries whereas in several former CMEA countries the secular ideological influence of Soviet sponsored Marxism-Leninism is still felt.

Cole and Cole (1993) contains a table (11.2, pp. 268–9), in which ten criteria were chosen to measure the eligibility of various countries for EU membership. Population, location, resources and GDP, political rights, health and education, and language, were all taken into account. In a system of lost points, out of a maximum of 50 that could be lost, the EFTA countries scored best, none losing more than 10. Czechoslovakia (now subdivided), Hungary and Poland were next, losing less than 20 points. In this second edition, data have been included to reflect various aspects of the countries of Central Europe, including ten of the 12 countries considered to be possible future EU members, but excluding Serbia and Bosnia. Peripheral countries are included mainly for comparative purposes: Russia, Belarus, Ukraine, the Maghreb countries and Turkey.

One of the main variables determining the direction of EU funding to backward and less favoured regions is GDP or GRP. Figure 13.1 illustrates vividly the economic abyss that lies beyond the EU and EFTA to the east and south, with background data provided in Table 13.2. Two different measures of GDP/GNP have been compared. In Figure 13.1 each country or group of countries is represented by a vertical bar or column proportional in (horizontal) width to population and proportional in height to GDP per capita. Real GDP has been chosen to represent economic differences because it shows what is actually produced and consumed in each country. The stretching or shrinking of GNP as converted to US dollars at current exchange rates is shown in the diagram for five single countries (letter d). The economies of Germany, France and Italy are flattered by exchange rates, while those of Russia and Turkey (weak currencies) are diminished greatly. The diagram shows that if what we have called the 'next Five' prospective joiners (Poland, the Czech Republic, Slovakia, Hungary and Slovenia) join the EU, the average real GDP per capita of 17,890 dollars for EUR 12 plus EFTA would be lowered to 16,080. Only Greece, Spain, Portugal (and Ireland) would remain below the new average. The example of changes in the level of infant mortality in selected EU and non-EU countries in Figure 13.2 shows that convergence has occurred in this area of healthcare in recent decades.

THE RESIDUE OF EFTA, MALTA AND CYPRUS

With the accession of Austria, Finland and Sweden to the EU in 1995, EFTA became the 'rump' of four countries, Norway, Switzerland, Liechtenstein and Iceland. All four countries satisfy the criteria necessary for consideration for EU membership. With 30,000 inhabitants,

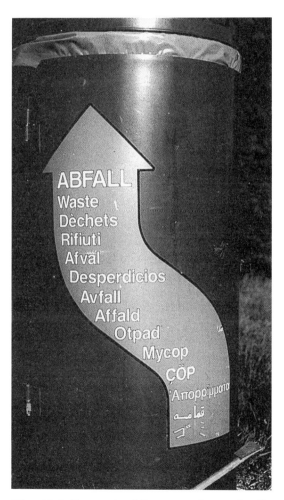

Plate 13.2 Crossroads of Europe: an internationally labelled waste bin in an Autobahn parking area, southern Germany. Nine of the eleven main EU languages are represented, together with five others

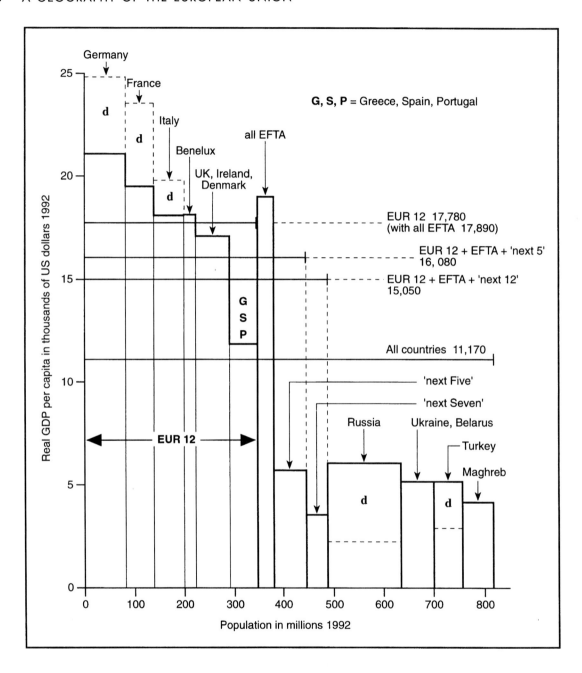

Figure 13.1 The economic 'abyss' beyond the European Union. Real GDP per capita income levels in the EU and EFTA compared with those of groups of countries in the 'near abroad'. For convenience Austria, Finland and Sweden are included with EFTA (as of 1992). Cyprus, Malta, the former Yugoslavia (except Slovenia) and Albania are not included. Note that Russia and Turkey extend well beyond the conventional confines of Europe and the Maghreb region is in Northwest Africa

Table 13.2 The background data for Figure 13.1

	Population in millions	Population – cumulative	Real GDP, billions of dollars	Real GDP in dollars per capita	Cumulative GDP
Germany	80.6	80.6	1,702	21,120	1,702
France	56.9	137.5	1,110	19,510	2,812
Italy	58.0	195.5	1,049	18,090	3,861
Belgium, Netherlands, Luxembourg	25.6	221.1	465	18,160	4,326
UK, Ireland, Denmark	66.5	287.6	1,136	17,080	5,462
Greece, Spain, Portugal	59.4	347.0	706	11,890	6,168
EFTA as of 1992	33.1	380.1	629	17,890	6,797
Original Visegrad plus Slovenia	66.3	446.4	380	5,730	7,177
Romania, Bulgaria, 3 Baltic States	40.1	486.5	145	3,620	7,322
Russia	149.3	635.8	917	6,140	8,239
Ukraine, Belarus	62.4	698.7	327	5,240	8,566
Turkey	59.2	757.4	310	5,240	8,876
3 Maghreb	60.6	818.0	258	4,260	9,134

Sources: WPDS 1992 (PRB 1992); UNDP 1995, Table 1
Note: The following countries in or near Europe are not included: former Yugoslavia except Slovenia, Albania, Moldova, Georgia, Armenia, Azerbaijan, Malta, Cyprus. The last two are included in Table 13.1

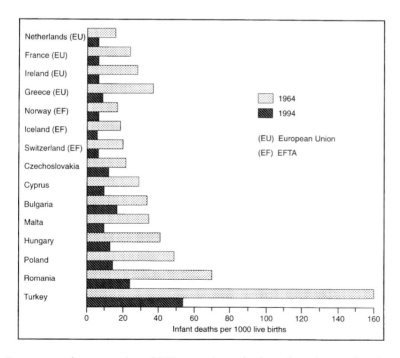

Figure 13.2 Convergence between selected EU countries and other selected countries, the example of infant mortality, 1964 and 1994. Compare with Figure 1.6

however, Liechtenstein is a handful of villages on behalf of which Switzerland carries out essential domestic and international functions and transactions. Table 13.1 contains data about Switzerland and Norway and includes two existing EU members for comparison, Denmark and Greece, which have the highest and lowest GNP per capita scores in the EU. In the rest of this section the residue of EFTA, together with Malta and Cyprus, will be examined with special reference to their 'qualifications' for EU membership. Other possible future members are discussed in the following section.

Norway

Negotiations for Norway's application for EU membership were completed in 1994, but as once already in a referendum in 1972, the majority of Norwegians voting on EU membership rejected the move to join. About 60 per cent of the exports of Norway go to just five EU countries, the UK, Germany, Sweden, France and Denmark. Assuming oil and natural gas continue to be in demand, Norway has enough reserves to last several decades at current production rates (more gas than oil). Its membership of the European Economic Area (EEA) gives it virtually all the advantages of EU membership while leaving it free of decisions that might be to its disadvantage, such as joining the Common Fisheries Policy. Most of its North Sea oil and gas reserves are close to the median line in the North Sea separating its economic zone from that of the UK. For technical and commercial reasons it is necessary to transport much of Norway's oil and gas to the UK and Germany (see Figure 5.1 in Chapter 5) rather than to Norway itself.

Iceland

Iceland has only about 260,000 inhabitants, fewer even than Luxembourg, which is by far the smallest country in the EU at present. Organisationally the presence of Iceland in the EU would be a nuisance rather than a problem, but economically its main industry is fishing, a sector that has caused many problems and several minor conflicts in Western Europe since the Second World War. Given the decline of the Soviet-Russian naval influence in the North Atlantic, Iceland's strategic location in NATO no longer ensures it special consideration.

Switzerland

Switzerland has so far distanced itself from the idea of EU membership, although its location, ethnic and linguistic background and trading links all effectively make close involvement with the Union unavoidable. Numerous road and rail links between major centres of population in the EU pass through Switzerland. The three main languages of Switzerland are German (spoken by almost 75 per cent), French (20 per cent) and Italian (5 per cent), while in religion the Christians are roughly equally divided into Roman Catholic and Protestant. About half of the exports of Switzerland go to four EU countries, Germany, France, Italy and the UK. To the outsider, Switzerland seems both cosmopolitan and parochial, but its neutrality no longer seems relevant either in Europe or globally. By staying out of the EU, Switzerland is able to control migration, to prevent the purchase of land, property and economic enterprises by foreigners, and to determine what foreign traffic can cross its territory, too many heavy goods vehicles being a prime concern. Switzerland could remain indefinitely a large 'onshore island' embedded in the EU.

Malta

After Malta became independent from the UK in 1964 its strategic location in the Mediterranean made it attractive to the USSR, seeking to strengthen its naval presence in the region.

Plate 13.3 A new road link in the transport network of Europe: the approach to a tunnel through the Alps to connect south Austria with the newly established country of Slovenia

Malta's maritime connections are maintained by the large number of ships registered there and by ship repairing, as well as by light industry, tourism and the growth of financial transactions. English and Maltese languages are used, the latter replacing Italian as an official language in 1994.

Cyprus

The situation in Cyprus is much more complex than that in Malta. One particular problem that has to be resolved before the country can become a full member of the EU is agreement between Greece and Turkey on the reunification of the southern, Greek-speaking and northern Turkish-speaking parts, partitioned following the Turkish invasion of 1974. Earlier there was pressure from Greek leaders to make the whole of the island part of Greece (enosis). Turkish intervention was justified as a move to protect the Turkish-speaking minority, accounting for less than a fifth of the total population. Militarily and strategically, Greece and Turkey are both key members of NATO, and southern Cyprus has two British air bases. The Cypriot economy is also highly dependent on tourism.

Malta, Cyprus, Norway, Iceland and Switzerland have a combined population of about 13 million inhabitants, little more than Portugal, rather less than the Netherlands. They are prosperous and stable economically. The three EFTA countries would be losers in relation to the EU budget if they joined the EU, contributing to the assistance increasingly spent in the poorest regions of the Union, although they might benefit in other ways. The arrival of Malta and Cyprus would make little difference to the budget. A comparison of the data in Table 13.1 shows that Switzerland and Norway are broadly comparable with Denmark, one of the most affluent EU countries. The next group of countries to be considered, the PHARE group, compare unfavourably economically in many respects even with Greece, one of the poorest EU members.

Figure 13.3 shows the broad features of the new Europe that has emerged since the end of the Cold War. Figure 13.4 shows one of the essential developments needed if Central Europe is to be integrated more closely with the EU and EFTA – the improvement to the existing transportation network of the region.

THE COUNTRIES OF THE PHARE PROGRAMME

The acronym PHARE stands for Poland–Hungary Assistance in the Restructuring of Economies. It has been extended from the two initial countries to embrace 13 altogether at the time of writing (see Table 13.1). It therefore includes all the countries of Central Europe except Serbia (with Montenegro) and Bosnia. By the time of the Madrid summit in December 1995, at which enlargement was discussed, ten of these countries had either applied for EU membership or were about to do so. Under the new Europe Agreements, the new generation of association agreements that recognise the aims of membership of the Union, there are ten countries: Bulgaria, the Czech Republic, Estonia, Hungary, Latvia, Lithuania, Poland, Romania, Slovakia and Slovenia.

Applying for EU membership and joining the EU are two different matters, as the UK found in the 1960s. Troev and Naudin (1995) assume that entry of some if not all the applicants will take place in due course, but in December 1995 Helmut Kohl was more spe-

cific, pointing to the Czech Republic, Hungary and Poland as the favourites. According to Felipe González, the Prime Minister of Spain, transitions of up to 15 years will be needed to offset the cost for EU farm and regional aid programmes, while the entry of other countries could take even longer. In the view of the authors, five countries stand the best chance of being in the first group to join, Helmut Khol's three, together with Slovakia and Slovenia. These five countries are described in more detail first, the remainder more briefly.

Hungary, Poland, the Czech Republic, Slovakia, Slovenia

With 67 million inhabitants, the five countries discussed in this subsection are together considerably larger in total population than the UK, Italy or France. The leaders of all five countries have expressed the hope that they can join the EU within a few years. Association agreements had already been made or were imminent at the time of writing. A decade ago, when the Cold War was still a dominant influence in determining the alignment of countries in Europe, the prospect of 'capturing' East Germany and the above countries would have seemed a significant gain in the context of world affairs. Although the official policy of the EU is to proceed with further enlargement, doubts are expressed elsewhere.

The economic consequences of the attainment of full EU membership by the four original Visegrad countries, Poland, the Czech Republic, Slovakia and Hungary, have been calculated by R. Portes of the Centre for Economic Policy Research (CEPR) in London, reported by Naudin (1994b):

> Full membership of Poland, Hungary and the former Czechoslovakia by the turn of the century is out of the question as the combined amounts of grants and subsidies would cost the EU two-thirds of its current budget . . . .

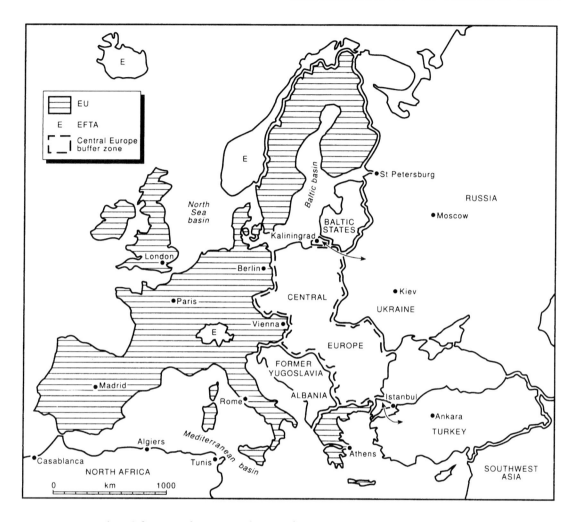

Figure 13.3 The broad features of the 'near abroad' of the European Union

Keen as they are to join the union as soon as possible, the four most advanced Central European countries are too poor, too populous and too agricultural to do so by the year 2000 . . . .

A senior official with EU Commission said that the resulting agricultural overproduction in the EU, if the four countries joined, would be such as to impose a sweeping reform of the common agricultural policy.

Clearly the above prospects are speculation and do not concern the EU Commission enough to prevent negotiations on enlargement.

There follows a discussion of the general features of the five countries (refer to Table 13.1) and then of specific features of relevance to their possible future EU membership.

Location

All five countries share a common boundary with at least one existing EU country (see Figure 1.5). If they join, then the centre of

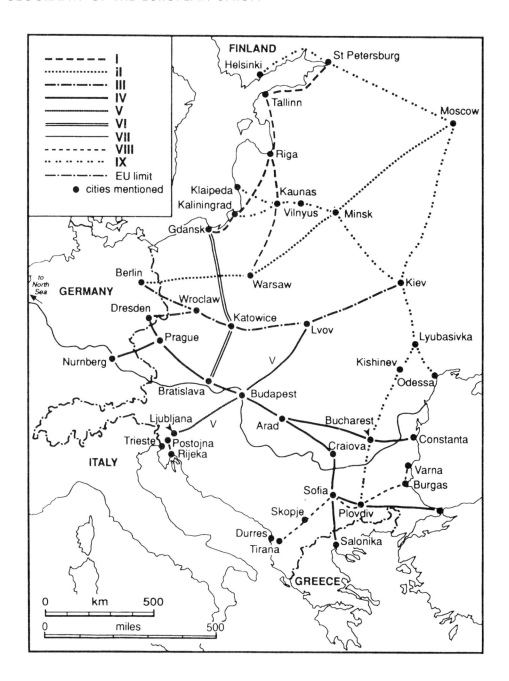

Figure 13.4 Proposed improvements to existing transportation links in Central Europe, and new links. Note that routeway VII is the Danube and its link with the Rhine. The other routes include both road and rail
Source: J. P. Cole 1996: 144

BOX 13.1 PHARE AND TACIS

The PHARE and TACIS programmes are EU initiatives that support the development of democracy and economic growth in the countries that used to make up the CMEA, i.e. the countries of Central and Eastern Europe and the former USSR.

- PHARE, which stands for Poland–Hungary Assistance in the Restructuring of Economies, began in 1990 and was aimed at the two countries in question. During the first five years of its existence, the programme was expanded to cover 11 partner countries listed in Table 13.3, and made 4.25 billion ECU available to them in the form of technical assistance.
- TACIS, in operation since 1991, stands for Technical Assistance for the CIS, and covers 12 of the former Soviet Republics, including Russia, plus the former non-Soviet Mongolian People's Republic shown in Table 13.3. It paid out 1.7 billion ECU for over 2,000 different projects between 1991 and 1994.

Both programmes are based on the principle of codetermination with the partner countries closely involved in deciding on how the funding is to be spent, within a framework agreed with the EU.

The programmes provide know-how from a wide range of non-commercial, public and private organisations to the partner countries, and act as a multiplier by stimulating investment. They also provide direct investment for infrastructure and other projects. The main priorities are common to all the countries concerned:

- Restructuring of state enterprises, including agriculture.
- Private sector development.
- Reform of institutions.
- Legislation and public administration.
- Reform of social services.
- Employment.
- Education and healthcare.
- Development of energy, transport and telecommunications infrastructure.
- Environment and nuclear safety.

In addition, major emphasis in TACIS has been placed on building an efficient food

Table 13.3 Partner countries of PHARE and TACIS

PHARE	TACIS
Albania	Armenia
Bulgaria	Azerbaijan
The Czech Republic	Belarus
Estonia	Georgia
Hungary	Kazakhstan
Latvia	Kyrgyzstan
Lithuania	Moldova
Poland	Mongolia
Romania	Russia
Slovakia	Tajikistan
Slovenia	Turkmenistan
	Ukraine
	Uzbekistan

Sources: COM (95) 366; COM (95) 349

production, distribution and processing system.

Within PHARE, for those countries which have signed Europe Agreements, funding is also focused on meeting the conditions required for membership of the EU, such as preparation for participation in the Single Market and development of infrastructure, especially in border regions. The 1997 EU draft budget provides for funding of over 1.8 billion ECU, a clear commitment to the future of PHARE and TACIS, which have become highly successful programmes following the initial difficulties encountered during the start-up period. The sums involved show the importance attached by the EU to having stable and prosperous neighbours in Europe, whether they eventually become members or not.

population of the enlarged EU would be near Strasbourg. All five countries are closer to the centres of population and of area of the EUR 15 than are several countries already in the EU. The Czech Republic and Slovenia are closer than the UK or Denmark, while the remaining three are much closer than Portugal, Greece or Finland. Great improvements to transportation links would, however, be needed to enable the new members to take advantage of their proximity.

Population

There is no 'right size' for an EU member, but Slovenia's small population size would add a less extreme case of the 'Luxembourg' problem, while Poland is almost as large as Spain in population, introducing a sixth 'larger' EU country. The population of the five countries is expected to grow by only 2–3 million by 2025, reaching about 70 million, a consideration of no more than minor significance.

Natural resources and production

By world standards all five countries are highly industrialised, but the industries are heavily dependent on home produced coal and lignite and on imports of oil and gas, as well as of most raw materials. Until the 1990s the USSR provided many of the primary products used by its CMEA partners in Central Europe. In the context of the present EU, most of the main industrial regions of the five countries would come into the category of Objective 2 regions (see Chapter 12), while large areas still heavily dependent on agriculture would be classed as Objective 1 regions. In Table 13.1, column (8) shows that about a fifth of the economically active population in Poland and Slovenia is still in agriculture, a level comparable to that in Greece. In all but Slovenia, a considerable portion of the total national area is classed as arable (field crops, permanent crops, or fallow), and the amount of cultivated land available per capita is higher than in most EU countries. Cereals are widely grown throughout Europe and cereal yields therefore serve as a broad guide to agricultural performance in general. Cereal yields (column (10)) are much lower in Central Europe than the highest achieved in the EU (e.g. Netherlands 7,530 kg per hectare in 1992–4, France and UK each around 6,500). While environmental and organisational conditions make it unlikely that Dutch or even French levels could be reached in the near future there appears to be scope for some increase soon, especially in Poland.

Plate 13.4 The Ignalina nuclear power station in Lithuania, one of several in Central Europe and the former USSR of increasing direct interest to the EU, in this case in view of Lithuania's application to join the Union

Economy

One of the main considerations is the transition from a centrally planned to a market economy. In this respect, according to Naudin (1994a), by mid-1994 the percentage share of GDP already accounted for by the private sector was 65 in the Czech Republic, 55 in Hungary, Poland and Slovakia, but only 30 in Slovenia.

Environment

As in East Germany, under communist governments little was done in Poland, Czechoslovakia and Hungary to protect the environment and prevent pollution. The burning of lignite, mainly in thermal power stations and in homes, was a particular contributor to pollution. Several suspect nuclear power stations in Central Europe are regarded as inadequately provided with safety facilities.

Ethnicity

Poland and the Czech Republic are among the few countries in Europe in which there is no sizeable ethnic minority. Slovakia, however, contains a considerable minority of Hungarians, while Hungary itself is concerned about the presence of Hungarians in Slovakia as well as in Romania and Serbia. Slovenia escaped the conflict in Yugoslavia in 1991–5 by keeping a low profile.

Languages

With the entry of three new countries in 1995, EUR 15 now has 11 main official languages, while several other languages are also used for some purposes. Each of the five countries described above would bring a different language, introducing for the first time four languages of Slavonic origin as well as Hungarian, which bears no resemblance to any other language in Europe, although it is similar in origin to Finnish and Estonian.

Institutions

The addition of at least five new commissioners (perhaps two for Poland) also raises a problem of finding portfolios for so many newcomers and the prospect that eventually countries might have to pair (e.g. Spain and Portugal, the UK and Ireland) to share such positions.

Much of the evidence presented in this section points to the emergence of new economic and organisational problems even if only 70 million people in five countries are added to the EU. In the sections that follow, other countries that have an interest in closer links with the EU will be described. All of them have many negative aspects, some far more grave than those noted in the five countries already described in this section.

Romania, Bulgaria, the Baltic States, the rest of Yugoslavia, Albania

The above countries, together with Serbia and Bosnia (not in PHARE at the end of 1995) have a total population of about 63 million, slightly larger than the UK or France. By current EU standards, all would qualify as Objective 1 regions although Croatia, Estonia and Latvia are more developed than the rest, while Albania is the least developed. All except the countries of former Yugoslavia and Albania have been strongly influenced by the USSR for four decades.

On strategic grounds, the Baltic States are the countries most likely to meet the strongest resistance from Russia if they are accepted into the EU. Like Ireland, Sweden and Austria, already members of the EU, they would not be obliged to join NATO or a specifically EU defence agreement. The idea in the early 1990s that they might become linked to the EU by the 'side-door' through an association with EFTA countries now sounds out of place since Finland and Sweden were the countries of EFTA closest in all respects to their Baltic neighbours.

The remaining countries, Romania, Bulgaria, the former Yugoslavia excluding Slovenia, and Albania, were in general less highly industrialised and less sophisticated than Hungary, Poland and the former Czechoslovakia and at the same time more rural and heavily dependent on agriculture. Croatia is the country culturally closest to the EU countries. Elsewhere, the legacy of the Ottoman Empire has left religious and ethnic problems, with an overlay of Soviet-style planning successes and failures to be sorted out. At the time of writing it was being argued that negotiations should begin with all ten countries with which there are Europe Agreements, since to create two groups would lead to instability in the countries left in second place, thereby contributing to worsening the situation in a region that is already far from secure. It is expected, therefore, that negotiations will begin with ten Central and East European countries as well as with Malta and Cyprus.

THE EURO-MED REGION

Four EU countries have a coastline on the Mediterranean. These countries have become increasingly concerned about developments along the rest of the Mediterranean coastlands, including the former Yugoslavia and Albania, the scene of the worst military conflicts in Europe since the Second World War, and the source of large numbers of refugees hoping to enter the EU. It was to address the growing problems of African and Asian countries bordering the Mediterranean, and to acknowledge their increasing impact on the EU itself, that the Euro-Med Conference was held in Barcelona in November 1995, leading to the adoption of the Barcelona Declaration, a far-reaching commitment to cooperation and mutual assistance in political, security, economic and financial affairs.

Of the so-called Mediterranean countries in question, four are in North Africa: Morocco, Algeria, Tunisia and Egypt. The rest are in Asia (the 'Near East'): Jordan (no coast on the Mediterranean), Israel, Gaza/West Bank, Lebanon, Syria, Turkey and Cyprus (discussed earlier in the chapter). Turkey has territory within the traditional limits of Europe, the population of

Cyprus is mainly Greek (and the Greeks 'invented' Europe), while even Israel (like Cyprus and Turkey) participates in European competitions such as the European Football Cup. All the countries except Israel and Cyprus are poorer than any of the EU countries, and all have fast growing populations. In all but Israel and Greek Cyprus Islam exerts varying degrees of influence on the political organisation, although Turkey is a non-secular state. In most of the other countries powerful dictators, civilian or military, have held power since independence from French or British influence.

Although Turkey and Cyprus are the only countries strictly eligible for EU membership, for a number of reasons the EU is increasingly involved in affairs in the region. Algeria and

Libya (not a participant in the Euro-Med meeting) are major suppliers of oil and gas to the EU. Morocco has half of the world's phosphate reserves, while it also 'exports' heroin to the EU. Migration, both legal and illegal, from Morocco, Algeria and Turkey, has been on a large scale for three decades. Illegal immigrants from the Maghreb countries mainly enter either through Ceuta or Melilla, Spain's territories in Morocco and a region of the EU (at NUTS level 1), or by landing undetected at places along Spain's long Mediterranean coast. Thanks to the Schengen Agreement they can then, at least in theory, without further barriers reach any of the countries of the EU that are signatories.

The latest approach of the EU to the Mediterranean problem is to give economic aid to

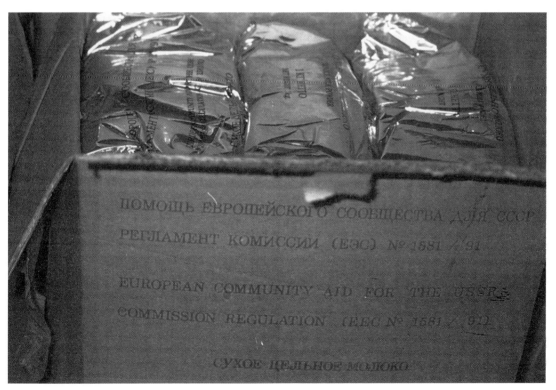

Plate 13.5 To Russia with love? An EEC consignment of powdered milk to the USSR, unthinkable assistance to a needy country, enough to make Stalin turn in his grave
Source: European Commission, Direction Générale X, Audiovisuel

various countries. In the Barcelona Declaration the EU allocated 4.7 billion ECU of aid to the region over the next five years through the MEDA programme, together with an equivalent amount of soft loans. A Mediterranean Free Trade Area is planned to be in place by 2010. It is too early to determine what effect this programme will have on the region, but given the size of the population of the countries involved, about 220 million people, almost two-thirds the population of the EU, the amount seems no more than a starter, a drop in the Mediterranean 'ocean'. Even more daunting is the prospect that this population is expected (PRB 1995) to reach about 365 million by 2025.

Apart from Cyprus, Turkey is the only country of those under discussion seriously considered as a candidate for EU membership. It qualifies on the grounds of having territory in Europe and a (mainly) market economy, but falls well short of EU expectations with regard to human rights and a democratic political system. Turkey's future in relation to the EU, Russia, and its various Islamic neighbours was uncertain at the time of writing. While it would be a great financial liability to the EU as a full Member State (its population is expected to grow from 62 to 98 million between 1995 and 2025), in comparison with the new countries that have emerged in the southern region of the former USSR it is well organised and relatively well developed.

THE COUNTRIES OF TACIS

TACIS stands for Technical Assistance to the Commonwealth of Independent States. While the Baltic States have with some careful manipulation been detached from Russian domination, the other 11 Republics still have links of varying strength with Russia, a situation implicitly recognised by the CIS part of TACIS. All, including the Baltic Republics, have consider-

able Russian minorities, all have strong economic ties with Russia, and in some there is a Russian military presence. Since Boris Yeltsin dissolved the USSR in 1991 and banned the Communist Party, the former USSR has been travelling, politically and economically speaking, in the opposite direction from the EU. Cohesion and convergence are no longer an element of policy, monetary union has been dissolved, and nationalistic sentiments have blossomed in Republics whose citizens were formerly coerced into accepting that cultural traditions took second place to economic considerations.

In the view of the authors it is unlikely that any of the former Soviet Socialist Republics have conditions that approach those required for EU membership, while in geographical location they are mostly remote from the present limits of the EU, several being located in Asia and therefore excluded automatically. In their turn, having recently gained political independence and a measure of economic independence, they are hardly likely to return to a strict central control imposed by Russia over any of their own affairs. Common sense should convince leaders of the new countries that the strongest reason from their point of view for joining the EU, the prospect of assistance on a large scale from EU Structural Funds, would not be forthcoming under present conditions.

While the former USSR is unlikely to produce any new fully integrated Member States for the EU for some decades to come, its natural resources are of great potential significance for the EU, as implied in the quotation at the head of the chapter. The former USSR, particularly Russia, is already one of the main sources of fuel for the EU, supplying both oil and natural gas, while it also exports various non-fuel minerals, timber and wood products. The former USSR has about 5 per cent of the total population of the world and the data in Table 13.4 show how generously it is endowed with bioclimatic resources, fossil fuels and many non-fuel minerals. Each of the former Soviet Socialist Republics

Table 13.4 Natural resources of the former USSR as a percentage of world resources

Bioclimatic resources			
Total land area	17%		
Arable and permanent crops	16%		
Permanent pasture	10%		
Forest	24%		
Mineral reserves			
Coal: anthracite and bituminous	20%		
Coal: sub-bituminous and lignite	26%		
Oil (4.9 per cent in Russia)	5.6%		
Natural gas (34 per cent in Russia)	40%		
Sulphur	26%	Manganese	14%
Potash	22%	Tungsten	14%
Silver	19%	Nickel	13%
Gold	17%	Lead	13%
Platinum	17%	Industrial diamonds	11%

Between 5 and 10 per cent: mercury, molybdenum, chromium, cobalt, copper, antimony, tin

Sources: FAOPY 1993 (FAO 1994); BP 1995; Bureau of Mines 1985
Note: Data for fossil fuels are for 1994, for other minerals for the mid-1980s

(SSR) or groups thereof will now be briefly considered.

Belarus

Apart from the Baltic Republics, Belarus is the closest former SSR to the EU. It lacks fuel and raw materials and has therefore come to depend heavily on Russia for supplies for its industry. The southeastern part has been blighted by the Chernobyl nuclear accident since 1986.

Ukraine

Ukraine is credited in popular publications with substantial natural resources. To be sure, it has some of the best agricultural land and bioclimatic conditions in the former USSR. On the other hand, its Donbass coalfield has high production costs, as does its iron and steel industry. The western part of Ukraine depends heavily on agriculture, the eastern part, with a large Russian element, specialises excessively in heavy industry, while the Crimea in the south is

regarded as vital for the maintenance of a strong Russian presence in the Black Sea area. The road network of Ukraine is poorly developed, while the main railways run north–south rather than east–west. Closer ties with Poland and the EU would require a great outlay on motorways into and across the country.

Transcaucasia

The three former Soviet Socialist Republics of Transcaucasia, Georgia, Armenia and Azerbaijan, are inside Europe according to some but part of Asia according to others. In the last few years they have been the scene of the worst conflicts anywhere in the former USSR except in Chechnya. Economically they are very backward by EU standards and are far more remote even than Ukraine, as the German armies found to their cost in the Second World War when, in 1942, they advanced towards the oilfields of the northern Caucasus and the Caspian Sea. The rejuvenation of oil exploration and production in Azerbaijan is a matter

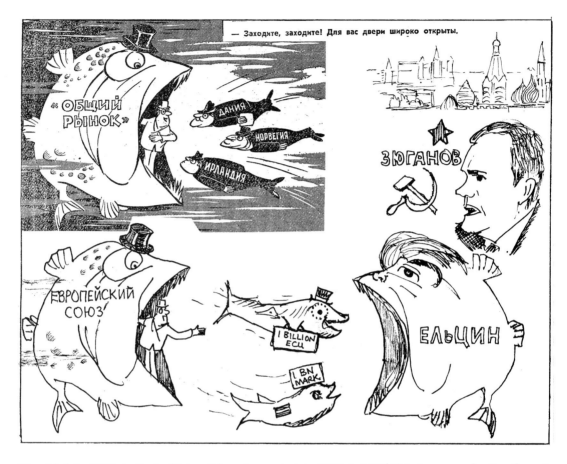

Plate 13.6 In 1970 the Soviet Union did not 'recognise' the EEC; in 1996 Yeltsin accepts funding from the EU through TACIS and from Member States

of interest to Western oil companies but hardly a strong enough card to help Azerbaijan into the EU.

Russia

Whether it stretches from the Baltic to the Pacific or merely from the Baltic to the Ural Mountains, Russia is too big and too unstable politically, ethnically and economically for the EU to swallow. On the other hand, some kind of association with the EU is gradually being worked out. On a visit to Moscow in 1991, one of the authors showed his EC watch to a Rus-

sian statistician. The dial of the watch is marked with the 12 flags of the EUR 12 countries. 'When we join the EC,' said the Russian, 'we can put our flag in the centre of the dial.' A good reason, perhaps, for thinking carefully about Russian entry.

FURTHER READING

Arter, D. (1993) *The Politics of European Integration in the Twentieth Century*, Aldershot: Dartmouth.
Commission of the EC (1991) *A Strategy for Enlargement*, Preliminary Report of the SG Study Group on Enlargement, Brussels, 14 Nov.
Donaldson, L. (1992) 'A twinning combination', *Geographical*, LXIV (12): 16–19. Twinning of EU towns

Plate 13.7 Jacques Delors is seen by Russia as an octopus capturing various parts of Europe through enlargement of the EU (*John Cole*)

with towns in Central Europe and the former USSR is on the increase.

Evans, R. (1995) 'From the Arctic to the Med', *Geographical*, LXVII (3): 28–31. Overview of current geography of the EU and points about possible future members.

Foucher, M. (ed.) (1993) *Fragments d'Europe mediane et orientale*, Lyon: Fayard, L'Observatoire européen de géopolitique. Definitive work on changes in Central Europe and the former USSR in the early 1990s. Detailed maps, five star presentation of material.

Gibb, R. and Michalak, W. (1993) 'The European Community and Central Europe: Prospects for Integration', *Geography*, 78 (1), January: 16–30.

McMillan-Scott, E. (1990) 'Greater Europe: 12 Plus 6 Plus 6', *Target Europe Papers* No. 4, Conservatives in the European Parliament.

Merritt, G. (1991) *Eastern Europe and the USSR: The Challenge of Freedom*, EC/Kogan Page.

Morgan, W. B. (1992) 'Economic reform, the free market and agriculture in Poland', *Geographical Journal*, 158 (2), July: 145–56.

Redmond, J. (1993) The Next Mediterranean, *Enlargement of the European Community: Turkey, Cyprus and Malta*, Aldershot: Dartmouth.

Sallnow, J. and Saiko, T. (1993) 'State of independence', *Geographical*, LXV (12), Dec.: 19–22. An economic crisis in Ukraine two years after its independence reveals its inescapable links with Russia.

Shaw, D. J. B (ed.) (1995) *The Post-Soviet Republics: A Systematic Geography*, Harlow: Longman.

Wessels, W. and Engel, C. (1993) *The European Community in the 1990s – Ever Closer and Larger*, Bonn: Europa Verlag.

14

THE EUROPEAN UNION AND THE REST OF THE WORLD

– Est-ce que la terre a diminué, par hasard?
– Sans doute, répondit Gauther Ralph. Je suis de l'avis de Mr. Fogg. La terre a diminué, puisq'on la parcourt maintenant dix fois plus vite qu'il y a cent ans. Et c'est ce qui, dans le cas dont nous nous occupons, rendra les recherches plus rapides

(Jules Verne, *Le Tour du Monde en Quatre-Vingt Jours*, 1873)

- Western Europe has had trade links with many parts of the world for up to five centuries and colonies at various times in all the continents. Globalisation is not therefore a new phenomenon.

- The movement of people, goods and information has become faster and cheaper in the twentieth century than ever before.

- With the USA, Japan and other developed regions the transactions of the EU are mainly in manufactured goods and investments.

- In the developing world the EU depends particularly on North Africa and the Middle East for imports of oil and natural gas.

- Many sectors of industry in the EU are threatened by competition from exports of manufactured goods from Latin America, southeast Asia and China.

- During the twenty-first century the relative share of world population, production and trade in the EU is likely to continue to diminish, as it has done for most of the twentieth century.

The relations of the EU with Central Europe, the former USSR and countries of the Mediterranean basin were discussed in Chapter 13. In 1995 the EU itself, together with those neighbouring areas, had a total population of some 930 million, about 16 per cent of the population of the world of 5,700 million. In this chapter the relations of the EU with the rest of the world are discussed. The remaining 84 per cent of the world's population includes the other main developed regions, North America, Oceania and Japan, as well as various newly industrialising countries, especially in Latin America and southeast Asia; China and India, each of which in its own way is rapidly becoming more integrated in the world economy; the Middle East with most of the world's oil reserves; and the poorest region of the world, Africa south of the Sahara. The EU has a particular commitment to assist the African, Caribbean and Pacific countries, which consist mainly of countries on mainland Africa and islands in the Caribbean Sea and the Pacific Ocean.

THE EU AND THE GLOBAL 'DEVELOPMENT GAP'

In this section 12 major regions of the world are compared with the help of data in Table 14.1. The first region listed, Western Europe, consists of the EU and EFTA. Central Europe and the former USSR have already been discussed in Chapter 13. Turkey and the Maghreb countries, also referred to in Chapter 13, are situated within the ninth region in the table, North Africa and southwest Asia. The EU itself is likely to be influenced in the future not only by its nearest neighbours but also, with varying degrees of intensity, by the remaining nine regions of the world. These are subdivided conventionally into three 'developed' regions, Japan and South Korea, the USA and Canada, and Oceania (regions 4–6 in Table 14.1), with a population of about 490 million, or 8.6 per cent of the total population of the world, and six 'developing' regions, with about 4,280 million inhabitants. More than half of the population of the world lives in regions 10–12, about 2,980 million people. The prospect that the population of the six developing regions could increase by about 60 per cent in the next three decades, adding about 2,570 million to the present size, cannot be ignored when the future of the EU is under consideration.

The countries and the major regions of the world differ in various ways. Since the end of the Second World War the gap between rich and poor regions and countries has become a central issue in world affairs. In Table 14.1, the data in columns (7) and (10), showing per capita energy consumption and GDP in the 12 regions, illustrate the gap in production and in consumption between different parts of the world. The implications of the fact that countries and regions of the world also differ greatly with regard to the natural resources available per inhabitant has received less publicity, but the prospect in the long run is that many countries that are poor in natural resources, including most in Europe, will depend increasingly on a few countries that are rich in natural resources.

The data in column (6) of Table 14.1 serve as a rough measure of the availability of natural resources in different regions. The indices are based on an estimate of the presence per inhabitant of five equally weighted natural resources: fresh water, bioclimatic resources (arable land, forest and pasture), fossil fuels (coal, oil, natural gas), non-fuel minerals and, to allow for future discoveries of minerals, finally total land area. The choice of criteria are justified in Cole (1996) and will not be elaborated here. It is evident from the data in column (6) of Table 14.1 that of the six developed regions (1–6), three are relatively poorly endowed with natural resources per inhabitant in relation to the world average level of 100: Western Europe, Central Europe, Japan and South Korea. Three developed regions are generously endowed: the former USSR, the USA and Canada, and Oceania. Similarly, of the six developing regions (7–12), three are relatively poorly endowed with natural resources per inhabitant: South Asia, Southeast Asia and China. Three developing regions are comparatively generously endowed: Latin America, Africa and southwest Asia.

Western and Central Europe and Japan have achieved and maintain their comparatively high

Table 14.1 Contrasts between 12 major regions of the world

	(1)	(2)	(3)	(4)	(5)	(6)	(7)	(8)	(9)	(10)
	Population in millions		Difference	% of world population in each region		Per capita natural resource score world = 100	Energy consumption tce per capita world = 100	Arable land, ha per capita		Real per capita GDP per capita world = 100
	1995	2025	1995–2025	1995	2025			1993	2025	
1 Western Europe¹	385	385	0	6.8	4.6	67	238	0.24	0.22	284
2 Central Europe²	130	134	4	2.3	1.6	63	221	0.44	0.42	108
3 Former USSR³	285	335	50	5.0	4.0	319	360	0.79	0.69	135
4 Japan and S Korea	170	177	7	3.0	2.1	16	184	0.04	0.03	261
5 USA and Canada	293	375	82	5.1	4.5	265	523	0.82	0.67	448
6 Oceania	28	39	11	0.5	0.5	1,040	280	1.79	1.28	260
7 Latin America	481	706	225	8.4	8.5	186	58	0.33	0.25	96
8 Africa south of the Sahara	614	1,348	734	10.8	16.2	137	16	0.28	0.12	25
9 North Africa and Southwest Asia	327	606	279	5.7	7.3	173	70	0.27	0.13	81
10 South Asia	1,221	1,899	678	21.5	22.9	24	14	0.17	0.11	22
11 Southeast Asia	485	704	219	8.5	8.5	63	19	0.17	0.13	46
12 China and immediate neighbours⁴	1,272	1,591	319	22.4	19.2	31	44	0.08	0.06	61
World	5,700	8,300	2,600	100.0	100.0	100	100	0.26	0.18	100

Sources: (1)–(5) WPDS 1995 (PRB 1995); (6)–(10) Cole 1996: (6) p. 64; (7), (10), p. 69; (8), (9), p. 413

Notes: tce – tonnes of coal equivalent
1 Western Europe = EU + EFTA
2 Central Europe is the rest of Europe including Baltic Republics but excluding former USSR
3 Former USSR excludes Baltic Republics
4 Includes Hong Kong, Taiwan, North Korea and Mongolia

levels of material production and development through being net importers from other parts of the world of primary products, especially oil and non-fuel minerals. Canada, Australia and the eastern regions of Russia, on the other hand, are very generously endowed with natural resources in relation to population and are net exporters of fuel, raw materials and food. Given their high levels of energy consumption and industrialisation per inhabitant, the six developed regions (1–6 in Table 14.1) use up non-renewable natural resources much more quickly than the six developing regions.

The developing regions have tended conventionally to be seen as net exporters of primary products, a feature of their colonial past, but a view now greatly oversimplified for two main reasons. First, the population of the six developing regions has roughly doubled in the period 1955–95, requiring more primary products simply to stay at the same level of development. Second, in many parts of the developing world considerable industrialisation has taken place since the Second World War, some even before it. The per capita consumption of energy and of raw materials has increased. Many developing countries are therefore now net importers of primary products and net exporters of manufactured goods. One implication for the leading market economy industrial regions of the world, the USA and Canada, the EU and Japan, is that there is a gradual shift in world trade patterns from a situation in which relatively few countries supplied most of the exports of manufactures to one in which relatively few countries will be supplying most of the exports of primary products.

The impact of the growing pressure of world population on natural resources may not be felt for some decades. The gap between rich and poor countries is therefore the issue and the problem that receives more publicity. Like North America and Japan, Western Europe is committed to providing development assistance to the poorer regions of the world. Official Development Assistance (ODA) from countries of the European Union is provided both by individual Member States and by the EU as a whole. Within each country ODA funding is in competition and conflict with other sectors of the budget. Within the EU as a whole, financial assistance with regional rather than sectoral destinations is now allocated to three competing regions: to poorer regions in the EU itself, to Central Europe and the former USSR to assist in the urgent need to transform their economies from centrally planned to market, and to the developing world, including in particular the African, Caribbean and Pacific (ACP) countries, which consist largely of former French and British colonies.

MEASURING THE DEVELOPMENT GAP

The development gap has been measured in various ways over the last few decades. Two measures are shown in Table 14.1, energy consumption and real Gross Domestic Product. Energy consumption reflects broadly the level of industrialisation of any economy and the use of natural resources, while Gross Domestic Product covers all types of production. Real GDP removes the possible distorting effect of the use of current exchange rates to convert other currencies to US dollars, since some are held or stay spontaneously artificially low (e.g. the Chinese) or high (e.g. the Japanese). Real GDP uses parity purchasing power to calculate and compare what can actually be purchased at home with given currencies.

In columns (7) and (10) of Table 14.1, the regional scores are expressed in terms of the amount consumed (energy) or produced (GDP) per inhabitant in relation to the world average of 100. Both criteria show a gap between the six developed and the six developing regions, although in terms of real GDP per inhabitant, Latin America is close to Central Europe and to the world average. There are marked differences also among the member regions of each group. Developed and developing regions do not fall into two discrete classes but rather form a continuum, a feature shown more clearly when individual countries are studied. In order to place the 15 Member States of the EU globally in the development gap continuum, four measures are shown in Table 14.2 for these countries and for 15 selected non-EU countries.

Per capita Gross National or Domestic Product is the single measure most widely used by the World Bank and many other bodies to indicate the level of development of the countries of the world. The modified real GDP index is shown in thousands of dollars per capita in column (1) of Table 14.2. Columns (2) and (3) show two measures that are combined in the United Nations Development Programme (UNDP) with real GDP to produce a synthetic human development index (HDI) (column (4)), intended to reflect more realistically the quality of life.

In the relevant table in *HDR 1995* (UNDP 1995) 174 countries are included. In Table 14.2 the ranking of each country on the HDI index is shown. All EU countries are in the class of 'high human development', the arbitrary cut-off score on the human development index, maximum possible 1,000, being 800. Of the top 20 countries among the 174 (score over 910), 12 are EU countries, 3 the rump of EFTA, together with Canada, the USA, Japan, Australia and New Zealand. Greece, Luxembourg (pulled down by a low score on educational attainment) and Portugal mingle in the next 20 with some mainly small developing countries and with the highest scoring countries in Central Europe. Below 500, the threshold into 'Medium human development', there are 47 countries. In World Bank (1995, Table 1), 132 countries are subjected to a similar screening, and all EU countries except Greece and Portugal come in the top, 'high-income

economies' class, which consists of the world's 24 richest countires.

One of the concerns of the EU institutions in the next century will be how to strengthen policy and performance towards reducing the global development gap or at least preventing it from continuing to widen. In the next section, international trade and development assistance will be discussed, followed by foreign investment in the following section. It has been widely argued that while international trade may bring benefit to all parties involved, it does not necessarily reduce global inequalities. Again, foreign investment may benefit recipient countries, but profits are extracted from them, usually to the benefit of richer countries. On the other hand, by definition, development assistance is a one-way transfer of resources, although even this transaction may bring problems, since such assistance may at times reach unintended destinations or increase environmental damage. One justification to the taxpayers in rich countries is to argue that they themselves also benefit as the economies of poorer countries grow.

THE EU AND INTERNATIONAL TRANSACTIONS

Foreign Trade

The EU as a whole is by far the largest trading bloc in the world. In 1993 EUR 12 accounted for about 35 per cent of all foreign trade (1,300 billion out of 3,687 billion dollars of exports, 1,295 out of 3,757 of imports). The remaining developed market economy countries of the world, EFTA members, Japan, the USA, Canada, Australia and New Zealand, accounted for almost another 30 per cent.

So great is the scale of the foreign trade between EU countries, however, that over 20 per cent of all international trade transactions worldwide took place between members of the

Table 14.2 A comparison of the human development indices of EUR 15 countries with those of 15 other selected countries

	HDI Rank in world	(1) Real GDP	(2) Life expectancy years	(3) Education index	(4) HDI		HDI Rank in world	(1) Real GDP	(2) Life expectancy years	(3) Education index	(4) HDI
Netherlands	4	17.8	77	95	936	Canada	1	20.5	77	99	950
Finland	5	16.3	76	98	934	USA	2	23.8	76	98	937
France	8	19.5	77	95	930	Japan	3	20.5	80	92	937
Spain	9	13.4	78	94	930	Argentina	30	8.9	72	90	882
Sweden	10	18.3	78	92	929	South Korea	31	9.3	71	91	882
Belgium	12	18.6	76	94	926	Czech Republic	38	7.7	71	89	872
Austria	14	18.7	76	94	925	Russia	52	6.1	68	89	849
Germany	15	22.1	76	93	921	Mexico	53	7.3	71	81	842
Denmark	16	19.1	75	94	920	Turkey	66	5.2	67	74	792
UK	18	17.2	76	92	916	China	111	2.0	69	71	594
Ireland	19	12.8	75	94	915	Morocco	117	3.4	63	41	554
Italy	20	18.1	78	88	912	India	134	1.2	60	52	439
Greece	22	8.3	78	88	907	Afghanistan	170	0.8	44	24	228
Luxembourg	27	21.5	76	85	893	Ethiopia	171	0.3	48	26	227
Portugal	36	9.9	75	83	874	Niger	174	0.8	47	13	207

Source: UNDP 1995, Table 1
Notes: (1) Real Gross Domestic Product per capita in thousands of ppp dollars, 1992
(2) Life expectancy at birth in years
(3) Education index, theoretical range of score 0–100, composed of adult literacy rate and combined first-, second-, and third-level gross enrolment rate, 1992
(4) Human development index (HDI), 1992, a composite index with theoretical limits of 1,000 and 0. GDP influences HDI very little above about 5,000 dollars per capita

[Plate 14.1] As yet an unusual sight 'Made in Europe', an EU photograph that will leave the reader guessing as to what has actually been made

Plates 14.1 and 14.2 Towards European Union?
Source: European Commission, Direction Générale X, Audiovisuel

[Plate 14.2] A little-known airline, Air Europe, parks near the giants

Union. Such trade is still classified as foreign or international. However, given the existence of the customs union, with internal free trade and the Single Market, it could be considered different from other international trade. The rest of this section deals with the 'external' trade of the European Union.

Table 14.4 shows the direction of the extra-EU merchandise trade of EUR 12 in 1994 (Austria, Finland and Sweden then still in

EFTA). The percentage of imports and exports originating in and going to major regions and groups of countries of the world is given in columns (1) and (2), the balance in billions of ECU in column (3). Over half of extra-EU trade is with other 'industrialised regions' of the world, the developed countries excluding Central Europe and the former USSR, classified separately in Eurostat (1996). China is similarly separated from the rest of the developing world,

BOX 14.1 THE 1996 OLYMPIC SHOP WINDOW

The EU can feel satisfaction from the staging of the original Olympic Games in ancient Greece, where also the first modern Games were hosted in 1896 in Athens. The scope of the various contests and the diffusion of information about proceedings and results through the media has made them one of the outstanding world events. An inspection of the Final Medals Table (see Table 14.3) shows sharp contrasts in the results among almost 200 participating countries and a feeling of failure in some, notably the UK (or Great Britain), that comparable countries did far better. Like so many human activities it is difficult to find a single objective measure to assess the performance of different countries.

The population size of countries favours smaller ones if medals are measured against population size. For example 13 medal-winning Central European countries could

Table 14.3 1996 Olympic medals won by major world regions

	(1) Population mlns	(2) Medals	(3) Medals per 100 million popn
1 Canada and Oceania	58	70	121
2 Central Europe	130	99	76
3 Western Europe	385	243	63
Developed World average	1,171	673	58
4 Former USSR	285	119	42
5 USA	263	101	38
6 Japan and South Korea	170	41	24
World average	5,771	839	15
7 Latin America	481	56	12
8 Southwest Asia and North Africa	327	17	5
9 China	1,272	58	5
10 Africa South of the Sahara	614	25	4
Developing World average	4,600	166	3.6
11 Southeast Asia	485	9	2
12 South Asia	1,272	1	0.1
EU 'Big Five' 1 Germany	82	65	79
2 France	58	37	64
3 Italy	57	35	61
4 Spain	39	17	44
5 UK	59	15	25

Source: Final Medals Table of 1996 Olympic Games in Atlanta

field far more competitors than Japan, yet they have roughly the same population size. The physical stature of the population favours participants in some sports such as basketball. The cost of training with horses and boats discourages entries from members of poor countries. In most if not all events the availability of training facilities is vital.

In the analysis of medals by regions shown in Table 14.3 three features stand out clearly. Comparability among regions differing markedly in population size (see Table 14.1 for details of the membership of the regions) is achieved in column (3) by calculating the number of medals (with no distinction between gold, silver and bronze) per 100 million people.

- Western Europe (the EU plus Norway and Switzerland) did very well on the criteria of performance used. Indeed about half of all the 839 medals won went to the countries of Europe, including the former USSR.
- The medals 'gap' between the rich and poor regions is enormous, the average won per population being more than 15 times as high for developed countries as for developing ones.
- An examination of the level of medals won by the five largest EU countries in population shows that the UK lagged far behind Germany, France and Italy and was also behind all other EU Member States except Portugal, Austria and Luxembourg.

which accounts for over a third of extra-EU trade. The most striking broad features of the situation illustrated in Table 14.4 are the enormous deficit in trade with Japan and the substantial surplus with some parts of the developing world. In relation to its population size, China's trade with the EU is small. The accession of Austria, Finland and Sweden to EU membership has greatly reduced the amount of trade between the EU and EFTA.

Official Development Assistance

Official development assistance refers to international net transfers of funds from rich to poor regions of the world. It therefore differs from international trade, in which flows in both directions roughly balance and cancel out, while in theory at least being mutually beneficial to the trading partners concerned. Until the late 1980s the principal donors of development assistance were the OECD market economy countries, the centrally planned countries of Soviet Union and Eastern Europe, and a few OPEC countries of Southwest Asia. Since the

break-up of the USSR in 1991 and the Gulf War in the same year virtually all official development assistance has come from OECD countries. In relation to the total world economy, development assistance to poorer countries has actually diminished both because two of the three donor groups have dropped out and because in OECD itself, foreign aid has been reduced, especially from the USA.

Table 14.5 shows the main sources of development assistance in 1993. In the interpretation of the data the great differences in the sizes of total population and of GNP should be taken into account. The countries are ranked in the table according to their 'generosity', as measured by the proportion of their total GNP allocated to foreign aid, which can be compared with a United Nations proposal that 0.7 per cent should be set aside. Only four of the 17 countries in the table exceed that level, the three Scandinavian countries and the Netherlands.

The EU is the largest single source of overseas development assistance. Since 1988 Japan has overtaken the USA to take second place, its

Table 14.4 EU trade balance and trade flows by main partners in 1994

	(1) Imports %	(2) Exports %	(3) Trade balance Bn ECU
All extra-EU trade	100.0[3]	100.0	−3.5
Developed countries[1]	55.6	53.1	−15.6
USA-Canada	18.8	19.3	2.1
Japan	9.0	4.9	−22.2
EFTA	22.7	22.1	−4.3
C. and E. Europe	9.0	9.0	−0.3
Developing countries[2]	29.5	34.0	23.3
Asian NICs	6.1	7.5	7.2
OPEC	7.5	6.9	−3.6
Mediterranean basin	7.8	10.2	12.3
ACP	3.4	2.8	−3.7
Latin America	5.0	5.2	1.3
China	4.2	2.3	−10.5

Source: Eurostat 1996: XIX
Notes: 1 The total for developed countries includes countries such as Australia not included in any of the sub-groups
2 The total for developing countries is somewhat smaller than the sum of totals of the sub-groups in the source
3 Total EU imports from all extra-EU trading partners were approximately 541 billion ECU, exports 538 billion ECU

Table 14.5 Net official development assistance (ODA) disbursed in 1993 by main contributors to aid flows

	Millions of dollars	As % of GNP			Millions of dollars	As % of GNP
1 Denmark	1,340	1.03	10 Australia		953	0.35
2 Norway	1,014	1.01	11 Switzerland		793	0.33
3 Sweden	1,769	0.98	12 UK		2,908	0.31
4 Netherlands	2,525	0.82	13 Italy		3,043	0.31
5 France	7,915	0.63	14 Austria		544	0.30
6 Canada	2,373	0.45	15 Japan		11,259	0.26
7 Finland	355	0.45	16 Spain		1,213	0.25
8 Belgium	808	0.39	17 USA		9,721	0.15
9 Germany	6,954	0.37	OECD[1]		55,960	0.30

Source: UNDP 1995, Table 29
Note: 1 Total includes small contributions from Portugal, New Zealand, Ireland and Luxembourg

contribution rising from 4.3 billion dollars in 1984 to 13.2 billion in 1994 compared with 10 billion from the USA in that year, an amount little changed since the mid-1980s. The OECD countries have a combined population of 830 million, a mere 15 per cent of the world's population and a proportion expected to decline further at least during the first half of the twenty-first century. With the Cold War over and the peace dividend materialising, if slowly, the 1990s may be a time in which a rethinking of development assistance takes place. Self-interest and altruism may combine to produce a greater effort on the part of the rich countries to help the poorer ones. The wasteful and mis-guided allocation of some development funds, which at times help the better-off developing countries and sectors of population or cause environmental damage, should be avoided.

Through the Lomé Convention, the EU has obligations to assist the 70 states forming the African, Caribbean and Pacific group. This group includes many of the poorest countries of Africa as well as many islands in the Caribbean and Pacific. In general the latter are more prosperous than Africa, but their economies suffer from very small size, and in the case of the Pacific islands, locations remote from any continents. According to *EP News* (European Parliament 1995a: 2), the final sum of 13.3 billion ECU has been allocated to the 8th European Development Fund for the period 1995 to 2000. Such sectors of agriculture as sugar and banana cultivation are supported, with preferential arrangements given to ACP countries, resulting in high prices for such foodstuffs in the EU. Germany and Spain, among other EU countries, are increasingly dissatisfied with this emphasis on only a small part of the developing world.

In order to review EU transactions with the rest of the world in more detail, the regions and countries in Table 14.4 have been subdivided into three groups: the rest of the developed world, Latin America/Africa/Southwest Asia and the rest of Asia. The developing world regions are thus divided into two groups. In the first group the natural resource endowment per inhabitant is fairly generous by world stan-

dards. In the second group the natural resource endowment per inhabitant is very limited.

THE EU AND NON-EU DEVELOPED REGIONS

For the purposes of this section, the non-EU developed countries are divided into four: EFTA (post 1994), the USA, Japan, and Canada/Australia/New Zealand. A combination of political, economic and social factors have delayed temporarily if not permanently the entry of Switzerland, Norway and Iceland into the EU. Switzerland's traditional neutrality and its strong financial position have enabled it to resist the possible effect of EU influence, as have the oil and gas reserves of Norway. The two countries trade very heavily with the EU, although the composition of their exports differs radically, most of Norway's consisting of primary products, most of Switzerland's manufactures and services (see Table 14.6). Their case for joining the EU in due course was discussed in Chapter 13. The three remaining single countries and groups of developed market economy countries will now be considered.

The USA

Relations between the EU and the USA can be divided into political and economic, the former governed to a large extent by the American

Table 14.6 Primary products as a percentage share of the merchandise exports of selected countries in 1993

Non-EU developed		Selected EU		Non-EU developing	
New Zealand	73	Netherlands	36	Saudi Arabia	91
Norway	69	Spain	22	Venezuela	84
Australia	65	France	22	Brazil	40
Canada	34	UK	19	India	25
USA	16	Sweden	15	China	19
Switzerland	6	Italy	10	South Korea	7
Japan	3	Germany	10	Hong Kong	7

Source: World Bank 1995, Table 15, Structure of merchandise exports

Table 14.7 US merchandise trade, 1993, in billions of dollars

	Exports	Imports	% of exports	% of imports
Total	464.8	580.5	100.0	100.0
Canada	100.2	110.9	21.6	19.1
Mexico	41.6	39.9	9.0	6.9
EUR 15	101.5	105.4	21.8	18.2
Germany	19.0	28.6	4.1	4.9
UK	26.4	21.7	5.7	3.7
France	13.3	15.2	2.9	2.6
Italy	6.5	13.2	1.4	2.3
Benelux	22.3	10.9	4.8	1.9
Japan	48.0	107.3	10.3	18.5
South Korea	14.8	17.1	3.2	2.9
China	8.8	31.5	1.9	5.4
Taiwan, Hong Kong, Singapore	37.9	47.5	8.2	8.2

Source: US Bureau of the Census 1994: 823–6
Note: Total also includes areas not listed

participation in the defence and security of Europe since the Second World War. In spite of the removal of the Berlin Wall and the Iron Curtain, NATO continues to play a strong role in the absence of any EU Common Foreign and Security Policy, and the USA provided the largest single military contingent for peace-keeping in the former Yugoslavia. In spite of efforts on both sides to reduce the need for US involvement in the defence of Europe, the USA is bound to continue to have a special geo-political role in Europe into the next millennium.

With regard to economic relations, in recent years there have been conflicts between the EU and the USA over trade and protectionism in industrial goods, settled amicably through the World Trade Organisation, which both blocs were instrumental in establishing through the GATT negotiations. Conflicts over the CAP and its disruption of world farm prices continue to arise, but it is hoped that reforms to the mechanisms of the CAP will ease relations and trade in agricultural goods.

The USA accounted in 1993 for 12.6 per cent of the exports and 16.1 per cent of the imports in total world merchandise trade. In

that year (see Table 14.7) the value of its imports therefore greatly exceeded the value of its exports. Primary products make up roughly the same proportion of imports and of exports (17–18 per cent), with oil and non-fuel minerals prominent among the imports, agricultural products among the exports.

The principal sources of US imports in 1993 were its NAFTA partners, Canada and Mexico (26 per cent), Japan (18 per cent) and the EU (including 1995 joiners, also 18 per cent). While the value of US trade with its NAFTA partners and with the EU as a whole is roughly in balance, its imports from Japan and, in recent years, also from China, greatly exceed its exports to these countries. If China and other countries of East and Southeast Asia such as South Korea and Malaysia continue to increase their exports and if they target the USA in particular, then the strong Japanese position with regard to its trade with the USA could be threatened. The more balanced EU position with regard to the USA could also be upset, always assuming the USA has products essential to the economies of EU countries.

The deterioration of the US balance of trade

Table 14.8 US direct investment position abroad, 1980 and 1992 in billions of dollars

	1980	1992	% 1992
Total	215.4	486.7	100.0
Canada	45.1	68.4	14.1
Europe	96.3	239.4	49.2
UK	28.5	77.8	16.0
Germany	15.4	34.0	7.0
Benelux	15.0	31.7	6.5
Switzerland	11.3	28.7	5.9
France	9.3	23.3	4.8
Italy	5.4	13.6	2.8
Japan	6.2	26.2	5.4
Latin America	39.6	88.7	18.2

Source: US Bureau of the Census 1994: 811

Table 14.9 Main sources of foreign direct investment in the USA in 1980 and 1992 in billions of dollars

	1980	1992	% of 1992
Total	83.0	419.5	100.0
Canada	12.2	39.0	9.3
Europe	54.7	248.5	59.2
UK	14.1	94.7	22.6
Netherlands	19.1	61.3	14.6
Switzerland	5.1	19.6	4.7
Germany	7.6	29.2	7.0
Japan	4.7	96.7	23.1

Source: US Bureau of Census 1994: 809

in the 1980s and early 1990s is matched by a change in its direct investment position in the world (see Table 14.8). In 1980 US cumulative investments abroad were 215 billion dollars, compared with 83 billion invested in the USA by foreign countries (see Table 14.9). The corresponding figures in 1992 were 487 billion and 420 billion respectively, the increase of the latter being far greater than that of the former, while the 'gap' narrowed very markedly. Tables 14.8 and 14.9 show

the position of Europe (in effect Western Europe) with regard to US investment.

Almost half of all US foreign investments are in Western Europe, another 14 per cent in Canada, both regarded as safe areas, albeit generally yielding lower returns than those obtained elsewhere. Only about 40 per cent of US investment in Europe is in manufacturing, while 28 per cent is in finance. In relation to their population size, Switzerland, the Benelux countries and the UK are the preferred destinations of US investment in Europe.

Japan

Like Switzerland, Japan has very few natural resources. Not surprisingly therefore primary commodities account for only 3 per cent of the value of its merchandise exports. In contrast, primary commodities account for 51 per cent of merchandise imports (food 18, fuels 21, other materials 13). Japan is even more dependent than the EU on imports of primary products. At first sight, therefore, the EU and Japan do not appear to be obvious trading partners since both are highly industrialised and are large importers of primary products. However, Japan must import large quantities of primary products from partners such as Saudi Arabia (oil) and Australia (non-fuel minerals). Such countries have relatively small markets and cannot absorb large quantities of Japanese manufactures. This is one reason why Japan has developed a strong and effective policy to establish a large surplus in its trade with the EU and the USA.

Relations with Japan are almost exclusively in the area of trade, with continuous concern in the EU over the imbalance in trade and accusations of protectionism on the part of the Japanese. In 1993 the total value of Japanese merchandise trade exports was 363 billion dollars (see Table 14.10). The EUR 12 countries took 56 billion, 15.5 per cent. In contrast, Japan's imports from the EUR 12 countries were only worth 30 billion, 12.5 per cent of Japan's total imports. Of the EU's imports from Japan, over 80 per cent consisted of machinery and equipment, virtually all the rest of other manufactured products. Of Japan's imports from EUR 12, manufactured goods accounted for about 90 per cent. Is this a case of 'taking in each other's washing', or is there some simple explanation?

Like the USA, Japan invests heavily in the EU. In 1993 its cumulative investment worldwide reached 423 billion dollars, not far short of the US total. Its 78 billion invested in EUR 12 in 1993 was 18.4 per cent of its total investment, contrasting with 42 per cent of all Japanese investment placed in the USA. In relation to the size of their populations, the Benelux countries and the UK are the most popular with Japanese investors. In spite of the publicity given to the manufacturing side of Japa-

Table 14.10 Japan's foreign trade with and investment in the EUR 12 countries

	(1) Exports to 1983	(2) 1993	(3) Imports from 1983	(4) 1993	(5) Investment 1993	(6) 1994
Germany	5.9	18.0	2.4	9.8	7.3	106
UK	5.0	12.0	1.9	5.0	31.7	206
Benelux	3.1	11.7	0.8	2.7	26.8	88
France	2.0	5.5	1.3	5.1	6.0	121
Italy	0.9	3.2	1.0	3.8	1.6	52
Other[1]	2.4	6.0	1.2	3.8	4.6	115
EUR 12	19.3	56.4	8.6	30.2	78.0	688

Source: JETRO 1995: 8
Notes: 1 Denmark, Greece, Ireland, Portugal, Spain. Note that in 1983 Portugal and Spain were not in the EU
(1)–(4)   Value of trade in billions of dollars
(5)   Cumulative investment 1951–93 in billions of dollars, 1993
(6)   Number of manufacturing facilities operated by Japanese affiliates, January 1994
          (no indication given of size)

nese investment, this sector accounts for only 24 per cent of all Japanese investment in the EU.

Canada, Australia and New Zealand

EU relations with Canada were soured in 1995 through a major dispute over fishing rights, a problem likely to resurface due to problems of overfishing throughout the North Atlantic, already discussed in Chapter 6. EU relations with Australia and New Zealand have suffered from the same conflicts over trade in agricultural goods as with the USA, with the EU accused of protectionism and dumping on the world markets.

Throughout the twentieth century the USA has been the main trading partner of Canada. In 1991 about 60 per cent of Canada's imports came from the USA which, in its turn, took over 70 per cent of Canada's exports. In contrast, the whole of the EU provided only about 10 per cent of Canada's imports and took less than 10 per cent of its exports.

The picture is similar now for Australia and New Zealand. Earlier this century the UK was the principal trading partner of both countries. In 1990 the whole of the EU took only 14 per cent of the exports of Australia and New Zealand, while providing 23 per cent of the imports of the two countries. Japan, however, took 25 per cent of the exports of Australia and New Zealand, while providing 18 per cent of their imports. The USA is also a major trading partner of the two countries, both of which mainly export primary products.

From the material presented so far in this chapter it is evident that in the 1990s the OECD countries dominate world trade and provide almost all development assistance. Virtually all the largest transnational companies are based in the USA, Japan or Western Europe. In Box 14.2 the mutual spatial relationships of these three dominant regions in the world economy

are considered, as are their positions in relation to the rest of the world.

EU TRANSACTIONS WITH DEVELOPING REGIONS WELL ENDOWED WITH NATURAL RESOURCES

The data in Table 14.3 show the percentage of EU imports and exports accounted for by non-EU countries, grouped as in an EU source. The developing countries, which account for roughly a third of EU trade with non-EU countries, are grouped into six major regions (as in Table 14.1), each of which is discussed in turn below.

Latin America

Culturally Latin America is a distinct region of the world, with up to three centuries of mainly Iberian colonial influence, but with independence from Spanish and Portuguese (Brazil) colonial rule achieved almost everywhere after about 1820. Until the last part of the nineteenth century, Latin American countries were almost exclusively exporters of primary products, especially non-fuel minerals and various food and beverage products. Modern industrialisation started on a small scale in the late nineteenth and early twentieth centuries and was further stimulated by shortages of manufactured goods in the two world wars and the policy of import substitution. During the second half of the twentieth century rapid industrialisation has taken place in the larger countries, albeit concentrated in a few small regions (e.g. around Mexico City, São Paulo, Buenos Aires), first of consumer manufactures then of metallurgical and engineering sectors. Initially intended to supply home markets, industries in the major Latin American economies now export manufactures not only to other countries within the region but also to North America and Western Europe. Such products

BOX 14.2 THE EU'S POSITION IN THE WORLD

In order to appreciate fully the global position of a place, whether a city, a country, or a group of countries like the European Union, it is advisable to consult a globe. Even then, it is impossible to see all the surface from a single viewpoint. To be printed in a book the spherical surface has to be flattened, in the process distorting either shape or scale or both. Two different types of map projection are used to illustrate the text in this Box. Figure 14.1 uses a familiar Mercator's projection, which in very simple terms is based on a cylinder touching the world round the equator and then 'unwrapped'. In Figures 14.2 and 14.3 zenithal projections are used. These are based on a flat surface touching the earth at one point and extending in all directions from that point.

In the 1980s and 1990s much has been said about the global system and the global 'village', in which communications interlink all places of importance, capital can be moved almost anywhere and goods and passengers flow freely over most of the earth's surface. Arguably the first global system was organised and used by Spain and Portugal when for 60 years from 1580 to 1640 they were united under a common crown (Times *Atlas of World History*: 158–9). Their combined trade routes extended round the globe, linking places in Iberia with Africa, Southern and Eastern Asia and Central and South America.

The world transport and communication networks, trade and flows of capital of today are dominated by North America, the EU and Japan. There are very strong links between these three parts of the developed world, and strong links between them and other parts of the

world. The links between pairs of developing regions (e.g. South America and Africa, southwest Asia and China) are mostly much weaker at present, although transactions are increasing. Most world maps in popular publications and more learned ones alike fail to make the best use of map projections, conventional or otherwise, to show where major regions are located on the earth's surface in relation to one another, as will be shown below.

Figure 14.1 shows the global position of each of the three major developed regions of the world. Look at panels A, B and C, and cover D and E in order to see the position of Japan in relation to the rest of the world. For the US position, look at panels B, C and D and cover A and E. Finally, the EU position can be seen in panels C, D and E, with A and B covered. None of the three world views gives any idea of the true positions of the three regions in relation to one another.

Figures 14.2 gives a better view of the situation of the EU, but gives little idea of the positions of other pairs of regions in relation to each other. In Figure 14.3 the zenithal projection is centred on the North Pole and from here distance outwards is correct in all directions. However, distance between other pairs of places not aligned along a meridian is distorted, becoming increasingly exaggerated away from the Pole, although not excessively so as far as the equator. The clear message from this map is that the USA, Japan and the EU occupy the three corners of a roughly equilateral (spherical-surface) triangle. Canada, Alaska and northern Russian, all thinly populated regions, lie within the triangle, while at

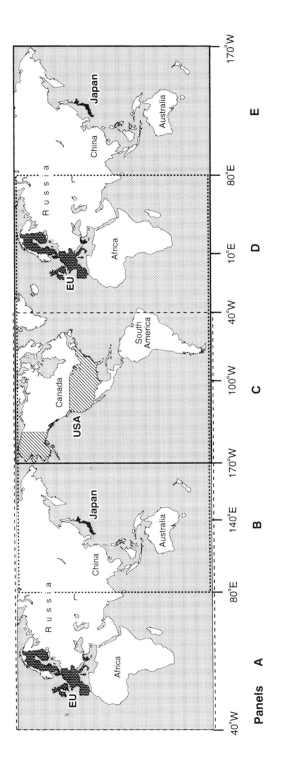

Figure 14.1 The conventional view of the world according to Mercator. To see the world with Japan at the 'centre', mask out panels D and E. For the EU, mask out A and B. See text for further explanation

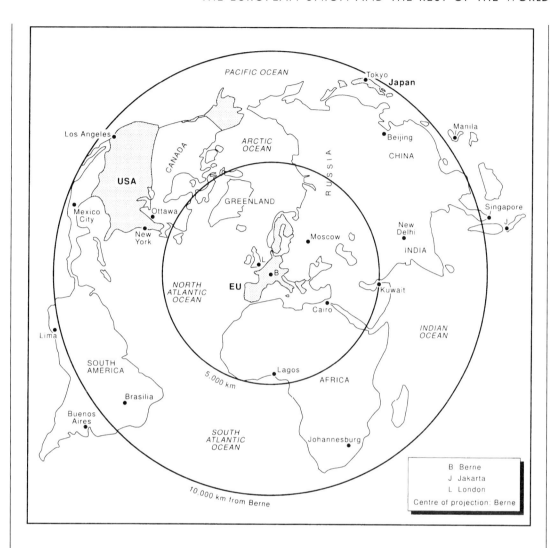

Figure 14.2 The EU at the centre of the world, and the 'hemisphere of Europe'. This projection is centred on Berne, Switzerland, Western Europe. Scale (i.e. distance) is consistent outward from the centre of the map which is an oblique zenithal equidistant projection. All other distances are not consistent in scale in other directions
Source: based on Spiessa 1993: 168

the centre of the triangle is the void of the frozen Arctic Ocean and the Greenland Ice Cap.

The EU and Japan are linked most directly by the Russian Trans-Siberian Railway, which until the late 1980s was usually 'out of bounds' for international traffic. The EU and eastern USA are linked by direct sea routes, as are Western USA and Japan. Two possible sea routes between the north Atlantic and north Pacific, the Northern Sea Route north of Russia and the sea route north of Canada and Alaska, are little used at present, but

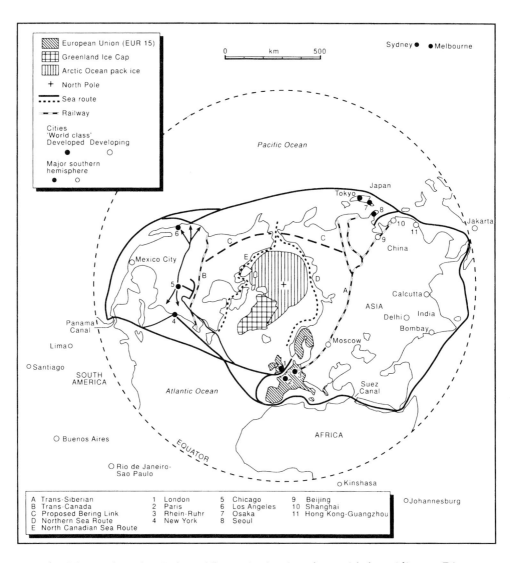

Figure 14.3 The northern hemisphere. The projection is polar zenithal equidistant. Distances are consistent outwards from the North Pole but not so in other directions. This view of the world gives a good idea of the true spatial relationships of the three major industrial regions of the world, the USA, the EU and Japan

in the next century means may be found to keep them open longer than the present short period in which they are not frozen. More futuristic and problematical is the proposal to link western North America and the Far East of Russia (close to Japan) by a railway, intended to carry goods rather than passengers, and to open up a large swathe of the world's northlands. Financially such a potential 'white elephant' could be a disaster that would make the Channel Tunnel financial situation seem a dazzling success.

'World class' cities in both the developed and developing world are shown on the map in Figure 14.3, together with some smaller cities in the southern hemisphere. All cities, including those south of the equator, are correct in their bearings and distances from the North Pole. Each of the three main developed regions has a

continent to its south: South America south of the USA, Africa south of the EU, Southeast Asia and Australia south of Japan.

Just as it is possible to distinguish central and peripheral parts of the EU, so in a global setting the North America, EU, Japan triangle may be regarded as 'central' in a spatial sense as well as an economic one.

are sometimes referred to as non-traditional exports.

Until the accession of Spain and Portugal to the EU, Latin America was a region largely ignored by the EC, whose former colonial ties were far closer to Africa and the Caribbean. In recent years, the isolated development projects financed by the EU have been overtaken by a more global policy for trade, given the increasing importance of the markets in South America and their exports to the EU. Relations are also characterised by concerns for human rights abuses in certain countries in the region and attempts to limit the production of hard drugs, which are the main source of supply to drug addicts in Europe. The EU is now actively financing programmes to encourage farmers in Colombia, Peru and Brazil to convert to other crops and to provide police training and cooperation to combat drug trafficking.

The data in Table 14.11 show that for the five largest Latin American countries in population – Brazil, Mexico, Colombia, Argentina and Peru – the percentage of all exports accounted for by manufactures has risen sharply over a very short period, 1970–93. Imports of manufactures still exceeded exports of manufactures in 1993, but if the trend in the 1970s continues, the gap could narrow or disappear early in the next century. Most of the exports of manufactures from Latin American countries

consist of consumer goods, while capital goods still form a large part of the imports.

In recent decades, intra-Latin American trade has gradually increased relative to total Latin American trade, with, for example, an increasing proportion of Venezuelan oil exports going to Latin American destinations, as also exports of manufactures from Brazil and agricultural products from Argentina. Latin America as a whole is still a net exporter of oil, non-fuel minerals and tropical agricultural products, and is therefore of interest to the EU in this respect. Political changes are, however, tending to emphasise existing divisions within Latin America. Now that it has joined NAFTA, the already large share of Mexico's foreign trade accounted for by the USA and Canada may increase. Although it would be 'geographically incorrect', Chile's application to join NAFTA underlines the new interest of the USA in its Latin American 'backyard' and there have already been discussions about an all-American free trade area. Such a move would leave only the Caribbean members of ACP still oriented explicitly towards the EU.

Africa South of the Sahara

This region consists of over fifty sovereign states of varying sizes, almost all of them colonies of West European powers between the 1880s and the 1960s. For some eighty years they were

Table 14.11 Value of exports and imports of manufactured goods as a percentage of total exports and imports in selected developing countries

Latin America and Africa	Exports 1970	1993	Imports 1993	Asia	Exports 1970	1993	Imports 1993
South Africa	41	74	90	South Korea	76	94	63
Brazil	15	60	66	Pakistan	57	85	62
Mexico	33	52	86	Bangladesh	57	81	41
Colombia	8	40	83	China	n.a.	81	85
Argentina	14	33	88	Philippines	8	77	75
Egypt	27	33	65	India	52	75	56
Peru	1	17	70	Thailand	8	73	81
Cameroon	9	14	78	Turkey	9	72	71
Ethiopia	2	4	82	Malaysia	8	65	84
Nigeria	1	2	n.a.	Indonesia	1	53	76

Source: World Bank 1995, Tables 14, 15
Note: n.a. not available

transformed with varying degrees of intensity into sources of primary commodities, almost entirely either agricultural products or non-fuel minerals. Western Europe was the destination of most of the exports, and the colonising powers exported manufactures to their colonies, while also placing limited numbers of settlers in some of them. The large number of settlers of European origin in South Africa and its exceptionally generous endowment of mineral resources have given it a more sophisticated and complex economic structure than that found elsewhere in Africa south of the Sahara.

The data for selected African countries in Table 14.10 show that manufactures still make up much of the value of the imports of the continent. Like all other countries of Africa south of the Sahara, however, the Cameroon, Ethiopia and Nigeria are essentially exporters of primary products. Most African countries have very small home markets. The idea that they should follow the path of the industrial countries and that Africa should have a heavy industrial base, proposed in the early 1960s in a United Nations Report, has not materialised. African countries have not even followed the path of many Latin American and Asian countries, which have developed light manufacturing and in some cases now also have impressive heavy industrial bases.

The EU countries that had colonies in Africa earlier this century carry some responsibility for the present state of the continent, both individually and now collectively under the Lomé Convention. Given the frequent ethnic conflicts, natural disasters and food shortages in Africa, much of the effort of the EU in improving conditions in the continent seems destined to be directed to coping with short-term problems and emergencies. In view of the rapid increase in population in Africa, the continued production of commercial crops for export seems unlikely since the cultivation of food crops will have to expand, while only Nigeria figures as a major exporter of oil. Non-fuel minerals are needed in the EU, the USA and Japan, but their production for export is highly localised and employs a comparatively small labour force. The export of minerals means that the chance to add value by using them in industry is lost except at the processing stage.

North Africa and Southwest Asia

This region is of particular interest to the EU because it has almost two-thirds of the world's proved oil reserves (at the end of 1994) and almost one-third of the world's proved natural gas reserves. At present rates of production the oil deposits have a reserves to production ratio of 'life' of almost 100 years. In 1994, Western Europe produced 287 million tonnes of oil, 256 million (almost 90 per cent) from Norway and the UK. It consumed 653 million tonnes of oil, having therefore in that year a net deficit of 366 million tonnes. In 1994 Western Europe imported 487 million tonnes of oil, of which 187 million came from the Middle East and 97 million from North Africa (mainly Algeria and Libya).

At 1994 dollar prices, the price of oil rose from under 10 dollars per barrel in 1972 to over 50 in 1980. In the 1980s it fell sharply to around 20 dollars, rose again following the Iraqi invasion of Kuwait, but had dropped to around 17 dollars by 1994. In real terms, therefore, the price of crude oil roughly doubled between the early 1970s and the mid-1990s. It is a matter of speculation whether or not the generally higher price of oil since 1973 stopped the rapid increase of oil consumption observed in the 1950s and 1960s from continuing indefinitely, especially in the developed regions of the world. Certainly the situation in the mid-1990s is such that the prediction that there would be an oil shortage by the year 2000 seems unlikely to prove correct. Even so, Western Europe is likely to face increasing competition for Middle East and other oil supplies on the part of the USA, Japan and a growing number of oil deficient industrialising countries in the developing world.

Southwest Asia and North Africa are of interest to the EU primarily as a source of oil and natural gas. The main reserves of non-fuel minerals of world class in North Africa are the phosphates of Morocco, of which it is credited with half of the world total. Only a very small proportion of the total area of the region is cultivated, and water supply is limited almost everywhere. Some countries are already

net importers of agricultural products and many other countries are likely to follow suit.

EU relations with North Africa and Southwest Asia are so complex that it would be difficult to develop a single EU policy towards the region as has been done for Africa south of the Sahara. As pointed out in Chapter 13, the Maghreb countries, which border the Western Mediterranean, are too near to the EU to be ignored. In the eastern Mediterranean, the conflicts between Israel and its Arab neighbours have made that area unstable, while Turkey has an ambivalent position, sharing an eastern border with Iraq, thereby projecting NATO into the Middle East, and there is a general lack of harmony among the Middle East countries themselves. Relations with North Africa have taken on a new dimension through the Barcelona Declaration and MEDA Programme aimed at cooperation in a wide range of activities in the Mediterranean Basin. As far as the Middle East is concerned, failure to establish a Common Foreign Policy has meant that the EU has contributed much less to the peace process than the USA, although it entertains close relations with the newly formed Palestinian Authority.

EU TRANSACTIONS WITH DEVELOPING REGIONS POORLY ENDOWED WITH NATURAL RESOURCES

South Asia

South Asia has almost a fifth of the total population of the world but the combined GDP of the countries in it is only a few per cent of the world total. The three largest countries of the region in population, India, Pakistan and Bangladesh, account for a very small percentage of total world trade and receive very little overseas development assistance. South Asia is still predominantly rural and the actual number of workers in the agricultural sector continues to grow in spite of migration from rural areas, while the cultivated area has hardly changed in recent decades. There is little surplus from the agricultural sector to provide exports and only a few items, such as cotton, jute and tea, are now exported in large quantities. The region is very poorly endowed with fossil fuels and non-fuel minerals, and is a net importer of oil.

The data in Table 14.11 show that in India, Pakistan and Bangladesh manufactures account for a large percentage of the value of exports. In all three countries this percentage grew substantially between 1970 and 1993. Even more surprisingly (at first sight), imports of manufactures account for a smaller percentage of the total than exports. The manufactures exported by Pakistan and Bangladesh consist almost entirely of textile fibres, textiles and clothing, whereas India's exports include in addition some capital goods (machinery and transport equipment) and a variety of other manufactures. Machinery and transport equipment figure prominently in the imports of all three countries, but in the 1990s the growth of the 'middle class' in India has led to an increase in the import also of consumer goods. Although India has achieved near if not total self-sufficiency in food production, in some years it is similar to Pakistan and Bangladesh in importing food as well as fuel and raw materials.

Southeast Asia

Until recently this region was seen as a source of primary products in the world economy. Rubber and other tropical agricultural products, tin, and more recently oil (mainly from Indonesia) were three of the principal items exported. Until the 1980s there was virtually no heavy industry and very limited light industry, processing being the main branch of industry apart from traditional crafts. The transformation from the dependence for exports on primary products to dependence on manufactures has taken place in some Southeast Asian countries more quickly than in any other major region of the world, as the figures in Table 14.11 show for the Philippines, Thailand, Malaysia and Indonesia. The exports of the first three countries include substantial amounts of machinery and transport equipment as well as the more familiar textile and clothing products. In contrast, Myanmar (formerly Burma) and Viet Nam, the other two large countries of Southeast Asia, are more self-sufficient and at the same time still export almost exclusively primary products.

Much of the manufacturing capacity in Southeast Asia is concentrated in a few large cities. Manila, Bangkok, Kuala Lumpur and Jakarta are now prominent industrial centres, producing manufactures for world markets with the help of cheap labour, including child labour. Their labour force is, however, competent and skilled enough to produce sophisticated engineering and electronic goods in factories recently established in the large cities referred to above as well as in special offshore zones such as Penang in Malaysia and Batam Island in Indonesia. Singapore is one of the main transport and financial centres of the region. There is less pressure on land and mineral resources in Southeast Asia than in South Asia, and little need to import primary products apart from fuel to Thailand and the Philippines. Manufactured goods still account for most of the imports.

China and Associated Countries

The political importance of China in East Asia is of concern to the EU, and the European Parliament has often expressed criticism of the Chinese authorities over human rights abuses, the situation of Tibet and sabre-rattling against Taiwan in order to subvert the democratic process in the strongly independent province. China has, however, constantly warned the

West not to interfere in its internal affairs and continues to enjoy privileged commercial relations with most Western European countries, which prefer to turn a blind eye to its political and social inadequacies due to its great economic potential.

South Korea (The Republic of Korea) resembles Japan as it was about three decades ago. It now has membership of OECD making it one of the first countries in the developing world to join. Its exports consist almost exclusively of manufactures and, like Japan, it imports food, fuel and raw materials, as well as manufactures, half of the value of which consists of machinery and transport equipment. It is increasingly investing abroad including in the EU. While South Korea's imports include various primary products, its exports consist almost entirely of manufactured goods, especially vehicles and electrical equipment. Japan and the USA each supply about a quarter of its imports and together take about 40 per cent of its exports, while the EU accounts for only about 15 per cent of South Korean foreign trade.

Like South Korea, Taiwan and Hong Kong both export almost exclusively manufactured goods, while importing both primary products and manufactures. Even when Hong Kong becomes part of China, its trading structure cannot change greatly, although the sources of its imports and the destination of its exports may change. Like South Korea, Taiwan trades heavily with the USA and Japan, while also in recent years supplying China with manufactures and placing capital there. The EU accounts for only about 15 per cent of Taiwan's foreign trade.

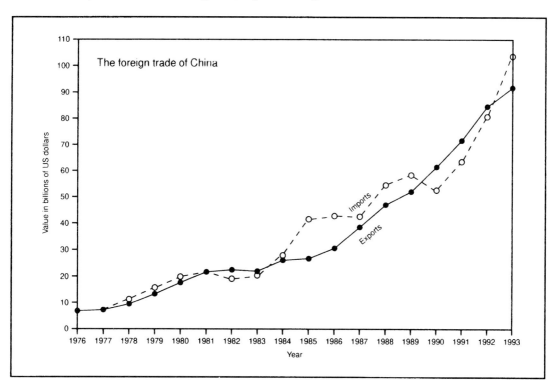

Figure 14.4 Changes in the value of Chinese foreign trade 1976–93

Table 14.12 Direction of China's foreign trade in 1993

	Billions of US $		Percentage	
	Exports	Imports	Exports	Imports
Hong Kong	22	11	24	11
Taiwan	1	13	1	13
Japan	16	23	17	22
European Union	12	16	13	15
Russia	3	5	3	5
N. America, Australia	19	14	21	13
All other partners	19	22	21	21
Total	92	104	100	100

Source: China Statistical Yearbook 1994, State Statistical Bureau of the People's Republic of China, Beijing, 1994

Since the Second World War, China's relationship to the rest of the world economy has changed markedly more than once. Between 1949 and the early 1960s most of China's foreign trade was with the USSR and Eastern Europe. In the 1960s and in the 1970s until the death of Mao Zedong in 1976, China's dependence on the Soviet bloc for development assistance in the form of capital manufactures and technical backup diminished sharply as relations deteriorated. Chinese leaders attempted to increase the influence of their country in various parts of the developing world, while for strategic reasons a policy of near economic self-sufficiency was pursued and foreign trade was very limited. With the policy of the four modernisations, introduced in the late 1970s, Chinese trade with Western countries began to increase.

Figure 14.4 shows that between 1976 and 1993 there was more than a tenfold increase in the value of China's foreign trade. Even when account is taken of the addition to the population of about 240 million people during that time, as well as some reduction in the real value of the US dollar, the scale of increase is paralleled in very few other countries. Understandably, a commensurate growth in interest in China both as a trading partner and as the location for investment has occurred in Japan, the USA and the EU.

Table 14.12 shows the direction of China's foreign trade in 1993. Hong Kong and Taiwan accounted for about a quarter, the developed countries of the world for about half, the rest of the world, including many developing countries and the former preferred partner, Russia, the remaining quarter. The EU supplied 15 per cent of China's imports and took 13 per cent of its exports. Like the EU, Japan and India, China is now essentially an exporter of manufactured goods, mainly textiles and machinery. These accounted for 82 per cent of the value of exports in 1993, primary goods for only 18 per cent. China is largely self-sufficient in food, fuel (mainly coal) and non-fuel minerals, and primary goods therefore make up only 14 per cent of all imports. Machinery and transport equipment account for about half of all imports.

Although the economy of China is theoretically still largely planned and managed by the state, foreign investment has been encouraged since the early 1980s. Whether the location of such investment is chosen or recommended by Chinese planners, or is determined by the foreign investors, there is a high degree of concentration in a few coastal municipalities and provinces. For example, Shanghai has little more than 1 per cent of China's population but it has more than 10 per cent of foreign investment, while Guangdong province, with which Hong Kong shares its only land bound-

Plate 14.3 'L'Europe dans le monde': unhindered by internal contradictions, Japan and the USA progress and flourish. Europe is beset by national squabbles, while Africa sinks
Source: 'Le douanier se fait la malle', cartoon by Plantu in *Le Monde Éditions*, Paris, 1992, p. 65

ary, has less than 6 per cent of the population of China yet gets a third of foreign investment. Most of the interior provinces of China are still heavily dependent on agriculture, have a poor infrastructure and are linked to the coastal cities and ports by long and overworked railways, features that make them unattractive to foreign investors. Hong Kong accounted in 1993 for 49 per cent of all non-Chinese investment in China, Taiwan for 8 per cent. In contrast, Japan (13 per cent), the EU (7 per cent) and the USA (5 per cent) together had a quarter of the total, as yet very small stakes in China's economy.

China's transactions with the EU are still modest in size. This subject has been covered here at some length to emphasise the difficult task that lies ahead if EU trade with and investment in China is to increase, always assuming that yet another catastrophic event on the human scene does not occur to set back economic development in China once again.

THE NEXT 50 YEARS

Many of the readers of this book may expect to live for another 50 years or more. In this section some prospects for the EU in the next century are outlined. It is said that there is nothing certain about the future except death and taxes. What follows is speculation rather than fact or opinion except that the choice of issues and problems discussed has been made subjectively by the authors and they are ones considered by the authors to merit consideration. The topic(s) covered in each chapter in the book will be discussed in turn. Each topic can be thought of as having ramifications at all or some of three levels: sub-EU, EU and global.

Table 14.1 serves as a starter to put the EU in a global perspective by comparing some of its salient features with those of 11 other 'major' world regions.

Columns (1)–(5) show how according to the Population Reference Bureau the population of the 12 regions is expected to change during 1995–2025. The share of world population in Western Europe (in effect EU plus EFTA) was 13.8 per cent in 1931, 11.4 per cent in 1958 and 6.8 per cent in 1995. It is expected to fall to less than 5 per cent in 2025 and by 2050 could reach less than 4 per cent of a total world population of 10–11 billion.

Columns (6)–(9) show that the natural resource endowment per inhabitant of Western Europe is below the world average but is more favourable than that of South, Southeast and East Asia (including Japan). Energy consumption per capita (column (7)) is well above the world average and since about half of the amount consumed is imported, the EU is likely to be competing with other parts of the world for oil and natural gas from the Middle East. On the other hand (columns (8), (9)), arable land per inhabitant in Western Europe is adequate and the amount is not likely to change much in the next few decades.

It is more difficult to speculate about the gap in GDP per inhabitant between rich and poor regions in the decades to come (column (10)). While the intention of EU leaders is to provide some development assistance to developing regions, especially the poorest, the implementation of policy on this issue is subject to delays and uncertainties. What has not yet been worked out is the extent to which the industrialising countries in the developing world will compete with the EU for fuel and raw materials nor whether migration from developing countries into the EU will increase as population grows in developing regions, especially Africa and Southwest Asia.

The topics and themes of each chapter in the book will now be reviewed.

Chapter 1 Introduction

Until the Second World War changes in the political map of Europe usually resulted either from merging, dividing or swapping territory by agreement of the rulers of states or from the acquisition of territory by the winners of wars. The European Union has developed through a peaceful process in which, in theory at least, the populations of distinct sovereign states have agreed to renounce some of their sovereignty to a supranational entity. Arguably CMEA was formed in a similar fashion, although the populations of the member countries had even less say about entry. The EU was initially formed through the initiative of French leaders, whose aim was to prevent yet another conflict with Germany. The Soviet threat to Western Europe in the 1950s also forced the leaders of the countries of the region to think seriously about strategic issues and collaboration in military matters. The situation is now entirely different. Unless EU leaders look to the future rather than the past they may continue to fight past battles and absorb a large number of new members from Central Europe and parts of the former USSR. The EU can trade with and invest in the rest of the countries of Europe as easily in the present situation as it could if they became full members of the EU. On the other hand, it is not at present obliged to commit a large slice of the EU budget to transforming and regenerating the economies of these countries (note the East Germany experience) as it would have to do if they joined.

Chapter 2 The Organisation of the EU

The future of the European Union under discussion in the mid-1990s has focused on two main issues, laid down on the agenda for the IGC, which began in March 1996 – enlargement to include the countries of Central and Eastern Europe, and further integration of its existing policies and mechanisms. Questions under integration include making the EU more efficient and accountable to the citizens, and developing policy areas that are still weak, in particular the Common Foreign and Security Policy, justice and home affairs, as well as the single currency and social policy. Some people view the structure of the EU in a multi-speed image, with certain Member States signing up to more integration ahead of or even without others. The prospect of enlargement itself gives rise to concerns over the impact on EU policies and institutions, and, in particular, how the EU budget will be able to finance the accession of 12 new countries, most of which will be net beneficiaries under the present arrangements. As so often with the EU, there is a strong political will and commitment to this new enlargement, without the practical, technical and financial issues being settled. Whether the EU will, as in the past, overcome these contradictions remains to be seen, given the scale of the difficulties it faces in the future.

Chapter 3 Population

It seems unlikely that fertility rates will change markedly in enough EU countries to produce a large increase or decrease in population in the next few decades. In developing countries the dependent population consists almost entirely of the young. In developed countries there has been a shift towards a preponderance of the elderly, with the need gradually to shift resources from education to healthcare. Such a trend is set to continue if current demographic trends do not change. Until the 1990s migration into the EU from Central Europe and the USSR was prevented by the Iron Curtain. With the expansion to 15 Member States and the removal of the Iron Curtain effect, the eastern 'frontier' of the EU has become much longer and also more leaky. Population growth in the Maghreb countries and Turkey means that more people from these areas are also likely to contemplate migrating to the EU. If it is deemed necessary to avoid an absolute decline in the population of the EU in the next few decades there are plenty of people ready to move in. A definitive immigration policy is needed in the EU.

Chapter 4 Transport and Communications

In the view of the authors there is a big difference between transport policy and expressed intentions on the one hand and reality on the other. The main dilemma is how to allocate the limited EU funds available to improve links between Member States and to produce an all-EU transport system. The main issue to be decided is whether to focus on road transport or on rail transport. In the view of the authors, a high-speed train network is too ambitious. Improvements in the rail network mainly affect passenger traffic rather than goods traffic yet rail is competitive for passenger traffic only over a limited range of distances.

Chapter 5 Energy and Water Supply

There is no common energy policy in the EU. Each Member State determines its priorities, Italy, for example, concentrating on the use of imported oil and natural gas, France on nuclear power. Apart from hydro-electricity, clean and renewable 'alternative' sources of power at present provide only a minute proportion of the energy consumed in the EU.

Chapter 6 Agriculture, Forestry and Fisheries

Although the EU is a net importer of agricultural products, it now broadly satisfies its consumption of food, a past battle fought and won, albeit resulting in the allocation of a large part of the EU budget to a small sector of the population. Issues in the next decades will probably include the question of how much higher yields can be pushed, the effect of the set-aside policy, how potential climatic change could affect different agricultural regions of the EU and whether in the longer term EU policy should be to increase agricultural production to assist those developing countries in which for political or environmental reasons food is in short supply.

Chapter 7 Industry

It was assumed by some in the 1980s that in 'the' post-industrial era a country could exist largely, if not entirely, on the strength of its service sector. In the 1990s there is widespread concern in the EU over both the loss of jobs in the industrial sector and a decline in output in many branches of manufacturing, if not in the industrial sector as a whole. In spite of being capable of making virtually every possible manufactured product, the EU is a large importer of manufactured goods. It faces competition from the USA and Japan in the manufacture of sophisticated capital and consumer goods and from many 'newly industrialising countries'.

Chapter 8 Services

Since the Second World War almost all the growth in employment in Western Europe has been in the service sector. Since the 1970s, however, the use of office machines such as electronic retail checkouts and computers has enabled shops, banks and other services to cut staff. Will new types of services in the next few decades produce new jobs, or will the reduction in jobs in the service sector continue to push up unemployment levels in the EU, as happened in the early 1990s?

Chapter 9 The Social Environment

Concern for the poor and the underprivileged has been expressed in EU policy statements about issues regarding ethnic minorities, gender disparities and low paid workers. As the movement of people, goods and information theoretically at least becomes unrestricted between EU Member States the problems of one country are increasingly being passed on to other countries, especially in such areas as healthcare (e.g. the spread of the HIV virus), crime (e.g. the movement of drugs), and illegal immigrants. The accession of Central European countries to full EU membership could aggravate social problems in the EU in view of differences in living standards and life-styles between the 'old' EU and former members of the Soviet bloc.

Chapter 10 The Environment

Issues of pollution and environmental degradation will not go away in the twenty-first century. Steps to prevent some kinds of pollution have already been taken, but issues such as measures to stabilise or even reduce the amount of fossil fuel consumed in the EU and indeed worldwide, while receiving frequent mention at conferences, are quickly shelved for later consideration. The idea that high levels of material consumption can somehow be maintained indefinitely through the application of sustainable development is very attractive but as yet without foundation in practice.

Chapters 11, 12 Regional Policies; Narrowing the Gaps

The extent to which there has been some narrowing of the differences in living standards between the richest and poorest parts of the EU since the 1950s depends to some degree on the evidence chosen to illustrate the problem. It was shown, for example, that over the last thirty years the infant mortality 'gap' has been greatly reduced between EUR 15 Member States (only six of which were originally members). On the other hand, for example, car ownership has increased in every region of the EU but, paradoxically, while the *relative* increase in the 1970s and 1980s was fastest in many of the poorer EU regions (e.g. threefold, from 50 to 150 per thousand population), the *absolute* increase tended to be fastest in richer regions (e.g. doubling from 150 to 300). In the assessment of any increase in production and consumption, care should be taken to discriminate between relative and absolute change. If the trends over the last four decades continue, then more of the same can be expected. To alter them dramatically, either a massive transfer of people from the least to the most favourable regions would have to be organised, or much future investment, whether sound economically or not, would have to be directed to the lagging regions. To achieve such transfers would require far more powers to be in the hands of the EU Commissioners, a much larger EU budget, and the presence of a superstate planning organisation reminiscent of GOSPLAN, which ran the experiment in the USSR for some seven decades.

Chapters 13, 14 Enlargement of the EU; the Rest of the World

Given its dependence on other parts of the world for both primary products and manufactured goods the EU cannot exist in isolation. For simplicity, the world can be subdivided into the rest of Europe together with certain nearby areas, and the rest of the world. Many countries in the rest of Europe are theoretically eligible for EU membership, while others, such as the three Maghreb countries of Northwest Africa, are so close to the EU in many senses that their future is tied to the fortunes of the EU itself. In the mid to late 1990s it is difficult to distinguish firm intentions about enlargement of the EU from the rhetoric aimed at keeping Central European governments happy and the use by politicians of enlargement to delay greater integration within the EU itself.

FURTHER READING

Edwards, G. and Regilsberger, E. (eds) (1990) *Europe's Global Links: The European Community and Inter-Regional Cooperation*, London: Pinter.

Featherstone, K. and Ginsberg, R. (1993) *The United States and the European Community in the 1990s*, London: Macmillan.

Grilli, E. R. (1993) *The European Community and the Developing Countries*, Cambridge: Cambridge University Press.

Nilsson, J.-E., Dicken, P. and Peck, J. (eds) (1996) *The Internationalization Process: European Firms in Global Competition*, London: Paul Chapman. Case studies of various European corporations, examining their fortunes in a changing world in which functional and geographical divisions of labour are related to forces operating both within and outside Europe.

Peterson, J. (1993) *Europe and America in the 1990s: The Prospects for Partnership*, Hampshire and Vermont: Edward Elgar.

United Nations Development Programme (1995) *Human Development Report 1995*, Oxford: Oxford University Press. Numerous international comparisons between the EU and both the other developed regions and the developing world.

Weidenfeld, W. and Janning, J. (eds) (1991) *Global Responsibilities: Europe in Tomorrow's World*, Gutersloh: Bertelsmann Foundation.

APPENDIX

Table A1 Data for NUTS level 1 regions of the EU, Part 1

	(1) Popn	(2) Popn change	(3) Agric.	(4) Industry	(5) Services	(6) Industrial change
BELGIUM	10.1	0.2	2.6	29.5	67.9	0.4
1 Vlaams Gewest	5.8	0.3	2.7	32.5	64.7	0.0
2 Région Wallonie	3.3	0.2	3.1	27.0	69.8	0.7
3 Bruxelles–Brussel	1.0	−0.4	0.2	16.7	83.1	2.5
4 DENMARK	5.2	0.1	5.1	26.0	68.8	−0.1
GERMANY	81.3	0.4	3.6	38.3	58.2	−2.5
5 Baden–Württemberg	10.2	0.8	3.2	45.3	51.5	−0.6
6 Bayern	11.9	0.7	5.8	39.4	54.8	−0.9
7 Berlin	3.5	1.2	0.7	27.6	71.7	−5.4
8 Brandenburg	2.4	−0.3	4.7	35.2	60.0	n.a.
9 Bremen	0.7	−0.1	0.8	31.2	68.0	−0.5
10 Hamburg	1.7	0.2	1.2	24.6	74.3	0.1
11 Hessen	6.0	0.5	2.6	35.7	61.7	−0.4
12 Mecklenburg–Vorpommern	1.8	−0.4	7.8	30.5	61.7	n.a.
13 Niedersachsen	7.6	0.4	4.8	36.7	58.5	−0.7
14 Nordrhein Westfalen	17.8	0.3	2.1	40.7	57.2	−0.4
15 Rheinland–Pfalz	3.9	0.6	3.5	38.8	57.6	−0.7
16 Saarland	1.1	0.1	0.7	38.6	60.7	0.6
17 Sachsen	4.6	−0.9	3.1	40.4	56.5	n.a.
18 Sachsen–Anhalt	2.8	−0.8	4.5	38.1	57.4	n.a.
19 Schleswig–Holstein	2.7	0.3	4.7	28.3	67.0	−0.1
20 Thüringen	2.5	−0.6	3.3	39.2	57.6	n.a.
ELLADA (GREECE)	10.4	0.6	21.3	24.2	54.5	0.6
21 Voreia Ellada	3.3	0.5	30.5	24.7	44.8	n.a.
22 Kentriki Ellada	2.5	0.6	38.5	20.3	41.3	n.a.
23 Attiki	3.6	0.6	1.1	27.9	70.9	n.a.
24 Nisia	1.0	0.7	28.2	17.1	54.7	n.a.
ESPAÑA (SPAIN)	39.1	0.3	10.2	30.8	59.0	1.1
25 Noroeste	4.4	0.0	24.7	26.4	48.8	−0.4
26 Noreste	4.1	0.0	7.3	37.3	55.3	0.4
27 Madrid	4.9	0.3	0.9	27.7	71.4	1.5
28 Centro	5.5	0.4	16.5	30.2	53.3	0.8
29 Este	10.5	0.2	5.2	37.9	56.9	1.5
30 Sur	8.2	0.8	13.3	23.6	63.1	1.9
31 Canarias	1.5	0.6	7.5	16.7	75.8	1.5
FRANCE	57.7	0.5	5.9	27.1	67.0	2.9
32 Ile de France	10.9	0.7	0.5	23.1	76.4	3.3
33 Bassin parisien	10.4	0.4	8.9	29.1	62.0	3.8
34 Nord-Pas-de-Calais	4.0	0.1	3.2	30.9	65.9	4.7
35 Est	5.1	0.2	3.7	35.0	61.3	2.3
36 Ouest	7.6	0.5	10.4	27.2	62.4	2.0
37 Sud-Ouest	6.0	0.5	11.0	25.9	63.1	0.8
38 Centre-Est	6.8	0.6	6.0	28.8	65.2	3.8
39 Mediterranée	6.8	1.0	4.4	22.8	72.8	1.5
40 Départements d'Outre-Mer	1.6	2.0	7.2	29.9	62.5	n.a.

Table A1 Part 1 (Cont.)

	(1) Popn	(2) Popn change	(3) Agric.	(4) Industry	(5) Services	(6) Industrial change
41 IRELAND	3.6	0.4	12.8	27.9	59.2	0.6
ITALIA (ITALY)	57.1	0.1	7.3	32.4	60.2	1.2
42 Nord Ouest	6.1	−0.4	5.5	35.2	59.3	2.5
43 Lombardia	8.9	0.0	2.8	42.9	54.3	1.3
44 Nord Est	6.5	0.1	6.4	37.9	55.7	−0.2
45 Emilia–Romagna	3.9	−0.1	6.9	34.8	58.3	0.8
46 Centro	5.8	0.0	6.1	34.8	59.2	1.5
47 Lazio	5.2	0.3	3.5	20.9	75.6	0.6
48 Campania	5.7	0.4	10.0	25.9	64.1	1.6
49 Abruzzi–Molise	1.6	0.2	12.5	30.6	56.9	−0.2
50 Sud	6.7	0.3	15.4	23.8	60.7	1.1
51 Sicilia	5.0	0.2	13.4	21.0	65.6	2.8
52 Sardegna	1.7	0.3	13.2	25.9	60.9	0.1
53 LUXEMBOURG (GRAND-DUCHÉ)	0.4	0.7	3.2	26.4	70.5	0.7
NEDERLAND (NETHERLANDS)	15.3	0.6	4.0	23.9	72.1	−1.0
54 Noord–Nederland	1.6	0.2	5.3	25.9	68.8	−0.7
55 Oost–Nederland	3.1	0.9	5.4	26.9	67.7	−2.3
56 West–Nederland	7.2	0.6	3.2	18.8	78.0	0.1
57 Zuid–Nederland	3.4	0.6	4.1	31.1	64.8	−1.6
ÖSTERREICH (AUSTRIA)	7.7	0.1	7.0	37.0	56.0	n.a.
58 Ostösterreich	3.2	0.0	n.a.	n.a.	n.a.	n.a.
59 Südösterreich	1.7	0.0	n.a.	n.a.	n.a.	n.a.
60 Westösterreich	2.7	0.4	n.a.	n.a.	n.a.	n.a.
PORTUGAL	9.9	0.1	11.6	32.9	55.6	n.a.
61 Continente	9.4	0.1	11.4	33.0	55.7	n.a.
62 Açores	0.2	−0.1	18.0	25.6	56.4	n.a.
63 Madeira	0.3	0.1	14.3	33.8	51.9	n.a.
SUOMI (FINLAND)						
64 Manner–Suomi	5.1	0.4	9.0	29.0	62.0	n.a.
65 Ahvenanmaa	0.03	0.7	n.a.	n.a.	n.a.	n.a.
66 SVERIGE (SWEDEN)	8.7	0.4	3.0	28.0	69.0	n.a.
UNITED KINGDOM	58.2	0.3	2.0	28.0	70.0	1.2
67 North	3.1	−0.1	1.8	31.0	67.2	2.1
68 Yorkshire and Humberside	5.0	0.1	1.7	33.7	64.6	1.0
69 East Midlands	4.1	0.5	2.4	36.8	60.8	0.7
70 East Anglia	2.1	0.8	4.5	28.5	66.9	−0.2
71 Southeast	17.8	0.3	1.2	23.6	75.2	1.9
72 South West	4.8	0.7	3.6	27.2	69.2	0.1
73 West Midlands	5.3	0.2	1.6	37.0	61.4	0.8
74 North West	6.4	−0.1	1.0	31.2	67.8	1.8
75 Wales	2.9	0.2	3.7	31.3	65.0	0.3

Table A1 Part 1 (Cont.)

	(1) Popn	(2) Popn change	(3) Agric.	(4) Industry	(5) Services	(6) Industrial change
UNITED KINGDOM Cont.						
76 Scotland	5.1	−0.1	3.1	29.8	67.0	0.9
77 Northern Ireland	1.6	0.4	4.8	27.4	67.8	0.4
EUR 15	369.7		5.6	31.6	62.8	

Source: *Regional Profiles* 1995
Notes: (1) Population in millions in 1993
(2) Annual population change, per cent 1993/1980
(3), (4), (5) Employment in agriculture, industry, services, as a percentage of total employed population, 1993
(6) Percentage variation in industrial employment 1993/1983

Table A1 Data for NUTS level 1 regions of the EU, Part 2

	(1) Infant mortality	(2) School	(3) Participation rate	(4) Over 64	(5) Av. GDP pps ECU	(6) Unempl. %
BELGIUM	6.9	90	40.7	15.2	16.1	8.2
1 Vlaams Gewest	6.7	82	41.6	14.7	16.3	6.1
2 Région Wallonie	7.3	93	38.7	15.5	13.0	11.2
3 Bruxelles–Brussel	7.0	130	41.5	17.5	25.7	10.8
4 DENMARK	6.6	81	56.0	15.6	16.3	10.3
GERMANY	6.2	94	48.8	15.0	16.6	7.5
5 Baden–Württemberg	5.1	93	49.8	14.3	19.8	4.2
6 Bayern	5.3	91	51.9	15.1	18.8	4.0
7 Berlin	6.3	91	52.8	14.0	15.3	10.4
8 Brandenburg	7.5	n.a.	51.3	12.3	6.2	14.0
9 Bremen	4.7	88	47.0	17.4	23.3	9.1
10 Hamburg	6.5	100	50.2	17.4	29.1	6.1
11 Hessen	5.8	93	48.3	15.5	22.1	4.9
12 Mecklenburg–Vorpommern	7.8	n.a.	51.5	11.1	5.6	16.9
13 Niedersachsen	6.1	92	47.4	15.8	15.6	6.7
14 Nordrhein Westfalen	7.0	99	45.9	15.1	17.1	6.9
15 Rheinland–Pfalz	6.7	89	47.1	15.9	15.5	4.9
16 Saarland	7.7	95	43.3	15.8	16.3	7.7
17 Sachsen	6.6	n.a.	48.3	15.9	5.8	13.7
18 Sachsen–Anhalt	7.9	n.a.	49.5	14.3	6.0	15.9
19 Schleswig–Holstein	5.7	91	49.5	15.9	15.5	5.5
20 Thüringen	7.8	n.a.	50.5	13.9	5.2	15.0
ELLADA (GREECE)	8.4	60	40.2	14.3	9.1	7.5
21 Voreia Ellada	8.0	n.a.	41.1	12.3	8.6	7.3
22 Kentriki Ellada	7.9	n.a.	40.1	17.8	8.7	7.9
23 Attiki	9.6	n.a.	39.4	12.9	10.0	10.4
24 Nisia	6.5	n.a.	40.4	17.5	8.2	4.3
ESPAÑA (SPAIN)	7.4	68	39.4	13.8	11.7	21.5
25 Noroeste	7.2	73	40.2	16.1	9.6	18.7
26 Noreste	8.0	79	40.0	14.6	13.5	18.8
27 Madrid	8.5	75	39.6	12.3	14.7	16.9
28 Centro	5.5	68	37.3	16.7	9.5	21.3
29 Este	6.6	63	42.0	14.1	13.5	19.2
30 Sur	8.8	63	36.8	11.7	9.2	30.1
31 Canaries	5.2	65	39.2	9.6	11.2	26.9
FRANCE	6.8	89	45.2	14.3	17.0	11.1
32 Ile de France	7.1	91	48.7	10.9	25.7	9.6
33 Bassin pariesien	7.2	85	43.6	14.5	15.5	11.6
34 Nord-Pas-de-Calais	7.1	91	39.0	12.2	13.5	14.5
35 Est	6.8	87	44.1	12.9	15.7	8.8
36 Ouest	6.3	91	43.7	15.7	14.1	10.8
37 Sud-Ouest	6.6	89	43.7	18.0	14.6	11.2
38 Centre-Est	6.1	90	45.5	14.1	16.2	10.8
39 Mediterranée	6.3	86	41.2	17.2	14.6	14.6
40 Départements d'Outre-Mer	9.0	n.a.	56.0	8.0	6.1	n.a.

Table A1 Part 2 (Cont.)

	(1) Infant mortality	(2) School	(3) Participation rate	(4) Over 64	(5) Av. GDP pps ECU	(6) Unempl. %
41 IRELAND	6.7	77	39.4	11.4	11.1	15.4
ITALIA (ITALY)	8.0	69	40.5	15.4	15.8	10.4
42 Nord Ouest	9.0	70	42.2	18.6	18.0	7.9
43 Lombardia	6.3	66	44.6	14.6	20.3	5.3
44 Nord Est	5.6	72	43.5	16.0	18.0	5.4
45 Emilia–Romagna	6.6	76	45.9	19.6	19.3	5.7
46 Centro	7.6	74	42.7	19.3	16.3	7.5
47 Lazio	9.7	73	40.4	14.2	18.0	10.2
48 Campania	9.6	67	35.2	11.2	10.9	20.2
49 Abruzzi–Molise	8.8	77	39.8	17.1	13.6	10.9
50 Sud	7.7	64	35.3	12.9	10.6	16.0
51 Sicilia	9.8	58	34.3	13.8	10.8	19.4
52 Sardegna	6.3	70	38.2	12.6	11.7	18.9
53 LUXEMBOURG (GRAND-DUCHÉ)	8.6	65	43.1	13.5	23.2	10.4
NEDERLAND (NETHERLANDS)	6.3	92	47.3	13.0	15.5	6.5
54 Noord-Nederland	6.5	92	44.5	14.1	15.6	8.3
55 Oost-Nederland	6.3	90	46.6	12.5	13.5	6.2
56 West-Nederland	6.3	93	48.4	13.5	16.6	6.3
57 Zuid-Nederland	6.2	90	47.1	11.7	14.7	6.1
ÖSTERREICH (AUSTRIA)	6.2	n.a.	n.a.	15.0	18.1	3.9
58 Ostösterreich	n.a.	n.a.	n.a.	n.a.	20.1	4.3
59 Südösterreich	n.a.	n.a.	n.a.	n.a.	14.4	4.1
60 Westösterreich	n.a.	n.a.	n.a.	n.a.	18.0	3.4
PORTUGAL	9.3	49	48.1	13.9	9.9	5.2
61 Continente	8.9	n.a.	48.3	14.0	10.0	5.3
62 Açores	16.3	n.a.	40.2	12.5	6.2	5.0
63 Madeira	11.2	n.a.	46.5	11.7	6.6	3.9
SUOMI (FINLAND)						
64 Manner-Suomi	4.4	n.a.	n.a.	14.0	14.6	16.4
65 Ahvenanmaa	n.a.	n.a.	n.a.	n.a.	20.7	5.1
66 SVERIGE (SWEDEN)	5.8	n.a.	n.a.	18.0	16.6	7.9
UNITED KINGDOM	6.6	66	49.5	15.7	14.9	10.0
67 North	7.1	65	46.9	15.9	13.2	11.3
68 Yorkshire and Humberside	6.7	68	49.3	15.9	13.5	9.9
69 East Midlands	6.9	62	50.7	15.7	14.1	8.8
70 East Anglia	4.6	61	51.8	17.0	15.1	7.7
71 Southeast	6.1	67	51.6	15.2	17.7	9.9
72 South West	5.7	65	50.0	18.6	14.2	9.0
73 West Midlands	8.2	69	49.1	15.3	13.7	10.4
74 North West	7.0	65	47.8	15.7	13.4	10.4
75 Wales	5.9	61	44.6	17.2	12.5	9.6

Table A1 Part 2 (Cont.)

	(1) Infant mortality	(2) School	(3) Participation rate	(4) Over 64	(5) Av. GDP pps ECU	(6) Unempl. %
UNITED KINGDOM (Cont.)						
76 Scotland	6.8	68	49.0	15.1	14.3	9.9
77 Northern Ireland	6.0	70	43.3	12.7	11.7	15.0
EUR 15	7.0	n.a.	45.2	14.8	15.1	10.4

Sources: Eurostat *Regions* 1994: variables (1), (2) and (4); *Regional Profiles* 1995: variables (3), (5) and (6)
Notes: (1) Infant mortality, 1991, deaths of infants under the age of 1 year per 1,000 live births
(2) Percentage of 16–18 year olds still in school, 1991
(3) Participation rate, 1993, employed population as a percentage of total population
(4) Population aged 65 and over as a percentage of total population
(5) Average GDP per inhabitant in ECU for the years 1990–91–92 in Purchasing Power Standards
(6) Average percentage unemployed, 1992–94

REFERENCES

Arkell, T. (1994) 'Crisis of confidence', *Geographical*, LXVI (9), Sept.: 14–15. The background to and implications of Italy's very low fertility rate.

Barney, G. O. (ed.) (1982) *The Global 2000 Report to the President*, Harmondsworth: Penguin Books.

Birrell, I. and Skipworth, M. (1991) 'Airlines warned their fares are far too high', *The Sunday Times*, 24 Nov., p. 15.

Blake, D. (1991) 'Time to abolish Luxembourg?', *The European*, 26–28 July, p. 10.

Bond, M. (1995) 'North Sea still awash with toxins', *The European*, 9–15 June, p. 4.

Born, M. (1995a) 'Hosts shamed by record on Co_2', *The European*, 17–23 March, pp. 1, 4.

Born, M. (1995b) '"Save our trees" plea to Brussels', *The European*, 10–16 Aug., p. 2.

Born, M. and Paterson, T. (1995) 'A drought and a deluge', *The European*, 24–30 Aug., p. 3.

Bounadonna, P. (1996) 'Europe's super rail plan hits the buffers', *The European*, 14–20 March, p. 6.

Bowditch, G. (1995) 'Scots defy drought by selling millions of gallons to Spain', *The Times*, 23 Aug., p. 7.

BP (1995) *BP Statistical Review of World Energy*, June, London: The British Petroleum Company plc, and earlier years.

Brierley, D. (1995) 'Rush for oil in the "new North Sea"', *The European*, 10–16 Aug., p. 17.

Bureau of Mines (1985) *Mineral Facts and Problems*, Bulletin 675, Department of the Interior, Washington.

Buscall, E. (1994) 'Tighten the net to save the needy tomorrow', *The European*, 11–17, Feb., p. 11.

Butler, C. (1995a) 'Single currency will close 24 Stock Exchanges', *The European*, 9–15 June, p. 17.

Butler, C. (1995b) 'The battle of the European bourses', *The European*, 14–20 Sept., p. 19.

Calendario Atlante De Agostini 1990 (1989) Novara: Istituto Geografico De Agostini.

Carazzi, M. and Segre, A. (1989) 'Città e industria: alcune linee interpretative a proposito di Torino e Milano', *Bollettino della Società Geografica Italiana*, Ser. XI, vol. VI., p. 197.

Casassus, B. (1991) 'Gulf war could fuel drive for the electric car', *The European*, 8–10 Feb., p. 25.

Clark, J. P. and Flemings, M. C. (1986) 'Advanced materials and the economy', *Scientific American*, 255 (4), October: 42–9.

Cole, F. J. (1996) 'The lesson of Babel', *European Voice*, 22–28 Feb., p. 14.

Cole, J. P. (1993) 'The Danish Connection: three new fixed links for Sjaelland?', *Focus,* Spring, pp. 2–7.

Cole, J. P. (1996) *Geography of the World's Major Regions*, London: Routledge.

Cole, J. P. and Cole, F. J. (1993) *Geography of the European Union*, London: Routledge.

Collett, B. (1996) 'Village life decline stunting growth', *The Times*, 16 July, p. 20.

COM 87–230 (1987) *Third Periodic Report from the Commission on the Social and Economic Situation and Development of the Regions of the Community*, Brussels: Commission of the European Communities, 21 May.

COM 89–564 (1990) *Communication on a Community Railway Policy*, Brussels: Commission of the European Communities, 25 Jan.

COM 90–609 (1991) *The Regions in the 1990s*, Fourth Periodic Report, Brussels: Commission of the European Communities, 9 Jan.

COM (95) 349 *Tacis Annual Report 1994*, Brussels: Commission of the European Communities, 18 July 1995.

COM (95) 357 *Report from the Commission to the Council, the European Parliament, the Economic and Social Committee and the Committee of the Regions on the state of health in the European Community*, Brussels: Commission of the European Communities, 19 July 1995.

COM (95) 366 *Phare 1994 Annual Report*, Brussels: Commission of the European Communities, 20 July 1995.

COM (95) 659 *For a European Union Energy Policy* (Green Paper), Brussels: Commission of the European Communities, 23 Feb. 1995.

COM (96) 300 European Commission Documents, *Preliminary draft, General Budget of the European Communities for the financial year 1997*, Vol. 0, General Introduction, 10 June 1996.

Coman, J. (1995) 'Caring falls victim to sick economies', *The European*, 16–22 Nov., p. 4.

Commission of the EC (1987a) *The European Community – Population*, 1:4 million map, Luxembourg: Office for Official Publications of the European Communities

Commission of the EC (1987b) *The European Community – Forests*, 1:4 million map, Luxembourg: Office for Official Publications of the European Communities.

Commission of the EC (1989) *European Regional Development Report*, Luxembourg: Office for Official Publications of the European Communities.

Commission of the EC (1991) *Europe 2000 – Outlook for the Development of the Community's Territory*, Directorate-General for Regional Policy, Brussels.

Commission of the EC (1992) *The Agricultural Situation in the Community*, 1991 Report, Brussels/Luxembourg.

Commission of the EC (1993a), *Our Farming Future*, Europe on the Move, Luxembourg: Office for Official Publications of the European Communities.

Commission of the EC (1993b) *Transport in the 1990s*, Europe on the Move, Luxembourg: Office for Official Publications of the European Communities.

Commission of the EC (1994a) *Competitiveness and Cohesion: trends in the regions*, EC Regional Policies, Luxembourg: Office for Official Publications of the European Communities.

Commission of the EC (1994b) *Panorama of EU Industry 94*, European Commission, Luxembourg: Office for Official Publications of the European Communities.

Commission of the EC (1994c) *Trans-European Networks*, Europe on the Move, Luxembourg: Office for Official Publications of the European Communities.

Commission of the EC (1995a) *Panorama of EU Industry 95–96*, Luxembourg: Office for Official Publications of the European Communities.

Commission of the EC (1995b) *Preliminary draft General Budget of the European Communities for the financial year 1996*, SEC(95) 850–EN, May 1995.

Conradi, P. (1995) 'MEPs try to fold their £140m travelling circus', *Sunday Times*, 25 June, p. 1.24.

Cornelissen, P. A. M. (1993) *Report of the Committee on Transport and Tourism in the approach to the year 2000*, European Parliament, Session Documents (DOC EN\RR\239\239180), 10 Nov.

CSE (95) 605 (1995) *Interim report from the Commission to the European Council on the effects of the policies of the European Union of enlargement to the associated countries of central and eastern Europe*, Brussels: Commission of the European Communities, 6 Dec. 1995.

CSO (1995) *Regional Trends*, 1995 edition, London: HMSO (Central Statistical Office, Editor J. Church).

Davies-Gleizes, F. (1991) 'US cleans up Europe's act', *The European*, 17–19 May, p. 24.

Dawe, T. (1995) 'Choking the life out of the countryside', *The Times*, 24 June, p. 24.

Dynes, M. (1992) 'Super-fast train shunted into sidings', *The Times*, 3 Jan., p. 5.

Dynes, M. and Brock, G. (1995) 'Britain's border controls face Euro court test', *The Times*, 15 Feb., p. 10.

Eason, K. (1994) 'Safety first for Britain', *The Times*, 2 June, p. 40.

Economic Commission for Europe (1991) *Transport Information 1991* (Geneva), New York: United Nations.

Endean, C. (1995) 'Immigration becomes Italy's hottest political issue', *The European*, 19–25 Oct., p. 9.

The European (1994) 'How motherhood is delayed', 28 Jan – 3 Feb, p. 6.

European Parliament (1990) *The Impact of German Unification on the European Community*, Directorate-General for Research, Working Document No.1, 6–1990.

European Parliament (1991) *On the State of the Environment in the Czech and Slovak Federative Republic* (CSFR), Delegation for Relations with Czechoslovakia. Committee on the Environment, Public Health and Consumer Protection, 26 March, SDI/CH/bs DOC-EN/ RESRCH/106892

European Parliament (1994a) *Report of the Committee of the Environment, Public Health and Consumer Protection on the Environmental Aspects of the Enlargement of the Community to include Sweden, Austria, Finland and Norway*, Session Documents A3–0008/94, 6 Jan.

European Parliament (1994b) *Fact Sheets on the European Parliament and the Activities of the European Union*, Directorate General for Research, Luxembourg: Office for Official Publications of the European Communities.

European Parliament (1995a) 'MEPs seek to counter aid cutbacks', *EP News*, 9–13 Oct., pp. 2–3.

European Parliament (1995b) 'All clear for Turkey deal', *EP News*, 11–15 Dec., p. 1.

European Parliament (1996) *The Regional Impact of Community Policies Executive Summary*, Regional Policy Series, W-16 External Study, Directorate General for Research, Luxembourg: Office for Official Publications of the European Communities.

Eurostat (1989) *Basic Statistics of the Community*, 26th edition, Luxembourg: Office for Official Publications of the European Community.

Eurostat (1991) *A Social Portrait of Europe*, Brussels/Luxembourg: Statistical Office of the European Communities.

Eurostat (1994a) *Regions statistical yearbook 1994*, Luxembourg: Statistical Office of the European Communities.

Eurostat (1994b) *Basic Statistics of the European Community*, 31st edition, Luxembourg: Office for Official Publications of the European Communities.

Eurostat (1995a) *Europe in Figures*, 4th edition, Luxembourg: Office for Official Publications of the European Communities.

Eurostat (1995b) 'Statistics in focus, regions', 8 Jan., Luxembourg.

Eurostat (1996) *External Trade*, Monthly statistics, 2, p. XIX.

Evans, J. (1994) 'Scandal and competition fail to dislodge London's crown', *The European*, 8–14 April, p. 17.

FAO (1994) *Food and Agriculture Organization Trade Yearbook 1993*, Rome: United Nations.

FAO (various years), *FAO Yearbook, Production*, Rome: Food and Agriculture Organization of the United Nations.

Foucher, M. (ed.) (1993) *Fragments d'Europe*, Fayard.

Froment, R. and Lerat, S. (1989) *La France a l'aube des années 90* (3 volumes), Paris: Bréal, Montreuil.

Gibson, M. (1996) 'Flying off the beaten track', *The European*, 4–10 July, p. 27.

Gilbert, S. and Horner, R. (1984) *The Thames Barrier*, London: Thomas Telford Ltd.

Gill, K. (1991) 'Conservation experts want 150,000 red deer culled', *The Times*, 17 Aug., p. 4.

Griffin, M. (1995) 'Hashish habit deepens farmers' dependency', *The European*, 9–15 Nov., p. 3.

Gunnemark, D. and Kenrick, D. (1985) *A Geolinguistic Handbook*, Gothenburg: Goterna.

Hallenstein, D. (1995) 'Citizens of the Sassi return to their underground homes', *The European Magazine (The European)*, 19–25 May, p. 3.

Helgadottir, B. (1994) 'Rich get richer, poor get poorer', *The European*, 4–10 Feb., p. 6.

Holloway, M. (1994) 'Trends in women's health: a global view', *Scientific American*, 271 (2), August: 66–73.

Hornsby, M. (1991) 'Farmers urge nitrogen cut', *The Times*, 14 May, p. 5.

ISTAT (1993) *Annuario statistico italiano 1993*, Rome: ISTAT (Istituto Nazionale di Statistica).

Jenkins, R. (1996) 'Ageing population causes huge rise in social workers', *The Times*, 9 Jan., p. 6.

Jensen, R. G., Shabad, T. and Wright, A. W. (1983) *Soviet Natural Resources in the World Economy*, Chicago: University of Chicago Press.

JETRO (1995) *Handy Facts on EU-Japan Economic Relations*, Tokyo: Japan External Trade Organization.

Jones, P. D. and Wigley, T. M. L. (1990) 'Global warming trends', *Scientific American*, 263 (2), August: 66–73.

Kuczynski, R. R. (1939) *Living-space and Population Problems*, Oxford Pamphlets on World Affairs, No. 8, Oxford: The Clarendon Press.

Langer, W. L. (1972) 'Checks on population growth: 1750–1850', *Scientific American*, 226 (2): 92–9.

Larson. E. D., Ross, M. H. and Williams, R. H. (1986) 'Beyond the era of materials', *Scientific American*, 254 (6): 24–31.

Lauby, J.-P. and Moreaux, D. (1994) *La France contemporaine*, Paris: Bordas, pp. 35–9.

Laurance, J. (1995) 'Two-tier NHS the way to meet overwhelming health demands', *The Times*, 19 Sept., p. 8.

Le Bras, H. and Todd, E. (1981) *L'Invention de la France*, Paris: Collection Pluriel.

Leonardi, R. (1993) 'Cohesion in the European Community: Illusion or Reality?', *West European Politics*, 16 (4), Oct.: 492–514.

Macintyre, B. (1995) 'French rail strikers fear dose of Dr Beeching's bitter medicine', *The Times*, 1 Dec., p. 16.

McIvor, G. (1993) 'Ghettos blast apart myth of the classless society', *The European*, 10–16 Dec., p. 13.

Mackinder, H.J. (1904) 'The geographical pivot of history', *Geographical Journal* XXIII (4): 421–2.

Mann, M. (1995a) 'Third of budget spent on regional aid', *European Voice*, 23–29 Nov., p. 14.

Mann, M. (1995b) 'Atlantic Arc redoubles efforts', *European Voice*, 23–29 Nov., p. 15.

Marchetti, R. (1985) *Quadro analitico complessivo dei risultati delle indagini condotte negli anni 1977–1980 sul problema dell' eutrofizzazione nelle acque costiere dell' Emilia-Romagna*, Regione Emilia-Romagna, Assessorato ambiente e difesa del suolo.

Martin, P. and Widgren, J. (1966) 'International Migration: A Global Challenge', *Population Bulletin*, 51(1), April, Washington: Population Reference Bureau.

Max, D. (1994) 'Park's science for creating jobs', *The European*, 2–8 Dec., p. 19.

Modern Teaching in the Senior School (undated, around 1930), London: The Home Library Book Co. (George Newnes).

Molle, W., van Holst, B. and Smit, H. (1980) *Regional Disparity and Economic Development in the European Community*, Westmead: Saxon House.

Mortished, C. (1995) 'BP close to $3.5 billion Algerian gas deal', *The Times*, Dec. 18, p. 40.

Musto, D. F. (1996) 'Alcohol in American History', *Scientific American*, 274 (4), April: 64–9.

National Audit Office (NAO) (1990) *Maternity Services*, London: HMSO.

Naudin, T. (1994a) 'Czech enterprise leads the way in former Soviet bloc', *The European*, 21–27 Oct., p. 18.

Naudin, T. (1994b) 'Cost will delay east joining EU', *The European*, 11–17 Nov., p. 17.

Naudin, T. (1994c) 'Frankfurt bids for the greater prize', *The European*, 22–28 Dec., p. 17.

Naudin, T. (1996) 'Capital investment', *The European*, 11–17 July, p. 22.

Nef, J. U. (1977) 'An early energy crisis and its consequences', *Scientific American*, 237 (5), Nov.: 140–51.

Nuttall, N. (1994a) '"Road tax" should be £630 a year', *The Times*, 24 Jan., p. 10.

Nuttall, N. (1994b) 'Three cities exceed EC limits on air pollution', *The Times*, 9 Dec., p. 7.

Nuttall, N. (1995) 'Expansion devours England's unspoilt havens', *The Times*, 1 Dec., p. 8.

Nuttall, N. (1996) 'Water firm ready to pipe supplies from France', *The Times*, 16 April, p. 10.

OECD (1991) *The State of the Environment*, Paris.

Olins, R. and Lorenz, A. (1992) 'BR Destination Unknown', *The Sunday Times*, 12 Jan, p. 3.3.

Palmer, R. and Rowland, J. (1990) 'North Sea chokes to death in its own filth', *The Sunday Times*, 4 March, p. A5.

Parry, M. (1990) *Climate Change and World Agriculture*, London: Earthscan Publications Ltd.

Plon, U. (1996) 'Dryshod across the Baltic', *The European Magazine*, 4–10 July, pp. 10–11.

Population Reference Bureau (PRB) (1987) *Population Bulletin*, 42 (1), March.

Population Reference Bureau (PRB) (various years) *World Population Data Sheet (WPDS)*, Washington D.C.

Ramesh, R. (1996) 'Passengers desert trains for inter-city minicabs', *The Sunday Times*, 11 Feb, p. 15.

Rand McNally (1987) *Goode's World Atlas*, Chicago: Rand McNally.

Read, J. (1995) 'Paris chokes with anger as toxic cloud thickens', *The European*, 26 Oct.–1 Nov., p. 3.

Regional Profiles (Profils régionaux), (1995) duplicated EU working paper, DGXVI-A4, September published only in French.

Ritchie, M. (1995) 'Scotland boasts unofficial EU embassy', *European Voice*, 23–9 Nov., p. 18.

Rollnick, R. (1995) 'Danger A – plant to reopen', *The European*, 31 Aug.–6 Sept., pp. 1,4.

Smith, S. (1994) 'Water war leaves regions feeling bitter as Spanish rivers run dry', *The European*, 26 Aug.–1 Sept., p. 4.

Smith, S. (1995) 'Will of iron saves blast of the past', *The European*, 20–26 Jan., p. 3

Spiessa, E. (1993) *Schweizer Weltatlas*, Zurich: Konferenz der Kantonalen Erziehungsdirektoren (EDK).

Stanners, D. and Bourdeau, P. (1995) *Europe's Environment: The Dobris Assessment*, Luxembourg: Office for Official Publications of the European Communities.

Thomas Cook (1991) *European Timetable*, July, Peterborough: Thomas Cook Publishing.

Thurow, L. C. (1987) 'A surge in inequality', *Scientific American*, 256 (5), May: 26.

Tillier, A. (1995) 'City of Light is the City of Conferences', *The European*, 25 April–1 May, p. 25.

Times Books (1979) *Times Atlas of World History* (ed. G. Barraclough), Times Books.

Times Books (1995) *The Times 1000, 1996*, London: Times Books.

Todd, E. (1987) *The Causes of Progress*, Oxford: Basil Blackwell.

Troev, T. and Naudin, T. (1995) 'Central Europe faces long wait at EU door', *The European*, 14–20 Dec., p. 20.

United Nations (1994) *1992 Energy Statistics Yearbook*, New York: United Nations, Table 13, pp. 156–68.

United Nations (various years) United Nations Statistical Yearbook, New York: United Nations.

United Nations Development Programme (1990) *Human Development Report 1990*, Oxford: Oxford University Press.

United Nations Development Programme (1993) *Human Development Report 1993*, Oxford: Oxford University Press.

United Nations Development Programme (1995) *Human Development Report 1995*, Oxford: Oxford University Press.

US Bureau of the Census (1994) *Statistical Abstract of the United States: 1994*, 114th edition, Washington DC.

Verchère, I. (1994a) 'Capital problems for the other cities', *The European*, 14–20 Jan., p. 19.

Verchère, I. (1994b) 'Airlines spiral into deeper debt', *The European*, 22–28 July, p. 23.

Vujakovic, P. (1992) 'Mapping Europe's myths', *Geographical Magazine*, LXIV(9), Sept.: 15–17.

Watson, R. (1995a) 'Bridging the gap between rich and poor regions', *European Voice*, 23–9 Nov., p. 13.

Watson, R. (1995b) 'Regions develop cross-border partnerships', *European Voice*, 23–9 Nov., p. 15.

Watt, N. (1993) 'Fearful Catholics plan to move their families from Ulster', *The Times*, 30 Oct., p. 2.

Wavell, S. (1995) 'Is the nuclear age over?', *The Sunday Times*, 17 Dec., p. 3.2.

Webster, J. (1995a) 'Who pays to clean the Mediterranean', *The European*, 9–15 June, p. 4.

Webster, J. (1995b) 'Parched Catalonia taps the Rhône for fresh water', *The European*, 26 Oct.–1 Nov., p. 2.

Webster, J. (1995c) 'Europe rebuilds the Wall in Africa', *The European*, 30 Nov.–6 Dec., p. 1.

The Week in Germany (1995) German Information Center New York, 24 Nov.

World Bank (1995) *World Development Report 1995*, Oxford: Oxford University Press.

Ziegler, C. E. (1987) *Environmental Policy in the USSR*, London: Frances Pinter.

GLOSSARY

Accumulated temperatures: a measure of the thermal resources for agriculture of different places, the addition of daily temperatures above the particular threshold at which different crops start to grow. Thus an average temperature on a given day of 18°C would count as 8 degrees for a plant with a threshold of 10°C.

Berlin Wall: a physical structure erected in 1961 to separate the Soviet occupied zone of Berlin from those of the Western allies and prevent the emigration of East German citizens to West Germany. Removed in 1989. The Iron Curtain already prevented movement elsewhere between East and West Germany.

Cabotage: a term referring to tramp shipping services, loading and unloading cargo at various ports, more recently referring to road haulage, when an HGV takes up a return cargo not necessarily on a set route.

Central Europe: since the late 1980s this term has increasingly replaced Eastern Europe when reference is made to countries located between the Western market economy countries and the former USSR. There is no definitive list of countries referred to as part of this area.

Cold War: confrontation without actual conflict between the Western capitalist countries and the Soviet communist bloc, late 1940s to late 1980s.

Customs union: in a customs union, customs duties between member countries are abolished. All goods can therefore be freely imported and exported without quotas, and a common external tariff is established for goods imported from non-EU countries.

Developing world: one of several terms used to refer to the poorer countries. Also Third World and South (as opposed to North).

East Germany: the Soviet occupied zone of post 1945 Germany, which became the German Democratic Republic until its unification with West Germany in 1990.

Europe Agreement: Association Agreement between EU and a potential member state in which the final objective of EU membership is explicitly cited.

European Agricultural Guidance and Guarantee Fund (EAGGF): established in 1962, the main agent of the Common Agricultural Policy. The 'Guidance' section is concerned with improvements in agricultural structures and marketing projects. The 'Guarantee' section ensures a minimum price to producers. It accounts for almost 90 per cent of the Fund's expenditure.

European Free Trade Association (EFTA): consisting after 1995 of Iceland, Liechtenstein, Norway and Switzerland.

Export refunds: subsidies granted to exporters of agricultural products to offset the difference between EU prices and the world prices of the products exported.

First World War, 1914–18: main participants, France, Britain, Russia, versus Germany, Austro-Hungary, Turkey, ending in victory for France and Britain, joined by Italy and the USA.

Free Trade Area: a set of countries that have entered into reciprocal free trade agreements within which goods can be traded free of quotas and customs duties or at reduced tariffs.

Import levies: duties charged on agricultural products imported from non-EU countries, the level determined by the difference between EU prices and world market prices.

Iron Curtain: refers to the barrier set up along the East–West German border to prevent the movement of people out of East Germany, and continuing southwards to isolate Czechoslovakia and Hungary from West Germany and Austria.

Limburg: a major administrative division of both Belgium (B) and the Netherlands (N).

Luxembourg: a sovereign Member State of the EU and also a major administrative division (province) of Belgium, distinguished sometimes by GD (Grand-Duché) and B (for Belgium).

Nordic countries: Denmark, Norway, Sweden, Finland and Iceland, of which the first three only form Scandinavia.

Second World War (in Europe) 1939–45: main participants, Germany, Italy versus France, Britain, and from 1941 the USSR and the USA, ending in defeat for Germany.

Single Market: an economic area within which persons, goods, services and capital have unrestricted freedom of movement. Customs barriers, as well as technical, tax and legislative obstacles are removed between member countries.

Transcaucasia: a major economic planning region of the former USSR comprising the now sovereign countries of Georgia, Armenia and Azerbaijan.

INDEX